UNITEXT

La Matematica per il 3+2

Volume 172

Editor-in-Chief

Alfio Quarteroni, Politecnico di Milano, Milan, Italy
 École Polytechnique Fédérale de Lausanne (EPFL), Lausanne, Switzerland

Series Editors

Luigi Ambrosio, Scuola Normale Superiore, Pisa, Italy

Paolo Biscari, Politecnico di Milano, Milan, Italy

Ciro Ciliberto, Università di Roma "Tor Vergata", Rome, Italy

Camillo De Lellis, Institute for Advanced Study, Princeton, USA

Lorenzo Rosasco, IDIBRIS, Università degli Studi di Genova, Genova, Italy
 Center for Brains Mind and Machines, Massachusetts Institute of Technology, Cambridge, MA, USA
 Istituto Italiano di Tecnologia, Genova, Italy

The **UNITEXT - La Matematica per il 3+2** series is designed for undergraduate and graduate academic courses, and also includes books addressed to PhD students in mathematics, presented at a sufficiently general and advanced level so that the student or scholar interested in a more specific theme would get the necessary background to explore it.

Originally released in Italian, the series now publishes textbooks in English addressed to students in mathematics worldwide.

Some of the most successful books in the series have evolved through several editions, adapting to the evolution of teaching curricula.

Submissions must include at least 3 sample chapters, a table of contents, and a preface outlining the aims and scope of the book, how the book fits in with the current literature, and which courses the book is suitable for.

For any further information, please contact the Editor at Springer: francesca.bonadei@springer.com

THE SERIES IS INDEXED IN SCOPUS

Marco Baronti · Enrico Calcagno ·
Filippo De Mari · Robertus van der Putten

Calculus Problems, II

Several Variables and Series

 Springer

Marco Baronti
Department of Mathematics
University of Genoa
Genoa, Italy

Filippo De Mari
Department of Mathematics
University of Genoa
Genoa, Italy

Enrico Calcagno
Department of Mathematics
University of Genoa
Genoa, Italy

Robertus van der Putten
Department of Mathematics
University of Genoa
Genoa, Italy

ISSN 2038-5714　　　　　　　ISSN 2532-3318　(electronic)
UNITEXT
ISSN 2038-5722　　　　　　　ISSN 2038-5757　(electronic)
La Matematica per il 3+2
ISBN 978-3-031-92912-0　　　ISBN 978-3-031-92913-7　(eBook)
https://doi.org/10.1007/978-3-031-92913-7

© The Editor(s) (if applicable) and The Author(s), under exclusive license to Springer Nature Switzerland AG 2025

This work is subject to copyright. All rights are solely and exclusively licensed by the Publisher, whether the whole or part of the material is concerned, specifically the rights of translation, reprinting, reuse of illustrations, recitation, broadcasting, reproduction on microfilms or in any other physical way, and transmission or information storage and retrieval, electronic adaptation, computer software, or by similar or dissimilar methodology now known or hereafter developed.
The use of general descriptive names, registered names, trademarks, service marks, etc. in this publication does not imply, even in the absence of a specific statement, that such names are exempt from the relevant protective laws and regulations and therefore free for general use.
The publisher, the authors and the editors are safe to assume that the advice and information in this book are believed to be true and accurate at the date of publication. Neither the publisher nor the authors or the editors give a warranty, expressed or implied, with respect to the material contained herein or for any errors or omissions that may have been made. The publisher remains neutral with regard to jurisdictional claims in published maps and institutional affiliations.

This Springer imprint is published by the registered company Springer Nature Switzerland AG
The registered company address is: Gewerbestrasse 11, 6330 Cham, Switzerland

If disposing of this product, please recycle the paper.

If people do not believe that mathematics is simple, it is only because they do not realize how complicated life is.

—*John von Neumann*

Preface

This book is the natural continuation of [1] and has the same features of the first volume. Globally, it contains 289 problems, either guided or suggested exercises. As far as the organization of the material is concerned, every chapter starts with a summary of the main results that should be kept in mind and used for the mathematics of that chapter and is followed by a selection of guided exercises. The theoretical preamble is meant to recapitulate the most relevant definitions and results and should also serve as a bird's-eye view on the topic treated in the chapter. Hence, the student can quickly review the pertinent theoretical facts and then, most importantly, "learn by examples", becoming acquainted with the specific techniques by seeing them applied directly to the problems. Each of the 152 guided exercises ends with a short comment which underlines the salient issues of that specific exercise, the leading ideas and the appropriate techniques. A selection of problems closes each chapter, the answers to which are all listed in Solutions, for a total of 137 problems in the book.

As in the case of the first volume, the scope of the book is to provide a practical working tool for students in Engineering, Mathematics, and Physics, or in any other field where rigorous Calculus is needed.

Perhaps the most distinctive feature of this book is that our approach is very direct and refers to a concrete experience. The material is in fact mostly taken from actual written tests that have been delivered in the years 2000–2020 at the Engineering School and at the Science School of the University of Genova. Literally thousands of students have worked on these problems, so our first and foremost acknowledgment goes to them, because they have helped us greatly over the years, tuning our views and making us see where the true difficulties really are, those that need both clear statements and specifically designed exercises. Their fellow colleagues, the present and future students, are of course our public and our intended readers.

The topics covered in this volume span approximately one half of the syllabus that in most undergraduate programs in Italy typically forms the core of the mathematical knowledge expected from Engineering or Physics students after their third year. More precisely, we have decided to cover, after an introductory chapter, three broad themes. The first, which is presented in Chaps. 2 and 3, is the basic apparatus concerning limits, continuity and differentiability issues of real valued functions of several variables (typically two or three) which is then applied in Chap. 4 to the study of unconstrained (and constrained) minima and maxima, and to implicit functions. The second theme, treated in Chap. 5, concerns multiple Riemann integrals, focusing again on the case of two or three variables. The third, developed in Chap. 6, includes sequences and series of functions.

We have chosen to postpone to a third volume the remaining somewhat more advanced topics, namely systems of ordinary differential equations, as well as special differential equations, and the vaste body of notions that might be collectively described as vector calculus on submanifolds of Euclidean space. This includes curves and line integrals, differential forms and vector fields, potentials, surfaces and surface integrals, namely the tools that are needed to develop the great achievements of the nineteenth century mathematics which revolve around the gradient, the curl, the divergence and the Laplacean, and culminate in the Gauss-Green formulae, the divergence theorem and Stokes' theorem.

We believe that anyone who can solve the suggested problems in our series of volumes with a reasonable degree of accuracy is in a safe position for positive results in most Italian universities. Our international experience also tells us that the same may be claimed for most universities around the world, for undergraduate Calculus or Advanced Calculus.

Genoa, Italy
November 2024

Marco Baronti
Enrico Calcagno
Filippo De Mari
Robertus van der Putten

Competing Interests The authors have no competing interests to declare that are relevant to the content of this manuscript.

Contents

1 **Structures and Functions on Euclidean Space** 1
   1.1   Vector Spaces 2
   1.2   Inner Product, Norm and Distance 4
   1.3   Topology in Euclidean Space 7
   1.4   Sequences 15
   1.5   Functions and Graphical Interpretations 17
       1.5.1 Functions of Two Variables 18
       1.5.2 Vector Fields, Complex Functions 19
   1.6   Coordinates 21
       1.6.1 Polar Coordinates 21
       1.6.2 Spherical Coordinates 23
       1.6.3 Cylindrical Coordinates 25
   1.7   Guided Exercises 26
   1.8   Problems 45
   References 46

2 **Limits and Continuity** 47
   2.1   Convergence Notions 47
   2.2   Results on Limits 50
       2.2.1 Uniformity and Polar Coordinates 52
   2.3   Continuity 53
   2.4   Topological Coda 55
       2.4.1 Curves 57
   2.5   Guided Exercises 59
   2.6   Problems 81
   Reference 83

3 Differentiation ... 85
3.1 Directional and Partial Derivatives ... 85
3.2 Differentiability ... 87
3.3 Higher Order Derivatives and Taylor Expansions ... 91
3.4 Differentiation Under the Integral Sign ... 95
3.5 Guided Exercises ... 98
3.6 Problems ... 141
Reference ... 145

4 Minima and Maxima, Implicit Functions ... 147
4.1 Unconstrained Minima and Maxima ... 148
4.2 Constrained Minima and Maxima ... 151
4.3 The Implicit Function Theorem ... 154
4.4 Guided Exercises ... 156
4.5 Problems ... 180

5 Multiple Integrals ... 183
5.1 Measure and Measurable Sets ... 183
5.2 Double Integrals ... 185
5.3 Triple Integrals ... 191
5.4 Barycenters and Moments of Inertia ... 194
5.4.1 Solids of Revolution ... 197
5.5 Guided Exercises ... 198
5.6 Problems ... 238
Reference ... 242

6 Sequences and Series of Functions ... 243
6.1 Pointwise and Uniform Convergence ... 243
6.1.1 Sequences ... 243
6.1.2 Series ... 246
6.2 Power Series ... 250
6.3 Taylor Series ... 252
6.4 Fourier Series ... 254
6.5 Guided Exercises ... 261
6.6 Problems ... 329
Reference ... 336

Solutions ... 337

Subject Index ... 351

Symbols

$\langle u, v \rangle$	Inner (or scalar) product between the vectors u and v
$\|u\| = \sqrt{\langle u, u \rangle}$	Norm of the vector u
$B(P, r)$	Open ball with center P and radius r
$C^0(\Omega)$	Continuous functions on $\Omega \subseteq \mathbb{R}^n$
$C^k(\Omega)$	Functions with continuous partial derivatives on $\Omega \subseteq \mathbb{R}^n$ up to order k
$C^\infty(\Omega)$	Functions with continuous partial derivatives on $\Omega \subseteq \mathbb{R}^n$ of all orders
$d(u, v)$	Distance between the vectors u and v
$\frac{\partial f}{\partial Q}(P)$	Directional derivative of f in the direction of the unit vector Q evaluated at P
$\frac{\partial f}{\partial x}(P)$	Partial derivative of f with respect to x evaluated at P
$\frac{\partial^2 f}{\partial x \partial y}(P)$	Second order derivative of f with respect to x and y evaluated at P
$\nabla f(P)$	Gradient of f evaluated at P
$\text{Dom}(f)$	Domain of the function f
$\partial \Omega$	Boundary of the set $\Omega \subseteq \mathbb{R}^n$
$Hf(P)$	Hessian matrix of f evaluated at P
$\iint_\Omega f$	Double integral of $f : \Omega \subseteq \mathbb{R}^2 \to \mathbb{R}$
$\iiint_\Omega f$	Triple integral of $f : \Omega \subseteq \mathbb{R}^3 \to \mathbb{R}$
$JF(P)$	Jacobian matrix of F evaluated at P
$\mathscr{P}_T(I)$	Periodic functions on I with period T
$PC(I)$	Piecewise continuous functions on I
PC_T	Piecewise continuous T-periodic functions on $\mathbb{R}$
$PS(I)$	Piecewise smooth functions on I
PS_T	Piecewise smooth T-periodic functions on $\mathbb{R}$
PS_T^e	Piecewise smooth even T-periodic functions on $\mathbb{R}$
PS_T^o	Piecewise smooth odd T-periodic functions on $\mathbb{R}$
Q_A	Quadratic form associated to the symmetric matrix A
$\sum_{n=0}^{\infty} f_n$	Series associated to the sequence of functions $(f_n)_{n \geq 0}$

χ_Ω	Characteristic funcion of the set $\Omega \subseteq \mathbb{R}^n$
$\|\Omega\|$	Peano-Jordan measure of the set $\Omega \subseteq \mathbb{R}^n$
$\overline{\Omega}$	Closure of the set $\Omega \subseteq \mathbb{R}^n$
$\mathring{\Omega}$	The interior points of the set $\Omega \subseteq \mathbb{R}^n$

Chapter 1
Structures and Functions on Euclidean Space

In classical Analysis, *Euclidean space* of dimension n refers to the n-fold Cartesian product $\mathbb{R}^n = \mathbb{R} \times \cdots \times \mathbb{R}$ endowed with its natural metric structure. This means that, inspired by Pythagoras' theorem, one defines the *length* of a *vector* in $\mathbb{R}^n$ and hence the distance between two points in $\mathbb{R}^n$ as the length of the segment joining them, the latter being just the "size" of the difference between the two vectors associated to the points. In order to do this, some basic notions need to be introduced. Most importantly, one introduces the structure of *vector space* on $\mathbb{R}^n$ and defines the *norm* of an element, namely its size. This, in turn, comes about from the notion of *inner product*. With these concepts at hand, Analysis in Euclidean space can be developed in every dimension n. Hence, from now on, n will always denote a positive integer, with the understanding that the most important cases that one should keep in mind are when $n = 2$ or $n = 3$. In these cases, the geometry that comes into play is that of the plane ($n = 2$) and of the ordinary space ($n = 3$), for which basic intuition is available and is in fact of great help. The reader is also invited to observe that when $n = 1$ the setup reduces to $\mathbb{R}$ and hence every relevant notion should be compared with its simple counterpart on the Euclidean line, a case that must always be understood first.

The notion of vector space, as those of inner product, norm and distance, allows for wide extensions, leading to very far-reaching generalizations that represent central areas of investigation in modern Analysis. For this reason they are best formulated with a certain degree of abstraction but the reader is invited to think very concretely, that is, to use the pictures that one can mentally draw in plane or in space and to develop a solid geometric intuition in these contexts.

In this chapter the basic metric and topological notions that are useful in Calculus are briefly recalled and the notation is fixed. A thorough treatment of metric or topological issues is beyond the scope of this book. The curious reader may consult for example [1].

1.1 Vector Spaces

Definition 1.1 A real vector space is a nonempty set V, the elements of which will be called *vectors*, on which two operations are defined: the sum of vectors, that is a map $V \times V \to V$

$$(u, v) \mapsto u + v \tag{1.1}$$

and the *scalar multiplication* by a real number, that is another map $\mathbb{R} \times V \to V$

$$(\alpha, u) \mapsto \alpha u, \tag{1.2}$$

satisfying the following properties:

(i) $u + v = v + u$ for every $u, v \in V$ (commutativity);
(ii) $u + (v + w) = (u + v) + w$ for every $u, v, w \in V$ (associativity);
(iii) there exists $0 \in V$ such that $0 + u = u$ for every $u \in V$ (identity for the sum);
(iv) for every $u \in V$ there exists $-u \in V$ such that $u + (-u) = 0$ (opposite elements);
(v) $\alpha(\beta u) = (\alpha \beta) u$ for every $\alpha, \beta \in \mathbb{R}$ and every $u \in V$ (compatibility of scalar and field multiplications);
(vi) $1u = u$ for every $u \in V$ (identity for the scalar multiplication);
(vii) $\alpha(u + v) = \alpha u + \alpha v$ for every $\alpha \in \mathbb{R}$ and every $u, v \in V$ (distributivity of scalar multiplication with respect to vector addition);
(viii) $(\alpha + \beta)u = \alpha u + \beta u$ (distributivity of scalar multiplication with respect to field addition).

The element $\alpha u + \beta v$ of V is called the *linear combination* of $u, v \in V$ with coefficients $\alpha, \beta \in \mathbb{R}$.

The basic example of vector space is of course $\mathbb{R}^n$. Vectors in $\mathbb{R}^n$ will often be denoted with capital letters like P or Q, which are reminiscent of classical notation in Euclidean Geometry. Thus, typically a vector is written $P = (x, y)$ in $\mathbb{R}^2$, $P = (x, y, z)$ in $\mathbb{R}^3$ and $P = (x_1, \ldots, x_n)$ in $\mathbb{R}^n$. Vectors are also called points, further stressing the geometry that lies behind.

The operation (1.1) in $\mathbb{R}^n$, that is, the sum of $(x_1, \ldots, x_n)$ and $(y_1, \ldots, y_n)$ is

$$(x_1, \ldots, x_n) + (y_1, \ldots, y_n) = (x_1 + y_1, \ldots, x_n + y_n)$$

and the multiplication (1.2) of $(x_1, \ldots, x_n)$ by the scalar $\alpha \in \mathbb{R}$ is given by

$$\alpha(x_1, \ldots, x_n) = (\alpha x_1, \ldots, \alpha x_n).$$

Geometrically, the sum of two vectors may be visualized as the diagonal of the parallelogram the sides of which are the original vectors. Positive scalars dilate or contract the vector, according as $\alpha > 1$ or $\alpha < 1$ (see Fig. 1.1), and negative scalars change the orientation.

1.1 Vector Spaces

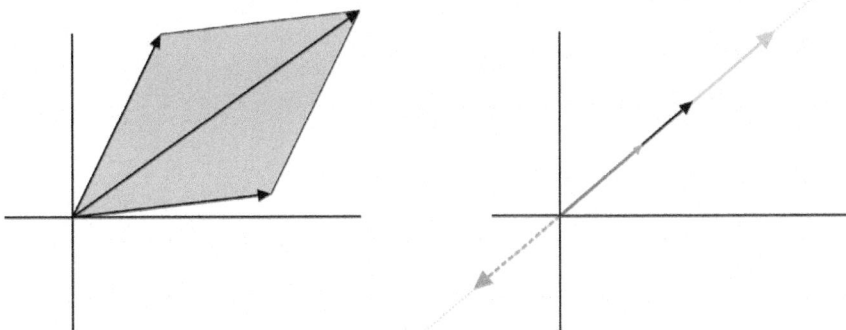

Fig. 1.1 Sum of vectors in $\mathbb{R}^2$ on the left, and multiplication of the black vector by positive scalars (grey) or negative scalars (dashed) on the right

Evidently, the zero vector is $0 = (0, \ldots, 0) \in \mathbb{R}^n$, often denoted O for short, and the opposite of $(x_1, \ldots, x_n)$ is the vector $(-x_1, \ldots, -x_n)$. The reader should check all the properties listed in Definition 1.1.

Of particular relevance are the *basis vectors* $e_1, \ldots, e_n$, defined by

$$e_j = (0, \ldots, 0, 1, 0, \ldots, 0), \qquad j = 1, \ldots, n.$$

Explicitly, all the entries of e_j vanish except the jth, which is 1. It is worth observing that each vector can be uniquely written as a linear combination of the basis vectors $e_1, \ldots, e_n$, namely

$$(x_1, \ldots, x_n) = \sum_{j=1}^{n} x_j e_j. \tag{1.3}$$

Without entering into a lengthy and detailed treatment of vector spaces, for which a vast literature is available [2], the above basic equality explains the reason why $\mathbb{R}^n$ is said to have *dimension n*.

An important example of real vector space of finite dimension, to wit $2n$, is $\mathbb{C}^n$, the elements of which are the n-tuples $(z_1, \ldots, z_n)$ of complex numbers, and where the same notions of sum and multiplication by real numbers as above may be given. In this case, a basis is

$$e_1, \ldots, e_n, i e_1, \ldots, i e_n,$$

where i is the imaginary unit, that satisfies $i^2 = -1$. Indeed, every element $(z_1, \ldots, z_n) \in \mathbb{C}^n$ may be uniquely written as

$$(z_1, \ldots, z_n) = \sum_{j=1}^{n} x_j e_j + i \sum_{j=1}^{n} y_j e_j, \tag{1.4}$$

where evidently $z_j = x_j + iy_j$ for $j = 1, \ldots, n$.

Many vector spaces of functions arise naturally in Analysis. For example, $C^0(\mathbb{R})$, $C^1(\mathbb{R})$ are real vector spaces (check this!), just like $C^\infty(\mathbb{R})$. A basic difference between $\mathbb{R}^n$ and the space $C^0(\mathbb{R})$ of continuous functions on $\mathbb{R}$ is that the latter is not finite dimensional, in the sense that there does not exist a finite set of functions for which an expansion like (1.3) holds. In other words, it is clear that any linear combination

$$\sum_{j=1}^{N} \alpha_j f_j$$

of functions $f_1, \ldots, f_N \in C^0(\mathbb{R})$ with real scalars $\alpha_1, \ldots, \alpha_N$ is again in $C^0(\mathbb{R})$, but one cannot express every continuous function as a linear combination of the same finitely many functions $f_1, \ldots, f_N$ fixed once and for all.

As a final comment in this section, observe that Definition 1.1 can be extended to cover the case of *complex vector spaces* by allowing for the scalars to be complex numbers. Of course, $\mathbb{C}^n$ is a complex vector space and as such it has dimension n. Sets of complex valued functions occur in many applications, many of which have the structure of complex vector spaces.

1.2 Inner Product, Norm and Distance

Definition 1.2 (*Inner product*) An *inner product* on the real vector space V is a map $\langle \cdot, \cdot \rangle \colon V \times V \to \mathbb{R}$ satisfying:

(i) $\langle u + w, v \rangle = \langle u, v \rangle + \langle w, v \rangle$ for every $u, v, w \in V$;
(ii) $\langle \alpha u, v \rangle = \alpha \langle u, v \rangle$ for every $\alpha \in \mathbb{R}$ and every $u, v \in V$;
(iii) $\langle u, v \rangle = \langle v, u \rangle$ for every $u, v \in V$;
(iv) $\langle u, u \rangle \geq 0$ for every $u \in V$ and $\langle u, u \rangle = 0$ if and only if $u = 0 \in V$.

A vector space V on which an inner product is defined is called an *inner product vector space*.

Properties (i) and (ii), together with (iii), say that $\langle \cdot, \cdot \rangle \colon V \times V \to \mathbb{R}$ is a symmetric *bilinear map* in the sense that by (i) and (ii) the map $u \mapsto \langle u, v \rangle$ is a linear map for any fixed $v \in V$ and, switching the roles of u and v using (iii) (that is, symmetry), such is also the map $v \mapsto \langle u, v \rangle$ for any fixed $u \in V$. Property (iv) is usually referred to by saying that the inner product is positive definite.

The basic example of inner product is the so-called *dot product* in $\mathbb{R}^n$, namely

$$\langle P, Q \rangle = P \cdot Q$$

where, for $P = (x_1, \ldots, x_n)$ and $Q = (y_1, \ldots, y_n)$, the right hand side is defined by

1.2 Inner Product, Norm and Distance

$$P \cdot Q := x_1 y_1 + \cdots + x_n y_n = \sum_{j=1}^{n} x_j y_j.$$

All the properties listed in Definition 1.2 are easily checked. In particular, it is worth remarking that

$$P \cdot P = x_1^2 + \cdots + x_n^2$$

which readily satisfies (iv).

Another important example of inner product vector space is obtained by considering on the space $C^0([a, b])$ of all continuous (hence Riemann integrable) functions on $[a, b]$ the inner product given by:

$$\langle f, g \rangle = \int_a^b f(t)g(t)\,dt, \qquad f, g \in C^0([a, b]).$$

On any inner product space it is customary to write

$$\|u\| = \sqrt{\langle u, u \rangle} \tag{1.5}$$

and to call $\|\cdot\|$ the *norm* associated with the inner product $\langle \cdot, \cdot \rangle$. The norm satisfies the basic properties described next.

Proposition 1.1 *In any inner product vector space V the following holds true for any $u, v \in V$ and $\alpha \in \mathbb{R}$:*

(i) $\|u\| \geq 0$, and $\|u\| = 0$ if and only if $u = 0$;

(ii) $\|\alpha u\| = |\alpha| \|u\|$;

(iii) $\|u + v\| \leq \|u\| + \|v\|$.

Furthermore, for every $u, v \in V$

$$|\langle u, v \rangle| \leq \|u\| \|v\|. \tag{1.6}$$

Inequality (iii) is called the triangle inequality *and (1.6) is the* Cauchy-Schwarz *inequality.*

The triangle inequality is actually equivalent to

$$\left| \|u\| - \|v\| \right| \leq \|u - v\|. \tag{1.7}$$

The norm of $P = (x_1, \ldots, x_n) \in \mathbb{R}^n$ associated with the dot product is called the *Euclidean norm* and is given by the non negative real number

$$\|P\| = \sqrt{x_1^2 + \cdots + x_n^2} = \left(\sum_{j=1}^{n} x_j^2\right)^{1/2}. \tag{1.8}$$

Observe that when $n = 1$ the norm of $x \in \mathbb{R}$ is just its absolute value: $\|x\| = |x|$. The norm of P is interpreted as the length of the segment joining the origin $0 \in \mathbb{R}^n$ to P. Indeed, upon identifying Euclidean space with $\mathbb{R}^n$ by means of orthogonal Cartesian coordinates (see Sect. 1.6), this is precisely the content of Pythagoras' theorem, classically stated when $n = 2$ (see Fig. 1.2). From this geometric viewpoint is then clear why item (iii) in Proposition 1.1 is called "triangle inequality": it states that the length of a side in a triangle must be smaller than the sum of the lengths of the other two (see also the leftmost picture in Fig. 1.1).

Any function $\|\cdot\| \colon V \to [0, +\infty)$ on a vector space V satisfying properties (i), (ii) and (iii) of Proposition 1.1 is called a *norm*, whether it actually comes from an inner product or not. Here are two important examples of norms on $\mathbb{R}^n$ that can be shown not to be associated with any possible inner product on $\mathbb{R}^n$:

$$\|P\|_1 = |x_1| + \cdots + |x_n| \tag{1.9}$$
$$\|P\|_\infty = \max\{|x_1|, \ldots, |x_n|\}. \tag{1.10}$$

Similarly, the quantities

$$\|f\|_1 = \int_a^b |f(t)|\,dt \tag{1.11}$$
$$\|f\|_\infty = \max\{|f(t)| : t \in [a,b]\} \tag{1.12}$$

define norms on $C^0([a,b])$. It is an instructive exercise to show that they are actually norms. These observations are formalized next.

Fig. 1.2 Pythagoras' theorem: in a right rectangle the square of the length of the hypotenuse, namely $\|(x,y)\|^2$, is equal to the sum $x^2 + y^2$ of the squares of the lengths of other two sides

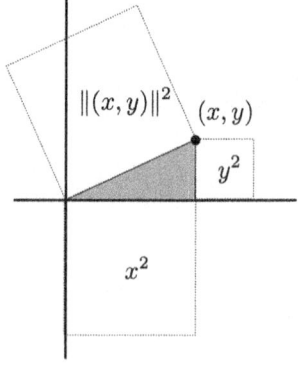

1.3 Topology in Euclidean Space

Definition 1.3 (*Normed space*) A vector space V is a *normed vector space* if it is equipped with a norm function $\|\cdot\|\colon V \to [0, +\infty)$ on V that satisfies properties (i), (ii) and (iii) of Proposition 1.1.

Definition 1.4 (*Distance*) The *distance* between the points u and v in the normed vector space V is

$$d(u, v) = \|u - v\| \qquad (1.13)$$

It is immediately verified that the distance function $d\colon V \times V \to [0, +\infty)$ satisfies:

(i) $d(u, v) \geq 0$ and $d(u, v) = 0$ if and only if $u = v$.
(ii) $d(u, v) = d(v, u)$
(iii) $d(u, w) \leq d(u, v) + d(v, w)$

for every $u, v, w \in V$.

The *Euclidean distance* between $P = (x_1, \ldots, x_n) \in \mathbb{R}^n$ and $Q = (y_1, \ldots, y_n) \in \mathbb{R}^n$ is therefore

$$d(P, Q) = \|P - Q\| = \sqrt{(x_1 - y_1)^2 + \cdots + (x_n - y_n)^2} \qquad (1.14)$$

and will play a central role in what follows. Observe again that for $n = 1$ the distance is simply $d(x, y) = |x - y|$. It is also worth observing that in $\mathbb{R}^2$ the distance between $P = (x, y)$ and $Q = (x_0, y_0)$ is

$$d(P, Q) = \|P - Q\| = \sqrt{(x - x_0)^2 + (y - y_0)^2}$$

and that in $\mathbb{R}^3$ the distance between $P = (x, y, z)$ and $Q = (x_0, y_0, z_0)$ is

$$d(P, Q) = \|P - Q\| = \sqrt{(x - x_0)^2 + (y - y_0)^2 + (z - z_0)^2}.$$

Here the reader should again notice the different choice of coordinates: (x, y) in $\mathbb{R}^2$, (x, y, z) in $\mathbb{R}^3$ and $(x_1, \ldots, x_n)$ in $\mathbb{R}^n$.

Distances can be defined abstractly on any set X, be it a normed vector space or not, by saying that $d\colon X \times X \to [0, +\infty)$ is a distance, or *metric*, if properties (i), (ii) and (iii) of Definition 1.4 are satisfied. A pair (X, d) where d is a metric on the set X is then called a *metric space*. In this monograph, unless otherwise stated, whenever the word "distance" appears, it is referred to the Euclidean distance (1.14).

1.3 Topology in Euclidean Space

The greek word τòπoς means "place". Topology is, roughly speaking, the study of the properties of sets that are independent of measurement but do depend on shape up to reasonable deformations. For example, two circles with different radii are geometrically different but topologically equivalent, and such are also a circle and an ellipse, because, at least abstractly, it is possible to shrink a circle into another or to bend a circle into an ellipse, but a circle and a segment are not topologically equivalent because cuts are not allowed and one must cut a circle to get something that can be bent into a segment. A deeper reflection reveals that the core of this train of thoughts resides in the notion of "close points", or "neighborhood".

The whole body of Topology (in fact its very definition, which will not be dealt with) is based on the notion of open set, and builds on it. Open sets should enjoy the following property: whenever a point P lies in the open set Ω, then all points sufficiently close to P are also in Ω. The concept of "all points sufficiently close to P" must of course be made precise. Whenever a notion of distance is available, this is an easy task. In this book, only Euclidean spaces are treated in detail, with simple hints to more general setups, but most notions may be widely generalized.

Definition 1.5 (*Ball*) The *open ball* centered at $P_0 \in \mathbb{R}^n$ of radius $r > 0$ is the set

$$B(P_0, r) = \{P \in \mathbb{R}^n : d(P, P_0) < r\}. \tag{1.15}$$

It should be clear from (1.15) that the notion of open ball can be given in any metric space. Notice that of course if $P_0 = (x_0, y_0) \in \mathbb{R}^2$, then

$$P = (x, y) \in B(P_0, r) \iff \|P - P_0\| = \sqrt{(x - x_0)^2 + (y - y_0)^2} < r$$

and similarly, if $P_0 = (x_0, y_0, z_0) \in \mathbb{R}^3$, then

$$P = (x, y, z) \in B(P_0, r) \iff \|P - P_0\| = \sqrt{(x - x_0)^2 + (y - y_0)^2 + (z - z_0)^2} < r.$$

Definition 1.6 (*Open and closed sets, neighborhoods*)

(i) A nonempty subset Ω of $\mathbb{R}^n$ is said to be *open* if for every $P \in \Omega$ there exists an open ball $B(P, r)$ centered at P and contained in Ω;
(ii) the complement $\mathbb{R}^n \setminus \Omega$ of an open set Ω is a *closed* set;
(iii) a subset $\mathcal{U}$ of $\mathbb{R}^n$ is a *neighborhood* of $P_0 \in \mathbb{R}^n$ if there exists an open ball $B(P_0, r)$ centered at P_0 contained in $\mathcal{U}$, that is $B(P_0, r) \subset \mathcal{U}$. If $\mathcal{U}$ is a neighborhood of $P_0 \in \mathbb{R}^n$, then $\mathcal{U} \setminus \{P_0\}$ is called a *punctured neighborhood* of P_0.

The empty set $\emptyset$ is defined to be both open and closed. This entails that the whole space $\mathbb{R}^n$ is also open and closed. In fact, $\mathbb{R}^n$ and $\emptyset$ are the only sets which are both open and closed. Whenever clear from the context, the complement $B \setminus A$ of a set A in B, is also denoted A^c. This is most common when $B = \mathbb{R}^n$.

Examples of open sets are of course the open balls. The upper-half plane in $\mathbb{R}^2$

$$\mathbb{R} \times \mathbb{R}_+ = \{(x, y) \in \mathbb{R}^2 : y > 0\}$$

is open because for any point $P = (x, y) \in \mathbb{R}_+^2$ the ball $B(P, y/2)$ is contained in $\mathbb{R} \times \mathbb{R}_+$. The set
$$C = \{(x, y) \in \mathbb{R}^2 : y \leq 0\}$$
is not open because $O = (0, 0) \in C$ but every open ball $B(O, r)$ with $r > 0$ contains the point $(0, r/2)$ which is not in C. The set C is closed because it is the complement of the upper-half plane. Although an open set is a neighborhood of all its points, neighborhoods need not be open, nor closed: a square that includes only two sides is neither open nor closed but it is a neighborhood, for example, of its center (prove these statements!). In Fig. 1.3 all these facts are illustrated.

Proposition 1.2 *(i) Arbitrary unions of open sets are open;*
(ii) finite intersections of open sets are open;
(iii) arbitrary intersections of closed sets are closed;
(iv) finite unions of closed sets are closed.

In Propostition 1.2, the word "arbitrary" means that collections with infinitely many elements can also be considered, regardless of cardinality issues. For example, consider the collection
$$\{B((3, y), r) : y \in \mathbb{Q}, 1 < r < 2\}$$
of open sets with rational centers along the vertical line $\{(3, y) : y \in \mathbb{R}\}$ and radii in the interval $(1, 2)$. This is an infinite, and uncountable, collection and statement (i) of Propostition 1.2 says that the union of its members
$$\bigcup_{y \in \mathbb{Q}, 1 < r < 2} B((3, y), r)$$

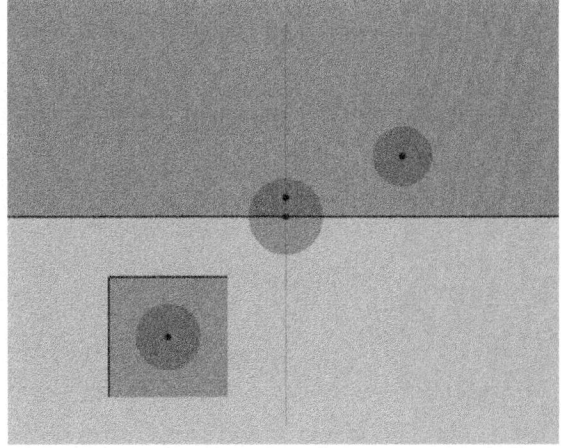

Fig. 1.3 Open and closed sets; a square neigborhood of $(-2, -2)$ which is neither open nor closed

is open. In fact, it is just the strip $\{(x, y) \in \mathbb{R}^2 : 1 < x < 5\}$, as the reader can check (see Fig. 1.4).

Definition 1.7 (*Interior point*) A point P is an *interior point* of $\Omega \subset \mathbb{R}^n$ if there exists an open ball centered at P and contained in Ω, that is, if for some $r > 0$ it is $B(P, r) \subset \Omega$. The union of all interior points of Ω is called the *interior* of Ω and is denoted $\overset{\circ}{\Omega}$.

The interior $\overset{\circ}{\Omega}$ is the largest open set contained in Ω. A set Ω is open if and only if $\Omega = \overset{\circ}{\Omega}$ or, which is the same, if and only if all its points are interior points. Nonempty sets may well have empty interior, for example a straight line in the plane: no open ball centered at any of its points is contained inside the line.

Definition 1.8 (*Closure, boundary*) Let Ω be a nonempty subset of $\mathbb{R}^n$.

(i) The *closure* of Ω is the smallest closed set containing Ω, and is denoted $\overline{\Omega}$;
(ii) the *boundary* of Ω is the set $\partial \Omega = \overline{\Omega} \cap \overline{(\mathbb{R}^n \setminus \Omega)}$.

A simple example illustrating these notions is given by $\Omega = B(O, 1)$, for which the closure

$$\overline{B(O, 1)} = \{P \in \mathbb{R}^n : \|P\| \leq 1\} = \left\{(x_1, \ldots, x_n) : \sum_{j=1}^n x_j^2 \leq 1\right\},$$

is the *closed unit ball*, and the boundary

$$\partial B(O, 1) = \{P \in \mathbb{R}^n : \|P\| = 1\} = \left\{(x_1, \ldots, x_n) : \sum_{j=1}^n x_j^2 = 1\right\}, \qquad (1.16)$$

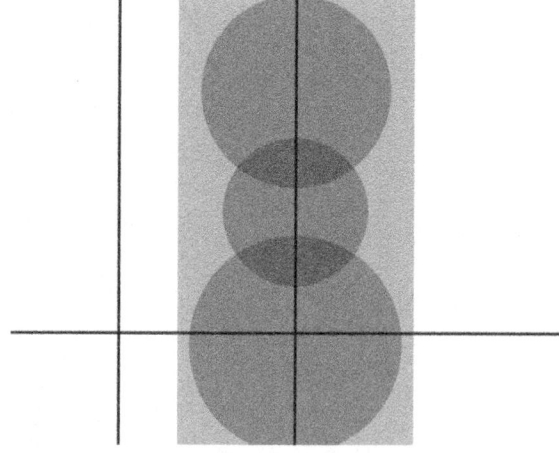

Fig. 1.4 Three open balls in the uncountable family $\{B((3, y), r) : y \in \mathbb{Q}, 1 < r < 2\}$ of open balls whose union is the open vertical strip $\{(x, y) \in \mathbb{R}^2 : 1 < x < 5\}$

1.3 Topology in Euclidean Space

is the *unit sphere*, also denoted S^{n-1}. When $n = 2$ these are depicted in Fig. 1.5. In this case one speaks of *disk* and *circle* instead of ball and sphere, respectively.

Observe that closed sets coincide with their closure and contain their boundary. The closure of a set is the intersection of all closed sets containing it. Also, being the intersection of two closed sets, any boundary $\partial \Omega$ is closed. An other way of describing a boundary point is by saying that $P \in \partial \Omega$ if and only if every neighborhood of P contains at least a point which is in Ω and one which is not in Ω. The boundary is what needs to be added to Ω to get its closure, namely

$$\overline{\Omega} = \Omega \cup \partial \Omega.$$

If one removes the interior from the closure what is left is the boundary:

$$\partial \Omega = \overline{\Omega} \setminus \overset{\circ}{\Omega}.$$

The above two equalities entail that some points of Ω, but none in $\overset{\circ}{\Omega}$, may well be boundary points. To see this, consider the set in $\mathbb{R}^2$ depicted in Fig. 1.6, namely

$$\Omega = B(O, 1) \cup \left\{(x, y) \in \mathbb{R}^2 : x^2 + y^2 = 1,\ x > 0,\ y > 0\right\},$$

which is obtained by adding to the unit open ball a piece of its boundary. The set Ω and $B(O, 1)$ are different but have the same closure, the same interior and the same boundary (prove these statements!). This seemingly strange fact happens because

$$\partial \Omega \cap \Omega = \left\{(x, y) \in \mathbb{R}^2 : x^2 + y^2 = 1,\ x > 0,\ y > 0\right\} \neq \emptyset$$

whereas $\partial B(O, 1) \cap B(O, 1) = \emptyset$.

A notion similar to that of boundary point is that of *limit point*, or *accumulation point*. It describes the property of being arbitrarily well approximated by (other) points of Ω.

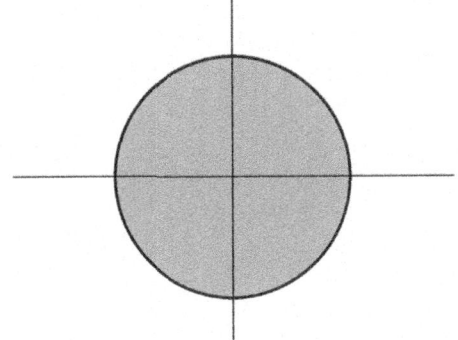

Fig. 1.5 In grey the open ball $\Omega = B(O, 1)$, in black its boundary $\partial B(O, 1)$; their union is the closed ball $\overline{B(O, 1)}$, the closure of $B(O, 1)$

Fig. 1.6 The set in this picture is not the open unit ball but has the same closure, interior and boundary as the open unit ball

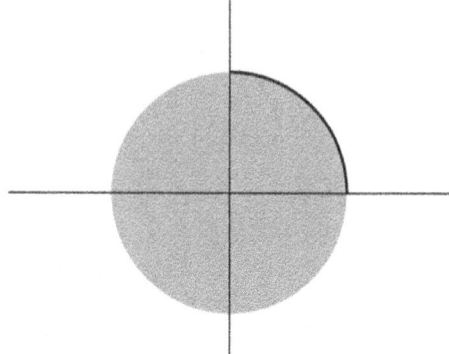

Definition 1.9 (*Limit point*) Take a nonempty subset Ω of $\mathbb{R}^n$. The point $P_0 \in \mathbb{R}^n$ is called a limit point, or an accumulation point, of Ω if every punctured neighborhood of P_0 intersects Ω in at least one element.

The above definition may be rephrased by saying that P_0 is a limit point of Ω if for every $r > 0$ there exists $P \in \Omega$ such that $0 < \|P - P_0\| < r$. Thus, a limit point of Ω cannot be separated from Ω by selecting a small enough radius r in such a way that in the ball $B(P_0, r)$ there are no points of Ω other than (possibly) P_0 itself.

The main difference between the definition of boundary point and that of limit point is that when deciding if P_0 is a limit point of Ω one must look at punctured neighborhoods of P_0, hence not at P_0 itself. For example, the point $P_0 = (1, 1)$ is a boundary point of $\Omega = B(O, 1) \cup P_0$ but is not a limit point of Ω because it cannot be arbitrarily well approximated by other points in Ω. In fact, see Fig. 1.7, the ball $B((1, 1), \varepsilon)$ contains no points of Ω except $(1, 1)$ if $\varepsilon > 0$ is small enough.

Definition 1.10 (*Isolated point*) The point $P_0 \in \Omega \subset \mathbb{R}^n$ is called an *isolated point* of Ω if there exists a neighborhood of P_0 that contains no other point of Ω.

Fig. 1.7 The point $(1, 1)$ is in the boundary of $\Omega = B(O, 1) \cup P_0$ but is not a limit point of Ω because sufficiently small balls centered at P_0 do not intersect $B(O, 1)$: it is an isolated point

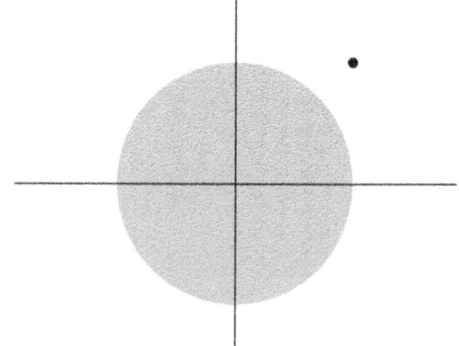

1.3 Topology in Euclidean Space

It should be clear that a boundary point is either a limit point or an isolated point. The latter ones are those that cannot be "reached" by other points in Ω (see Sect. 1.4).

Definition 1.11 (*Dense set*) A subset Ω of $\mathbb{R}^n$ is *dense* in $\mathbb{R}^n$ if $\overline{\Omega} = \mathbb{R}^n$.

Dense sets should be thought of as sets that fill out almost the whole space, in the sense that the smallest closed set that contains a dense set is the whole space $\mathbb{R}^n$. Examples of dense sets in $\mathbb{R}^2$ are $\mathbb{R}^2 \setminus \{P_1, \ldots, P_N\}$ or $\mathbb{R}^2 \setminus \{\text{axes}\}$. A subtler example in the plane is $\mathbb{Q}^2$, the set of points with rational coordinates. The reader is refered to Sect. 1.4 for better understanding.

Definition 1.12 (*Connected set*) Take a nonempty subset Ω of $\mathbb{R}^n$. Suppose that there exist nonempty sets $U_1, U_2 \subset \mathbb{R}^n$ such that:

(i) U_1 and U_2 are both open or both closed;
(ii) $U_1 \cap U_2 = \emptyset$;
(iii) $V_1 = U_1 \cap \Omega \neq \emptyset$ and $V_2 = U_2 \cap \Omega \neq \emptyset$;
(iv) $\Omega = V_1 \cup V_2$.

Then Ω is *disconnected*. If Ω is not disconnected, then it is *connected*.

Informally, a connected set is a set Ω that consists of "one piece". If Ω is not connected, then it is possible to find two different boxes (the two nonempty disjoint open or closed subsets U_1 and U_2 of $\mathbb{R}^n$) each of which contains different pieces of Ω (the piece V_1 and the piece V_2). Clearly, if Ω is the union of two nonempty open disjoint subsets $U_1, U_2 \subset \mathbb{R}^n$, then it is clearly disconnected because $V_i = U_i \cap \Omega = U_i$, for $i = 1, 2$. Similarly, Ω is disconnected if it is the union of two disjoint nonempty closed sets. The sets U_1 and U_2 are said to be the *disconnecting subsets* for A. See Fig. 1.8 for an example.

The connected subsets of $\mathbb{R}$ ($n = 1$) are precisely the intervals.

A quick way to check if a set is connected is to join any two points of the set with a polygonal chain. This is formally given by a finite set of points $\{P_1, \ldots, P_m\}$ which label the vertices of the polygonal chain, which consists of the segments $\overline{P_1 P_2}$, $\overline{P_2 P_3}, \ldots, \overline{P_{m-1} P_m}$.

Definition 1.13 (*Polygonally connected set*) A nonempty subset Ω of $\mathbb{R}^n$ is *polygonally connected* if for any pair of points A and B of Ω there is a polygonal chain entirely contained in Ω joining A and B, that is, points $A = P_1, \ldots, P_m = B \in \Omega$ such that $\overline{P_j P_{j+1}} \subset \Omega$ for $j = 1, \ldots, m-1$.

It is quite clear that a polygonally connected set is necessarily connected. The converse is not true: a circle in the plane, for example, is connected but not polygonally connected because a circle contains no segments. Using the idea that chains of connected sets should produce connected sets, it is not difficult to prove the following result, which is useful in some instances.

Proposition 1.3 *Suppose that each set in the family $\{C_j : j = 0, 1, 2, \ldots\}$ is connected, or polygonally connected, and assume further that $C_{j+1} \cap C_j \neq \emptyset$ for every $j \geq 0$. Then the union*

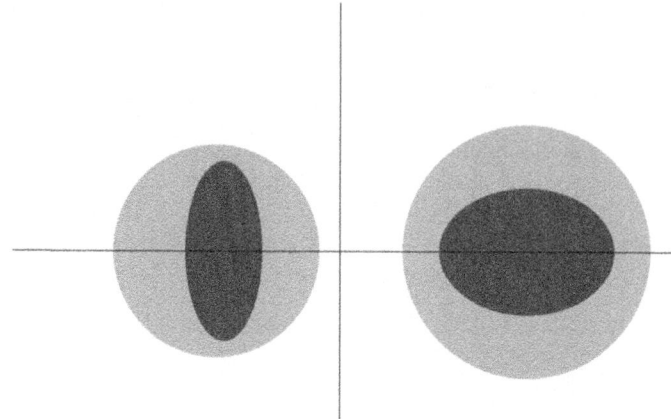

Fig. 1.8 The darker set (union of two elliptic regions) is disconnected: there are two disjoint open balls that cover the set, each containing a portion of it

$$C = \bigcup_{j \geq 0} C_j$$

is also connected, or polygonally connected, respectively.

The rightmost picture in Fig. 1.9 displays the situation considered in Proposition 1.3: four (polygonally) connected sets each intersecting the next one.

Definition 1.14 (*Convex set*) A nonempty subset Ω of $\mathbb{R}^n$ is *convex* if for any pair of points A and B of Ω the segment $\overline{AB}$ is contained in Ω.

Clearly, convex sets are polygonally connected, hence connected. The converse, however, is not true: the polygonally connected set in Fig. 1.9 is not convex because the segment joining the first and the last point of the polygonal chain is not contained in Ω. In Fig. 1.10 are depicted two convex sets: all ponts in a convex set "see eachother".

Definition 1.15 (*Bounded set*) A nonempty subset Ω of $\mathbb{R}^n$ is *bounded* if there exists $r > 0$ such that $\Omega \subset B(O, r)$.

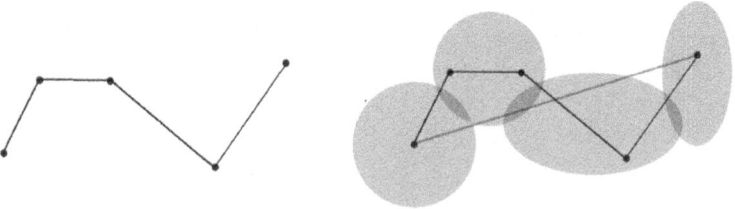

Fig. 1.9 A polygonal chain joining five points (left) and a polygonally connected set (right)

1.4 Sequences

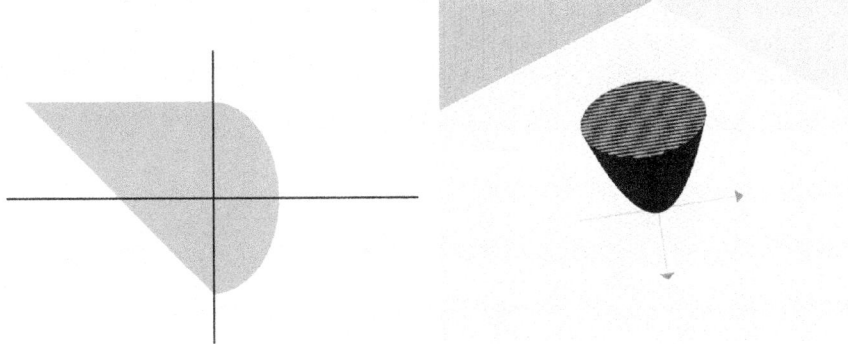

Fig. 1.10 A convex set in the plane (left) and in space (right)

The following famous theorem plays a crucial role in Analysis in Euclidean space.

Theorem 1.1 (Heine-Borel) *For a nonempty subset Ω of $\mathbb{R}^n$ the following are equivalent:*

(i) Ω is closed and bounded;
(ii) for any collection $\{U_i : i \in I\}$ of open sets such that $\Omega \subset \bigcup_{i \in I} U_i$ there exists a finite subcollection $U_{i_1}, \ldots, U_{i_N}$ such that $\Omega \subset (U_{i_1} \cup \cdots \cup U_{i_N})$.

Definition 1.16 (*Compact set*) A nonempty subset Ω of $\mathbb{R}^n$ is *compact* if it satisfies one, hence both, of the conditions in Theorem 1.1.

Property (ii) in Theorem 1.1 is not elementary in nature but allows for a very wide generalization: whenever a sensible notion of open set is available (to wit, in any topological space), it is possible to introduce the notion of compact set by means of this covering property. Indeed, a collection $\{U_i : i \in I\}$ of open sets such that $\Omega \subset \bigcup_{i \in I} U_i$ is called an *open cover* and hence the general definition of compact set goes as follows: "a set K is compact if every open cover of K admits a finite subcover". The power of the Heine-Borel theorem resides in the fact that it gives an easy characterization of compacta in Euclidean space: they are precisely the bounded and closed sets. The reader is invited to exhibit examples of sets that are compact, sets that are closed but not bounded, and sets that are bounded but not closed.

1.4 Sequences

Some of the topological issues discussed in the previous section can be formulated by means of *sequences* of points in $\mathbb{R}^n$. Recall that a sequence of points in the set X is a map from the natural numbers $\mathbb{N}$ into X, traditionally denoted $(P_n)_{n \geq 0}$, whereby P_n stands for the image of $n \in \mathbb{N}$. In the present context, the sequences of interest are sequences in $\mathbb{R}^n$. When $n = 1$ sequences are usually denoted $(a_n)_{n \geq 0}$.

Definition 1.17 (*Convergent sequences*) The sequence $(P_n)_{n\geq 0}$ of points in $\mathbb{R}^n$ is said to *converge* to $P \in \mathbb{R}^n$ if for any $\varepsilon > 0$ there exists $N \in \mathbb{N}$ such that for $n > N$ it holds $d(P_n, P) = \|P - P_n\| < \varepsilon$. In this case it is customary to write

$$P_n \to P, \quad \lim_n P_n = P, \quad \lim P_n = P$$

and to say that the limit of $(P_n)_{n \geq 0}$ is P.

The sequence $(P_n)_{n\geq 0}$ converges to P if and only if it converges componentwise. This means, for example, that $P_n = (x_n, y_n) \in \mathbb{R}^2$ converges to $P = (x, y) \in \mathbb{R}^2$ if and only if $x_n \to x$ and $y_n \to y$, and analogously in $\mathbb{R}^n$.

It should be clear from the definition that the notion of convergent sequence can be given in any metric space, that is, whenever a sensible concept of distance is available. In particular, whenever a norm is defined. In what follows, only Euclidean distance is considered, but occasional reference to other setups will be made.

Proposition 1.4 *Take Ω be a nonempty subset of $\mathbb{R}^n$. Then:*
 (i) *P_0 is in the closure $\overline{\Omega}$ if and only if there exists a sequence $(P_n)_{n\geq 0}$ of points in Ω such that $P_n \to P_0$;*
 (ii) *P_0 is in the boundary $\partial\Omega$ if and only if there exist a sequence $(P_n)_{n\geq 0}$ of points in Ω and a sequence $(Q_n)_{n \geq 0}$ of points in the complement $\mathbb{R}^n \setminus \Omega$ such that $\lim P_n = \lim Q_n = P$;*
 (iii) *P_0 is a limit point of Ω if and only if there exists a sequence $(P_n)_{n\geq 0}$ of points in $\Omega \setminus \{P_0\}$ such that $P_n \to P_0$.*

By item (i) in Proposition 1.4 it follows that a set Ω is dense in $\mathbb{R}^n$ if every point $P \in \mathbb{R}^n$ is the limit of a sequence of elements in Ω. Hence $\mathbb{Q}^n$ is dense in $\mathbb{R}^n$. By items (i) and (iii), any isolated point is in $\overline{\Omega}$ but is not a limit point (prove this fact).

Recall that a *subsequence* of the sequence $(P_n)_{n\geq 0}$, usually denoted $(P_{n_k})_{k \geq 0}$, is a sequence obtained from $(P_n)_{n\geq 0}$ by operating a strictly increasing selection $k \mapsto n_k$, that is, by pre-composing the sequence map with a strictly increasing map $\mathbb{N} \to \mathbb{N}$. If $P_n \to P$, then $\lim_k P_{n_k} \to P$ for every subsequence $(P_{n_k})_{k \geq 0}$. There are, however, sequences that do not converge but have convergent subsequences. For example, when $n = 1$ the sequence $a_n = (-1)^n$ does not converge but if $n_k = 2k$, then clearly $a_{n_k} = (-1)^{2k} = 1$ converges.

Theorem 1.2 (*Bolzano-Weierstrass*) *Every bounded sequence in $\mathbb{R}^n$ has a convergent subsequence.*

It is worth observing that if $(P_n)_{n \geq 0}$ is a sequence of points in a compact set Ω, then it is bounded and hence it admits a convergent subsequence $(P_{n_k})_{k \geq 0}$, whose limit P is in the closure $\overline{\Omega} = \Omega$. Thus every sequence of points in a compact set has a subsequence that converges to some point in the set.

Definition 1.18 (*Cauchy sequences*) The sequence $(P_n)_{n \geq 0}$ of points in $\mathbb{R}^n$ is a *Cauchy sequence* if for every $\varepsilon > 0$ there exists $N \in \mathbb{N}$ such that if $n, m > N$, then $d(P_n, P_m) = \|P_n - P_m\| < \varepsilon$.

Again, Cauchy sequences can be defined in any metric space. It is immediate to check that any convergent sequence is Cauchy. The converse statement is true in $\mathbb{R}^n$, but not for every metric space. A paramount example is $\mathbb{Q}$, where many Cauchy sequences do not converge (to rational numbers).

Definition 1.19 (*Completeness*) A nonempty subset Ω of $\mathbb{R}^n$ is said to be complete if every Cauchy sequence of points in Ω converges to a point in Ω.

Observe that every complete subset in $\mathbb{R}^n$ is necessarily closed.

Theorem 1.3 (Compactness and completeness) *Every compact subset in $\mathbb{R}^n$ is complete.*

The converse of Theorem 1.3 is false: the unbounded interval $[0, +\infty)$ is complete.

1.5 Functions and Graphical Interpretations

The main object of study of this book is the analysis of functions between Euclidean spaces, namely maps of the form

$$F \colon A \subset \mathbb{R}^n \to \mathbb{R}^m \tag{1.17}$$

where in most cases $n = 2, 3$ and $m \geq 1$. If $m = 1$ then one speaks of *functions of several variables* or, borrowing ideas and notation from Physics, *scalar fields*. When $m > 1$, then F is called a *vector valued* map. If $n = m$ the map is also called *vector field*, that is $F \colon A \subset \mathbb{R}^n \to \mathbb{R}^n$, often with some additional regularity assumptions. Clearly, a vector valued map is just a collection of m scalar functions defined in one and the same subset A of $\mathbb{R}^n$. Thus, the standard way of denoting a vector valued map as in (1.17) is

$$F(x_1, \ldots, x_n) = (f_1(x_1, \ldots, x_n), \ldots, f_m(x_1, \ldots, x_n)),$$

but shorter and more geometric notation such as

$$F(P) = (f_1(P), \ldots, f_m(P))$$

is also used. It is quite common to denote scalar fields by lower case letters like f, g, h, and vector valued maps by capital letters like F, G, H.

Observe that, upon identifying the complex field $\mathbb{C}$ with the Euclidean plane $\mathbb{R}^2$, a complex valued function $f \colon \Omega \subset \mathbb{C} \to \mathbb{C}$ can be thought of as a two dimensional real vector field. Explicitly, if one writes both the complex variable z and the complex image $f(z)$ in terms of real and imaginary parts, that is

$$z = x + iy, \quad f(z) = (\Re f)(z) + i(\Im f)(z)$$

and hence puts
$$(\Re f)(z) = u(x, y), \qquad (\Im f)(z) = v(x, y),$$
then the equality
$$f(x+iy) = u(x, y) + iv(x, y)$$
leads to the natural identification
$$f \longleftrightarrow F = (u, v)$$
between complex valued functions of a complex variable and two-dimensional vector fields.

Drawing pictures that represent functions of several variables or vector valued functions is not always an easy task. Some standard examples are illustrated in the next section.

1.5.1 Functions of Two Variables

The *graph* of a funcion $f : A \subset \mathbb{R}^2 \to \mathbb{R}$ is of course
$$\Gamma(f) = \{(x, y, z) \in \mathbb{R}^3 : z = f(x, y)\} \subset \mathbb{R}^3$$
and as such it represents some kind of surface in three dimensional space. If the function is not too complicated, it is actually possible to attempt a picture (see Fig. 1.11). Computers are of great help in this endeavour.

A good alternative idea comes from cartography, and consists in drawing the *level sets* of a function of two variables. These are the sets
$$L_\lambda = \{(x, y) \in \mathbb{R}^2 : f(x, y) = \lambda\}$$
as λ ranges in $\mathbb{R}$. Clearly, the level set L_λ will be empty whenever λ is not in the image of f. If $\lambda \in f(A)$, then L_λ is some region in the plane, in many cases a nice curve, possibly a bunch of points, or a region, or a combintaion of all these. Real life maps are very often drawn precisely in this way, that is, by indicating the curves along which a terrain maintains the same height (see the third picture in Fig. 1.12). This is achieved by slicing the profile with planes at constant height.

Sometimes graphical rendering is very effective and intuitively clear such as that in Fig. 1.11, other times much harder to produce and not immediate to read, such as that the second picture of Fig. 1.12.

1.5 Functions and Graphical Interpretations

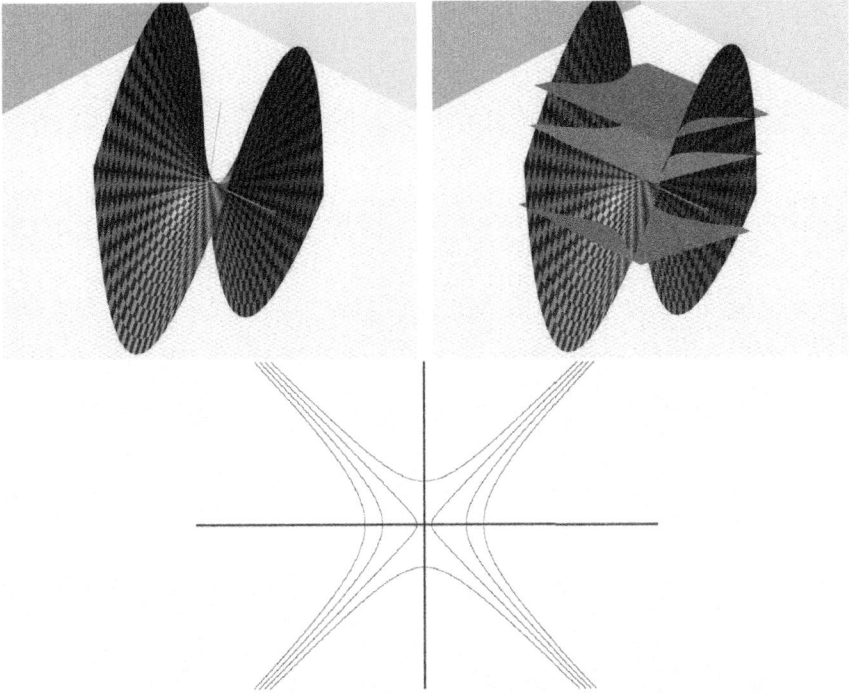

Fig. 1.11 The graph of the function $f(x, y) = x^2 - y^2$ (above left), its intersection with three horizontal planes (above right), and the outcoming set of level lines (below)

1.5.2 Vector Fields, Complex Functions

A useful graphical rendering of vector fields in the plane is obtained by selecting a grid of parallel lines and then drawing its image. In Fig. 1.13 this technique is illustrated for the vector field

$$F(x, y) = (x^2 - y^2, 2xy),$$

which corresponds to the complex valued map $f(z) = z^2$, because

$$z^2 = (x + iy)^2 = (x^2 - y^2) + i(2xy) \quad \longleftrightarrow \quad F(x, y) = (x^2 - y^2, 2xy)$$

Notice incidentally that the parabolae on the right meet at right angles, just like the straight lines on the left. It is an instructive exercise to write down the equation of the image of a horizontal or vertical line.

A different approach for representing vector fields in the plane, particularly useful when dealing with differential equations, is to draw at each point $P \in \mathbb{R}^2$ the vector $F(P)$, which is again an element in $\mathbb{R}^2$, as an arrow issuing from P as if P were the

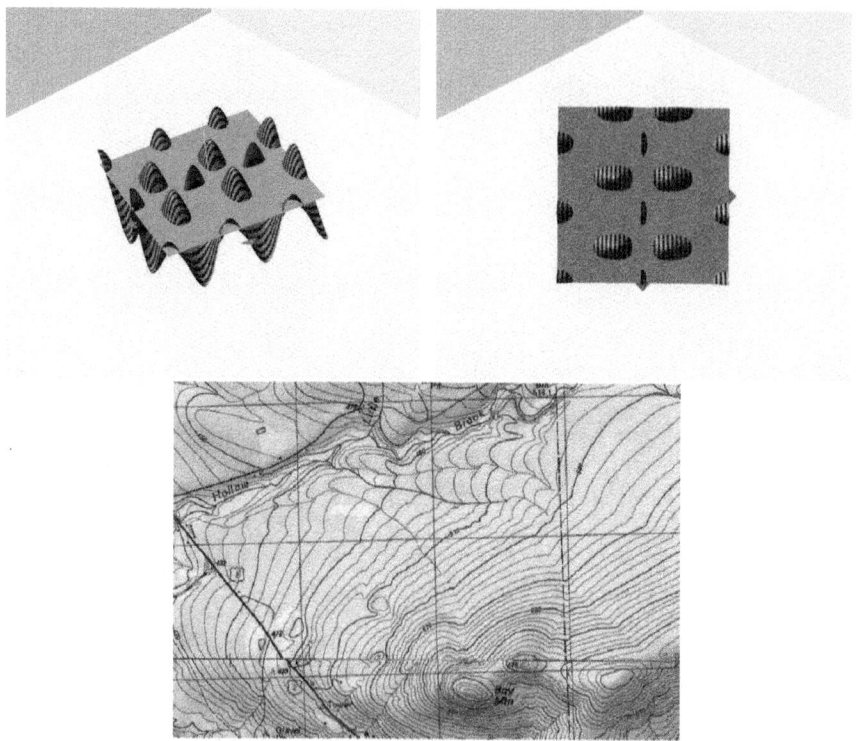

Fig. 1.12 Two versions of the graph of $f(x, y) = 4 \sin x \cos(2\sqrt{|y|})$, indicating what the level sets might look like; real life level curves on the picture below

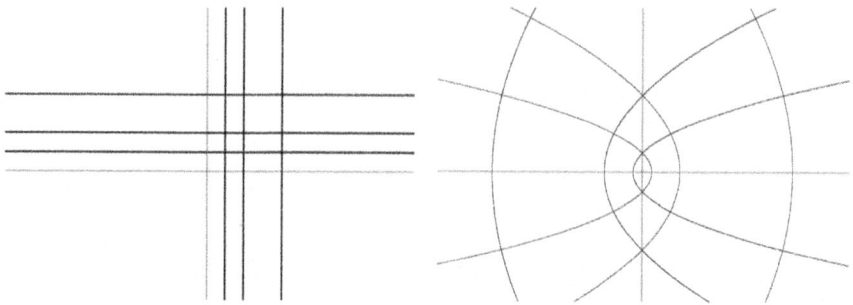

Fig. 1.13 Straight lines parallel to the axes (left) are sent into parabolae (right) by the vector field $F(x, y) = (x^2 - y^2, 2xy)$, which corresponds to the complex valued map $f(z) = z^2$

origin of a Cartesian plane. In Fig. 1.14 there is an example of a vector field defined outside a circular region. The picture should clarify the reason of the word "field", because in this representation one is led to think about a grass field, each single grass blade representing the vector assigned at that point.

1.6 Coordinates

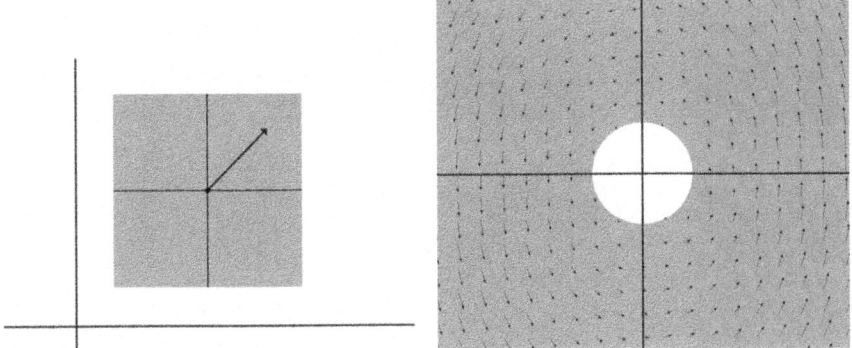

Fig. 1.14 Vector fields as fields of arrows; on the left the representation of $F(P)$ as an arrow issuing from P pointing at the point $F(P)$ in the plane with origin in P, on the right a vector field

1.6 Coordinates

Loosely speaking, a *coordinate system* in $\mathbb{R}^n$, or in an open subset Ω thereof, is a way of assigning n numbers to each point $P \in \Omega$ that uniquely identifies it. More formally, it is an injective mapping $F : \Omega \subset \mathbb{R}^n \to \mathbb{R}^n$ where Ω is an open subset of $\mathbb{R}^n$. The most obvious coordinates are the Cartesian coordinates, given by the identity map that assigns to each $P = (x_1, \ldots, x_n) \in \Omega = \mathbb{R}^n$ the vector $(x_1, \ldots, x_n)$ in $\mathbb{R}^n$. Below is a list of the most commonly used coordinate systems.

1.6.1 Polar Coordinates

The geometric idea behind *polar coordinates* is very simple: given the point P in the punctured plane $\mathbb{R}^2 \setminus \{O\}$ just draw the circle centered at the origin and passing through P. Then, identify P by means of the radius ρ of that circle and by the angle θ determined by moving counterclockwise from the positive x-ray to the segment $\overline{OP}$ (see Fig. 1.15).

Formally, the polar coordinates in the plane are given by the map

$$F: [0, 2\pi) \times (0, +\infty) \to \mathbb{R}^2, \qquad (\theta, \rho) \mapsto (\rho \cos \theta, \rho \sin \theta), \tag{1.18}$$

a bijection of the strip $[0, 2\pi) \times (0, +\infty)$ onto the punctured plane $\mathbb{R}^2 \setminus \{(0, 0)\}$, as shown in Fig. 1.16. It is also customary to indicate the map by writing

$$\begin{cases} x = \rho \cos \theta \\ y = \rho \sin \theta. \end{cases}$$

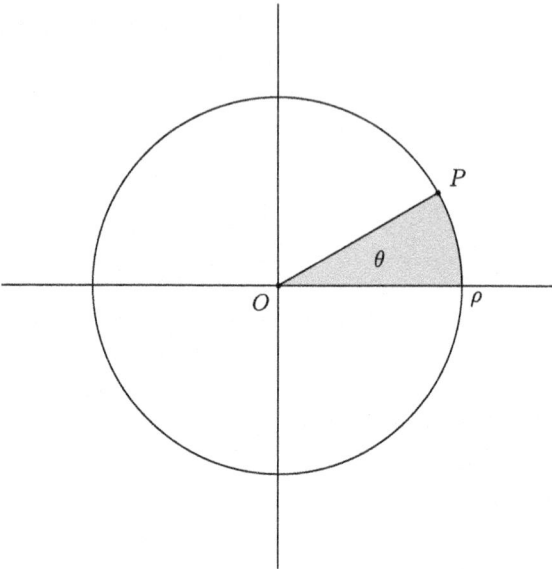

Fig. 1.15 A point in the punctured plane is uniquely determined by the radius of the circle through the origin on which it lies and the angle between the ray through the point and the horizontal one

Fig. 1.16 Polar coordinates: two points with the same polar angle $\pi/4$ and different radii

Clearly, the polar coordinate map identifies uniquely any point in the plane except the origin. The coordinate θ is called the polar *angle* whereas ρ is the polar *radius*.

The inversion of F is as follows: first of all $\rho(x, y) = \sqrt{x^2 + y^2}$ and then $\theta(x, y)$ is simply the unique angle in $[0, 2\pi)$ that identifies the point corresponding to (x, y) in the circle of radius $\rho(x, y)$. In formulæ:

1.6 Coordinates

$$\theta(x, y) = \begin{cases} \arccos\left(\dfrac{x}{\sqrt{x^2+y^2}}\right) & y \geq 0 \\ 2\pi - \arccos\left(\dfrac{x}{\sqrt{x^2+y^2}}\right) & y < 0 \end{cases}$$

$$\rho(x, y) = \sqrt{x^2 + y^2}.$$

Polar coordinates can of course be defined with respect to any *pole* $P_0 = (x_0, y_0)$ other than the origin. This is done by a simple translation, that is

$$\begin{cases} x = x_0 + \rho \cos\theta \\ y = y_0 + \rho \sin\theta. \end{cases} \tag{1.19}$$

These coordinates may be useful, for example, when computing limits as $P \to P_0$ (see Chap. 2).

1.6.2 Spherical Coordinates

The *spherical coordinates* are the three-dimensional analogue of polar coordinates: given the point $P \in \mathbb{R}^3 \setminus \{O\}$ just take the sphere centered at the origin passing through P and identify P by means of the radius ρ of that sphere and by the pair $(\theta, \varphi) \in [0, \pi] \times [0, 2\pi)$ representing the altitude and the azimuth of P on the sphere, respectively. Thus, unravelling the language of Astronomy, θ is the angle between the z-axis and the segment $\overline{OP}$, and φ is the value of the angular variable in polar coordinates of the orthogonal projection Q of the point P onto the xy-plane. Borrowing from Geography this time, $\theta \in [0, \pi]$ represents the colatitude, whereas $\varphi \in [0, 2\pi)$ is the longitude. Thus, colatitude is zero at the north pole and 90^0, that is $\pi/2$, at the equator, whereas longititude is zero on the meridian plane "$y = 0$" and $\pi/2$ on the meridian plane "$x = 0$". The geometry is illustrated in Fig. 1.17.

The spherical coordinate map is $F: [0, \pi] \times [0, 2\pi) \times (0, +\infty) \to \mathbb{R}^3$, given by

$$(\theta, \varphi, \rho) \mapsto (\rho \sin\theta \cos\varphi, \rho \sin\theta \sin\varphi, \rho \cos\theta),$$

a bijection of the cylinder $(0, \pi) \times [0, 2\pi) \times (0, +\infty)$ onto the slit space $\mathbb{R}^3 \setminus \{z\text{-axis}\}$. It is also customary to indicate the map by writing

$$\begin{cases} x = \rho \sin\theta \cos\varphi \\ y = \rho \sin\theta \sin\varphi \\ z = \rho \cos\theta. \end{cases}$$

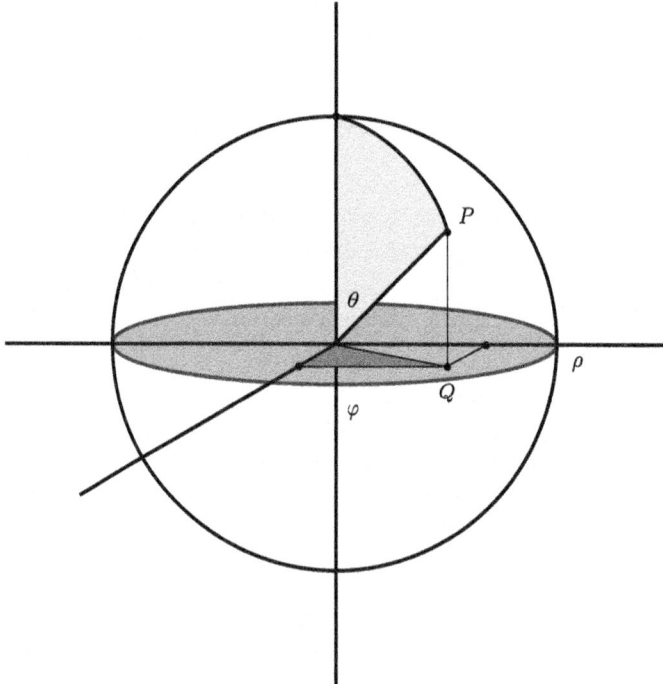

Fig. 1.17 A point in the punctured space is uniquely determined by the radius ρ of the sphere through the origin on which it lies and its "geographical" coordinates (θ, φ) on that sphere

Spherical coordinates are inverted by

$$\theta = \arccos \frac{z}{\sqrt{x^2 + y^2 + z^2}}$$

$$\varphi = \begin{cases} \arccos\left(\dfrac{x}{\sqrt{x^2 + y^2}}\right) & y \geq 0 \\ 2\pi - \arccos\left(\dfrac{x}{\sqrt{x^2 + y^2}}\right) & y < 0 \end{cases}$$

$$\rho = \sqrt{x^2 + y^2 + z^2}.$$

Just as for polar coordinates in the plane, spherical coordinates can of course be defined with respect to any pole $P_0 = (x_0, y_0, z_0)$ other than the origin. This is done by a simple translation, that is

$$\begin{cases} x = x_0 + \rho \sin\theta \cos\varphi \\ y = y_0 + \rho \sin\theta \sin\varphi \\ z = z_0 + \rho \cos\theta. \end{cases} \tag{1.20}$$

They may be useful when computing limits as $P \to P_0$ (see Chap. 2).

1.6.3 Cylindrical Coordinates

These are coordinates in $\mathbb{R}^3 \setminus \{z-\text{axis}\}$ obtained by combining Cartesian and polar coordinates. Take a point $P \in \mathbb{R}^3 \setminus \{z-\text{axis}\}$ and describe it by its z-Cartesian coordinate and by the polar coordinates of its projection onto the xy-plane. In formulæ

$$F: [0, 2\pi) \times (0, +\infty) \times \mathbb{R} \to \mathbb{R}^3, \qquad (\theta, \rho, z) \mapsto (\rho \cos\theta, \rho \sin\theta, z).$$

Geometrically, one takes the unique cylinder centered around the z-axis passing through P and then identifies P by the z-coordinate of the unique horizontal circle of the cylinder to which P belongs, and then by its coordinates on the circle, given as before by polar radius and polar angle. The geometry is illustrated in Fig. 1.18.

Cylindrical coordinates can be defined with respect to any pole

$$\begin{cases} x = x_0 + \rho \cos\theta \\ y = y_0 + \rho \sin\theta \\ z = z_0 + z. \end{cases}$$

They may be useful when computing limits as $P \to P_0$ (see Chap. 2).

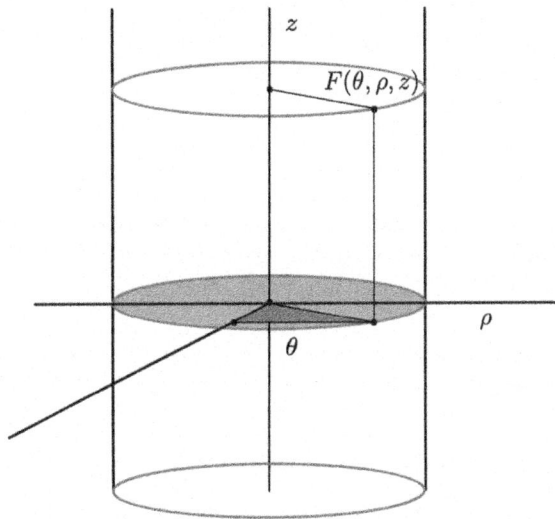

Fig. 1.18 A point in the slit space (i.e. $\mathbb{R}^3$ without the z-axis) is uniquely determined by the distance ρ from the z-axis, its longitudinal polar angle coordinates and its height

1.7 Guided Exercises

1.1 Describe the domains of the following functions:

(a) $f_1(x, y) = \log(xy)$

(b) $f_2(x, y) = \dfrac{1}{\sqrt{1-x^2}\sqrt{1-y^2}}$

(c) $f_3(x, y) = \log(x^2 - y - 1)$

(d) $f_4(x, y) = \sqrt{\dfrac{y^2-1}{1-x^2}}\log(1+xy)$

(e) $f_5(x, y) = \sqrt{\dfrac{x}{e^x - y}}$.

Answer. (a) In order for xy to be positive, x and y must have the same sign. Hence $\mathrm{Dom}(f_1) = \{(x, y) \in \mathbb{R}^2 : x > 0, y > 0\} \cup \{(x, y) \in \mathbb{R}^2 : x < 0, y < 0\}$, see Fig. 1.19, top left.
(b) It must be that both $1 - x^2 > 0$, namely $-1 < x < 1$, and $1 - y^2 > 0$, namely $-1 < y < 1$. Thus $\mathrm{Dom}(f_2) = (-1, 1) \times (-1, 1)$, the open unit square, see Fig. 1.19, top right.
(c) Evidently, it must be $x^2 - y - 1 > 0$, that is $\mathrm{Dom}(f_3) = \{(x, y) \in \mathbb{R}^2 : y < x^2 - 1\}$, see Fig. 1.19, second row, left.
(d) Clearly $(x, y) \in \mathrm{Dom}(f_4)$ if and only if

$$\begin{cases} 1 + xy > 0 \\ \dfrac{y^2 - 1}{1 - x^2} \geq 0. \end{cases}$$

The first equation is the unbounded open region Ω of the plane that contains the axes and has as boundary the two branches of the hyperbola $1 + xy = 0$, which are not part of Ω. The second equation is equivalent to the systems

$$\begin{cases} y^2 - 1 \geq 0 \\ 1 - x^2 > 0 \end{cases} \cup \begin{cases} y^2 - 1 \leq 0 \\ 1 - x^2 < 0 \end{cases} \iff \begin{cases} |y| \geq 1 \\ |x| < 1 \end{cases} \cup \begin{cases} |y| \leq 1 \\ |x| > 1. \end{cases}$$

The points satisfying the conditions of the first system form two "vertical" strips centered around the y-axis, with $x \in (-1, 1)$ and $y \geq 1$ or $y \leq -1$, respectively. Similarly, the points satisfying the conditions of the second system form two "horizontal" strips centered around the x-axis, with $y \in [-1, 1]$ and $x > 1$ or $x < -1$, respectively. Thus $\mathrm{Dom}(f_4)$ is obtained by intersecting Ω with the union of the four strips. Notice that the two sets $\{(x, 1) \in \mathbb{R}^2 : x > -1, x \neq 1\}$ and $\{(x, -1) \in \mathbb{R}^2 : x < 1, x \neq -1\}$ belong to $\mathrm{Dom}(f_4)$ whereas the points along the hyperbola $1 + xy = 0$ do not. See Fig. 1.19, second row, right. The two half lines have been depicted in black together with the portions of the axes that lie in $\mathrm{Dom}(f_4)$.

1.7 Guided Exercises

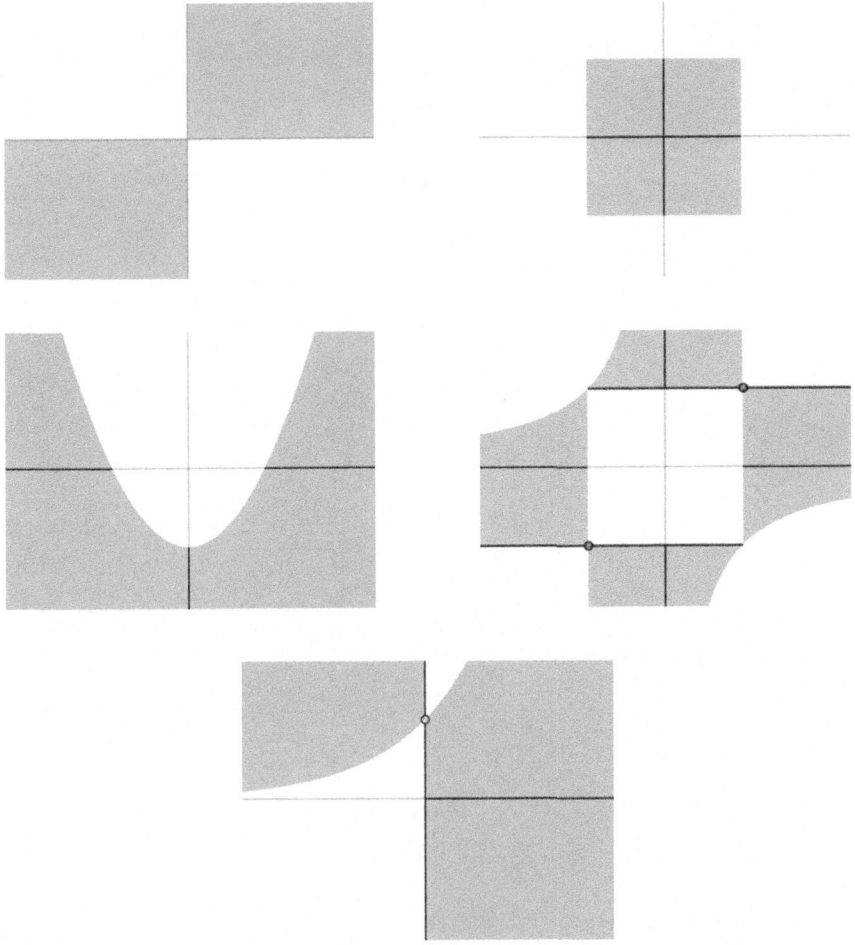

Fig. 1.19 The domains of the functions in Exercise 1.1 are in grey; significant segments or lines that are part of the domains are in black

(e) The points $(x, y) \in \text{Dom}(f_5)$ must satisfy the conditions

$$\begin{cases} x \geq 0 \\ e^x - y > 0 \end{cases} \cup \begin{cases} x \leq 0 \\ e^x - y < 0. \end{cases}$$

The first system singles out the region below the graph of $x \mapsto e^x$ for $x \geq 0$, the second the region above the graph of $x \mapsto e^x$ for $x \leq 0$. Thus, the y-axis belongs to $\text{Dom}(f_5)$ except the point $(0, 1)$, whereas only the right half of the x-axis does. See Fig. 1.19, third row.

This exercise is meant to show in very simple cases that describing the domain of a function of two variables requires a little geometric thinking, translating equations and inequalities into geometric facts. In particular, it requires having a clear understanding of how systems of equations amount to finding intersections of subsets (of the plane).

1.2 Describe the indicated level sets.

(a) The level set of $f_1(x, y) = \dfrac{1+x}{\sqrt{y+1}}$ passing through $(0, 1)$;

(b) all level sets $f_2(x, y) = \lambda$ as λ varies in $\mathbb{R}$, where $f_2(x, y) = \dfrac{\log(xy)}{x^2 - 1}$;

(c) all level sets of $f_3(x, y) = \begin{cases} \dfrac{x+y}{x-y} & x \neq y \\ 0 & x = y; \end{cases}$

(d) all level sets of $f_4(x, y) = \begin{cases} xy + \dfrac{x}{y} & y \neq 0 \\ 0 & y = 0. \end{cases}$

Answer. (a) The domain of f_1 is $\text{Dom}(f_1) = \{(x, y) \in \mathbb{R}^2 : y > -1\}$. Furthermore, $f(0, 1) = 1/\sqrt{2}$. Hence the level set is

$$L_{1/\sqrt{2}} = \{(x, y) \in \mathbb{R}^2 : y > -1,\ f_1(x, y) = 1/\sqrt{2}\}.$$

Now, since $x > -1$

$$f_1(x, y) = \frac{1}{\sqrt{2}} \iff 2(1 + x) = \sqrt{2(y + 1)}$$
$$\iff 4(x^2 + 2x + 1) = 2y + 2$$
$$\iff y = 2x^2 + 4x + 1.$$

Therefore $L_{1/\sqrt{2}}$ is the branch of the parabola depicted in Fig. 1.20, top left. The point $(-1, -1)$ is not part of $L_{1/\sqrt{2}}$.

(b) The domain of f_2 is $\text{Dom}(f_2) = \{(x, y) \in \mathbb{R}^2 : xy > 0,\ x^2 - 1 \neq 0\}$, depicted in grey in Fig. 1.20, top right. The two half-lines do not belong to the domain, which is contained in the first and third open quadrants. Fix now $\lambda \in \mathbb{R}$. Then

$$f_2(x, y) = \lambda \iff \log(xy) = \lambda(x^2 - 1)$$
$$\iff xy = e^{\lambda(x^2-1)}$$
$$\iff y = \frac{1}{x} e^{\lambda(x^2-1)}.$$

The function $x \mapsto e^{\lambda(x^2-1)}/x$ is odd. Hence it is enough to study it for $x > 0, x \neq 1$. Suppose first $\lambda > 0$. Evidently

$$\lim_{x \to 0^+} \frac{1}{x} e^{\lambda(x^2-1)} = +\infty, \quad \lim_{x \to +\infty} \frac{1}{x} e^{\lambda(x^2-1)} = +\infty$$

1.7 Guided Exercises

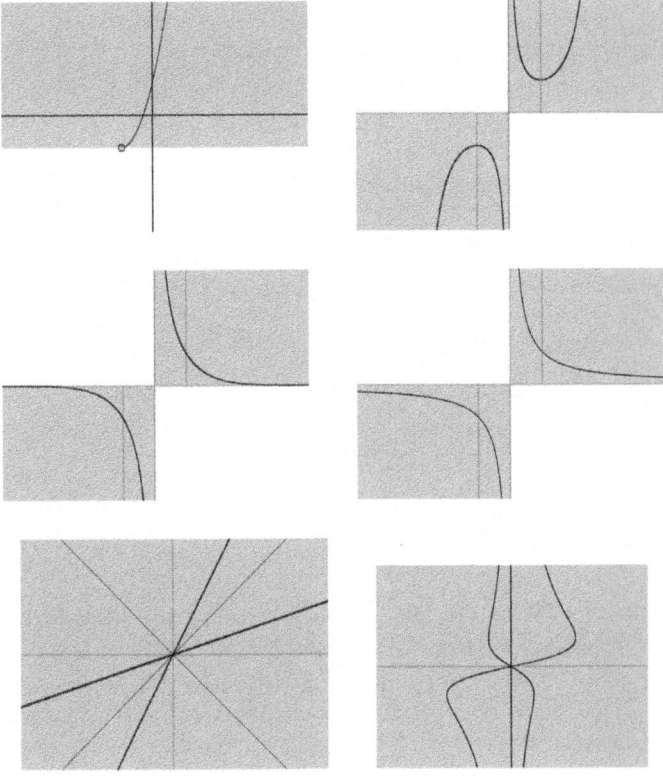

Fig. 1.20 Level sets of the functions in Exercise 1.2 in black; in grey their domains

and
$$\frac{d}{dx}\left(\frac{1}{x}e^{\lambda(x^2-1)}\right) = \left[-\frac{1}{x^2} + 2\lambda\right]e^{\lambda(x^2-1)} = \frac{2\lambda x^2 - 1}{x^2}e^{\lambda(x^2-1)},$$

which has the sign of the numerator. Therefore f_2 is decreasing in $(0, 1/(2\lambda)^{1/2})$ and increasing in $(1/(2\lambda)^{1/2}, +\infty)$. For fixed $\lambda > 0$, the level set is the graph of $x \mapsto e^{\lambda(x^2-1)}/x$, that has just been analyzed. In Fig. 1.20 top right it is shown the case $\lambda = 1/2$. Suppose next $\lambda < 0$. In this case

$$\lim_{x \to 0^+} \frac{1}{x}e^{\lambda(x^2-1)} = +\infty, \qquad \lim_{x \to +\infty} \frac{1}{x}e^{\lambda(x^2-1)} = 0$$

and the derivative is negative, so that f_2 is decreasing both in $(0, 1)$ and in $(1, +\infty)$. In Fig. 1.20, second row, left, it is shown the case $\lambda = -0.3$. Finally, if $\lambda = 0$ then the level set is just the hyperbola $xy = 1$ depicted in Fig. 1.20 second row, right.
(c) By definition, Dom(f_3) = $\mathbb{R}^2$. Consider the level sets L_λ where $f_3(x, y) = \lambda$. Suppose first $\lambda = 0$. On both lines $y = \pm x$ the function vanishes, so that the union of

them is L_0. Suppose next $\lambda \neq 0$. Then $f_3(x, y) = \lambda$ if and only if $x + y = \lambda(x - y)$. If $\lambda \neq -1$ then the equation $f_3(x, y) = \lambda$ is equivalent to

$$y = \left(\frac{\lambda - 1}{\lambda + 1}\right) x$$

with $(x, y) \neq (0, 0)$. The real number $(\lambda - 1)/(\lambda + 1)$ takes on all possible values except 1. If $\lambda = -1$ the equation $f_3(x, y) = \lambda$ is equivalent to $x = 0$ except the origin. Summing up, the level sets are all possible straight lines through the origin (without the origin itself) except the lines $y = \pm x$ which together form a single level set.

(d) Again, by definition, $\text{Dom}(f_4) = \mathbb{R}^2$. Clearly, the x-axis is contained in the level set L_0. Furthermore, for fixed $\lambda \in \mathbb{R}$ and $y \neq 0$ one has

$$f(x, y) = \lambda \iff xy + \frac{x}{y} = \lambda$$
$$\iff xy^2 + x = \lambda y$$
$$\iff x = \lambda \frac{y}{1 + y^2}.$$

Therefore, also the y-axis is contained in the level set L_0, which is actually the union of the two coordinate axes. When $\lambda \neq 0$ the corresponding level set is given by the graph of the odd function $x = \lambda y/(1 + y^2)$ defined for $y \neq 0$. Two such sets are drawn in Fig. 1.20, third row, right. In particular, for $\lambda > 0$ it is the curve in the first and third quadrants, and for $\lambda < 0$ it lies in the second and fourth quadrants.

This exercise combines some work along the same lines as that of the Exercise 1.1, that amounts to finding domains, together with more refined analysis on solving equations that single out one-dimensional curves. Indeed, under favourable circumstances, an equation like $f(x, y) = \lambda$ may be solved explicitly, in the sense that it is easily shown to be equivalent to another equation of the form $y = \varphi(x)$ or $x = \psi(y)$ whereby φ or ψ are explicit. This is by no means always the case, as will be discussed later.

1.3 Show that $A = \{(x, y) \in \mathbb{R}^2 : x^2 + y^2 \leq 2x, \ y \leq x\}$ is closed and bounded.

Answer. Observe that

$$x^2 + y^2 \leq 2x \iff (x - 1)^2 + y^2 \leq 1$$

so that $\{(x, y) \in \mathbb{R}^2 : x^2 + y^2 \leq 2x\} = \overline{B((1, 0), 1)}$. It follows that

$$A \subset \{(x, y) \in \mathbb{R}^2 : x^2 + y^2 \leq 2x\} \subset B((1, 0), r)$$

for every $r > 1$, and hence A is bounded. Next, observe that $A = A_1 \cap A_2$, where

$$A_1 = \{(x, y) \in \mathbb{R}^2 : x^2 + y^2 \leq 2x\} = \overline{B((1, 0), 1)}$$
$$A_2 = \{(x, y) \in \mathbb{R}^2 : y \leq x\}.$$

1.7 Guided Exercises

Showing that A is closed is equivalent to showing that its complement $\mathbb{R}^2 \setminus A = A^c$ is open. Now, $A^c = A_1^c \cup A_2^c$ and since A_1^c is obviously open because A_1 is closed, the matter reduces to proving that A_2^c is open. Take then $P_0 = (x_0, y_0) \in A_2^c$. This entails that $y_0 > x_0$. Set $\delta = (y_0 - x_0)/2$. If $P = (x, y) \in B(P_0, \delta)$, then

$$x - x_0 \leq |x - x_0| = \sqrt{(x - x_0)^2} \leq \sqrt{(x - x_0)^2 + (y - y_0)^2} < \delta$$
$$y_0 - y \leq |y - y_0| = \sqrt{(y - y_0)^2} \leq \sqrt{(x - x_0)^2 + (y - y_0)^2} < \delta.$$

Therefore

$$x < \delta + x_0 = \frac{y_0 - x_0}{2} + x_0 = \frac{y_0 + x_0}{2} = y_0 - \frac{y_0 - x_0}{2} = y_0 - \delta < y.$$

This shows that $x < y$, so that $(x, y) \in A_2^c$. Therefore if $P_0 = (x_0, y_0) \in A_2^c$ then the whole open ball $B(P_0, \delta)$ is contained in A_2^c, which is therefore open, as desired.

Here it is asked to show in full detail that the intersection of a closed ball with a closed half-plane is closed and bounded, that is, compact. While boundedness is obvious, one possible way to prove that the intersection of these two sets is closed is to show that both their complements in the plane are open. The details are depicted in Fig. 1.21.

1.4 Find the boundary of $A = \{(x, y) \in \mathbb{R}^2 : y \leq |x|\}$.

Answer. It is quite intuitive that $\partial A = \{(x, y) \in \mathbb{R}^2 : y = |x|\}$. To show this, it is enough to take $P_0 = (x_0, y_0)$ with $y_0 \neq |x_0|$ and show that $P_0 \notin \partial A$, because the inclusion $\partial A \supset \{(x, y) \in \mathbb{R}^2 : y = |x|\}$ is clear. Suppose for instance that $y_0 < |x_0|$. As in Exercise 1.3, it is possible to find $\delta > 0$ such that

$$B(P_0, \delta) \subset \{(x, y) \in \mathbb{R}^2 : y < |x|\} \subset A,$$

so that $B(P_0, \delta) \cap A^c = \emptyset$. This contradicts the fact that any ball centered at a boundary point must intersect non trivially both A and its complement A^c, as observed in

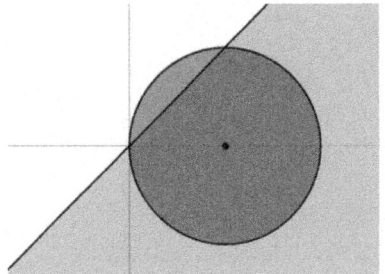
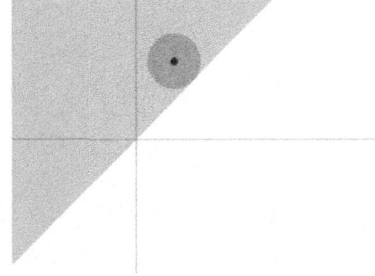

Fig. 1.21 The set A, left, as the intersection of a closed ball and a closed half-plane; the set A_2^c, right, is open: any point in A_2^c is surrounded by an open ball contained in A_2^c

the remarks following Definition 1.8. The argument when $y_0 > |x_0|$ is analogous. Therefore $P_0 \notin \partial A$.

This elementary exercise indicates that the boundary of the epigraph (or the ipograph) of a continuous function is just the graph itself. The direct technique displayed here is quite easy, and is illustrated in Fig. 1.22.

1.5 Using either of the formulae

$$\partial \Omega = \overline{\Omega} \cap \overline{\Omega^c} = \overline{\Omega} \setminus \mathring{\Omega}, \tag{1.21}$$

find the boundary of the following subsets of $\mathbb{R}$

(a) $(0, 1)$

(b) $\mathbb{Z}$

(c) $\left\{ \dfrac{1}{n} : n = 1, 2, 3, \ldots \right\}$

(d) $\mathbb{Q} \cap [0, 1]$

Answer. (a) Clearly $\overline{(0, 1)} = [0, 1]$. This may be seen in several ways, one is to observe first that $0 = \lim_n (1/n)$ and $1 = \lim_n (1 - 1/n)$ so that by Proposition 1.4 both 0 and 1 are in the closure of $(0, 1)$. This implies $[0, 1] \subset \overline{(0, 1)}$. But $[0, 1]$ is closed because its complement $(-\infty, 1) \cup (1, +\infty)$ is manifestly open. Therefore $[0, 1]$ is a closed set that lies between $(0, 1)$ and its closure, so it is its closure. Finally, $(0, 1)$ is open and hence the second formula in (1.21) entails that $\partial (0, 1) = [0, 1] \setminus (0, 1) = \{0, 1\}$.

(b) Being the union of open sets

$$\mathbb{Z}^c = \bigcup_{n \in \mathbb{Z}} (n, n+1)$$

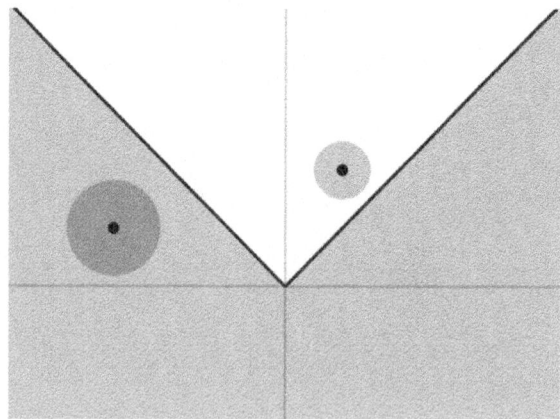

Fig. 1.22 The set ∂A is the graph of $x \mapsto |x|$. Any point in the region below or above the graph is surrounded by a ball that does not intersect the other region, hence is not on the boundary

1.7 Guided Exercises   33

is open, and hence $\mathbb{Z}$ is closed. The previous formula also shows that the closure of $\mathbb{Z}^c$ is $\mathbb{R}$ because any integer n is the limit $n = \lim_m (n - 1/m)$ of points in $\mathbb{Z}^c$, hence in the closure of $\mathbb{Z}^c$. The first formula in (1.21) becomes then $\partial \mathbb{Z} = \mathbb{Z} \cap \mathbb{R} = \mathbb{Z}$.

(c) Clearly
$$\Omega \cup \{0\} := \left\{ \frac{1}{n} : n = 1, 2, 3, \dots \right\} \cup \{0\} :$$

is closed because its complement

$$\bigcup_{n>0} \left(\frac{1}{n+1}, \frac{1}{n} \right) \cup (-\infty, 0) \cup (1, +\infty)$$

is open. Since $0 = \lim_n (1/n)$, the set $\Omega \cup \{0\}$ is contained in the closure of the original set Ω, and it is closed. Therefore $\overline{\Omega} = \Omega \cup \{0\}$. However, no point of Ω is interior because any point $1/n$ is an isolated point. The second formula in (1.21) gives $\partial \Omega = (\Omega \cup \{0\}) \setminus \emptyset = \Omega \cup \{0\}$.

(d) It is clear that
$$\overline{\mathbb{Q} \cap [0,1]} = [0,1].$$

Indeed, all points in $[0,1]$ are limits of rational sequences in $[0,1]$ and hence are in the closure of $\mathbb{Q} \cap [0,1]$. Therefore $[0,1]$ is a closed set beween $\mathbb{Q} \cap [0,1]$ and its closure, so it is in fact its closure. Now, no point of $\mathbb{Q} \cap [0,1]$ is interior because any ball centered at any of its points meets irrational numbers, so the second formula in (1.21) gives

$$\partial(\mathbb{Q} \cap [0,1]) = \overline{\mathbb{Q} \cap [0,1]} \setminus \emptyset = [0,1].$$

This exercise requires to explain in detail a number of more or less obvious facts about elementary sets, mostly having to do with finding their closure. It is useful to observe that by taking limits of points in Ω one gets points in $\overline{\Omega}$, hence if adding some limit points one obtains a closed set, then this is necessarily $\overline{\Omega}$, because it is a closed set that lies between Ω and $\overline{\Omega}$, and the latter is the smallest closed set containing Ω.

Observe that (a) is an example of a set the boundary of which is disjoint from it, in (b) the set coincides with its own boundary and in (c) one has to add just a point to the set to find its boundary. Finally, (d) is an example of a somewhat surprising fact: contrary to intuition, a boundary may well be a rather "fat" set.

1.6 Describe the topological properies of $\Omega = \{a_n : n \geq 1\}$, where

$$\begin{cases} a_{n+1} = \sqrt{a_n + 2}, & n = 1, 2, 3, \dots \\ a_1 = 1. \end{cases}$$

That is, is Ω open? Closed? Bounded? What are $\overline{\Omega}$ and $\partial \Omega$?

Answer. Observe that $a_n > 0$ for all n, as easily established by an induction argument. Next, inspecting the monononicity of the sequence it turns out that

$$a_{n+1} > a_n \iff \sqrt{a_n + 2} > a_n \iff a_n + 2 > a_n^2 \iff -1 < a_n < 2.$$

Now, indeed $a_1 = 1 < 2$ and if inductively $a_n < 2$, then

$$a_n + 2 < 4 \implies a_{n+1} = \sqrt{a_n + 2} < \sqrt{4} = 2.$$

Hence $(a_n)_{n \geq 1}$ is a positive increasing sequence, bounded above by 2. Hence it converges to some $\lambda > 0$. Moreover, since

$$\lambda = \lim_n a_n = \lim_n a_{n+1} = \lim_n \sqrt{a_n + 2} = \sqrt{\lambda + 2},$$

necessarily $\lambda = 2$. Since $2 \notin \Omega$, the set Ω is not closed. It is bounded because $\Omega \subset [1, 2]$ and it is not open because for example $a_1 = 1 \in \Omega$ is an isolated point. Indeed, for any sufficiently small r the ball centered at a_1 with radius r meets Ω only at a_1 itself. Since $a_2 = \sqrt{3}$ it is enough to take $r < \sqrt{3} - 1$. The same property actually holds for any other point of Ω which are thus all isolated. In particular, Ω has no interior points. Finally, $\overline{\Omega} = \Omega \cup \{2\} = \partial \Omega$, the latter being a consequence of the second formula in (1.21).

In short, all this exercise is about is to prove that the sequence $(a_n)_{n \geq 1}$ is strictly increasing and bounded above, hence convergent to a real number λ which is therefore *not* in the set Ω. In fact, all sequences $(a_n)_{n \geq 1}$ that are strictly increasing and bounded above, hence convergent to a real number which is therefore *not* one of the sequence values, give rise to value sets $\Omega = \{a_n : n \geq 1\}$ that enjoy exactly the same properties: they are neither open nor closed and have no interior points; their closure is obtained by adding the unique limit point and thus coincides with the boundary. Similar considerations apply to strictly decreasing sequences.

1.7 Is the set $\mathbb{Q}$ connected in $\mathbb{R}$?

Answer. No, it is not: indeed, for any $z \in \mathbb{R} \setminus \mathbb{Q}$, the two sets

$$U_1 = \{x \in \mathbb{R} : x < z\}, \qquad U_2 = \{x \in \mathbb{R} : x > z\}$$

are both nonempty open sets of $\mathbb{R}$, they satisfy $U_1 \cap U_2 = \emptyset$ and the sets $V_1 = U_1 \cap \mathbb{Q}$ and $V_2 = U_2 \cap \mathbb{Q}$ are both nonempty, and $\mathbb{Q} = V_1 \cup V_2$.

This is a straightforward application of the requirements in Definition 1.12 for being disconnected, the negation of connected.

1.8 Show that the set of points in $\mathbb{R}^2$ that have at least one rational coordinate is connected.

Answer. Let Ω be the set consisting of those points $P = (x, y)$ such that either $x \in \mathbb{Q}$ or $y \in \mathbb{Q}$, or both $x, y \in \mathbb{Q}$. It is enough to prove that Ω is in fact polygonally

1.7 Guided Exercises

connected. Actually, two points in Ω may be joined by a finite number of segments each parallel to the coordinate axes and contained in Ω. Indeed, let $P = (x_0, y_0) \in \Omega$ and $P_1 = (x_1, y_1) \in \Omega$. At least one of the following cases certainly arises:

(i) $x_0 \in \mathbb{Q}$ and $x_1 \in \mathbb{Q}$;
(ii) $x_0 \in \mathbb{Q}$ and $y_1 \in \mathbb{Q}$;
(iii) $y_0 \in \mathbb{Q}$ and $x_1 \in \mathbb{Q}$;
(iv) $y_0 \in \mathbb{Q}$ and $y_1 \in \mathbb{Q}$.

Assume without loss of generality that $x_0 < x_1$ and $y_0 > y_1$ (Fig. 1.23).

In case (i), choose $Y \in \mathbb{Q}$ larger than y_0. The vertical segments $\{(x, y) \in \mathbb{R}^2 : x = x_0 \in \mathbb{Q}, \ y_0 \leq y \leq Y\}$ and $\{(x, y) \in \mathbb{R}^2 : x = x_1 \in \mathbb{Q}, \ y_1 \leq y \leq Y\}$ together with the horizontal segment $\{(x, y) \in \mathbb{R}^2 : y = Y \in \mathbb{Q}, \ x_0 \leq x \leq x_1\}$ are completely contained in Ω and join P_0 to P_1.

In case (ii), the vertical segment $\{(x, y) \in \mathbb{R}^2 : x = x_0 \in \mathbb{Q}, \ y_1 \leq y \leq y_0\}$ and the horizontal segment $\{(x, y) \in \mathbb{R}^2 : y = y_1 \in \mathbb{Q}, \ x_0 \leq x \leq x_1\}$ are completely contained in Ω and join P_0 to P_1.

In case (iii), the vertical segment $\{(x, y) \in \mathbb{R}^2 : x = x_1 \in \mathbb{Q}, \ y_1 \leq y \leq y_0\}$ and the horizontal segment $\{(x, y) \in \mathbb{R}^2 : y = y_0 \in \mathbb{Q}, \ x_0 \leq x \leq x_1\}$ are completely contained in Ω and join P_0 to P_1.

In case (iv), choose $X \in \mathbb{Q}$ larger than x_1. The vertical segment $\{(x, y) \in \mathbb{R}^2 : x = X \in \mathbb{Q}, \ y_1 \leq y \leq y_0\}$ together with the horizontal segments $\{(x, y) \in \mathbb{R}^2 : y =$

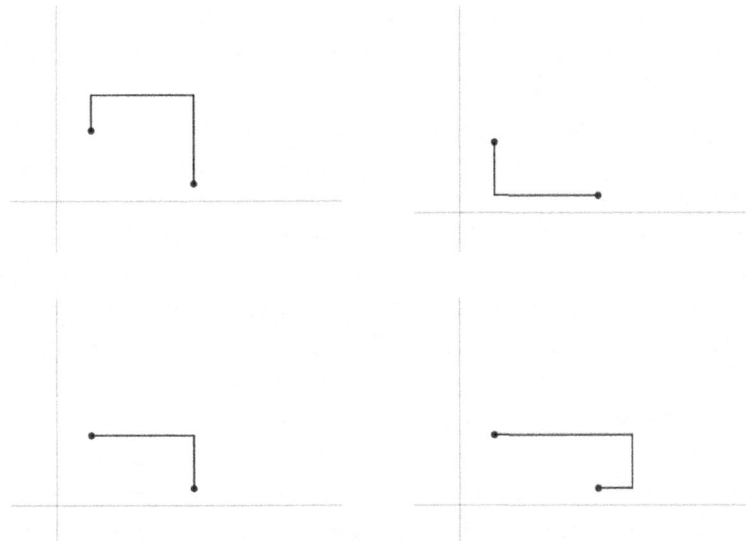

Fig. 1.23 Suppose $P = (x, y)$ and $P_1 = (x_1, y_1)$ are such that $x_0 < x_1$ and $y_0 > y_1$; the four pictures describe, in the same order, the polygonal paths explained in Exercise 1.8 for the situations (i), (ii), (iii) and (iv), respectively

$y_0 \in \mathbb{Q}$, $x_0 \leq x \leq X\}$ and $\{(x, y) \in \mathbb{R}^2 : y = y_1 \in \mathbb{Q}, x_1 \leq x \leq X\}$ are completely contained in Ω and join P_0 to P_1.

One way to think of this exercise is as follows. If $P \in \Omega$, then at least one of its coordinates is rational, hence at least one, or both, the coordinate lines passing through P are completely contained in Ω. Therefore, Ω determines a grid of horizontal and vertical lines along which it is possible to travel, going from any point in Ω to any other point in Ω.

1.9 Consider the function

$$f(x, y) = \begin{cases} \dfrac{x^2 + y^2}{|x| + |y|} & (x, y) \neq (0, 0) \\ 0 & (x, y) = (0, 0). \end{cases}$$

Establish if $A = \{(x, y) \in \mathbb{R}^2 : f(x, y) \geq 1\}$ is open, closed, bounded.

Answer. Observe first that $O = (0, 0) \notin A$ and in fact

$$A = \{(x, y) \in \mathbb{R}^2 \setminus \{(0, 0)\} : x^2 + y^2 \geq |x| + |y|\}$$
$$= \left\{(x, y) \in \mathbb{R}^2 \setminus \{(0, 0)\} : \left(|x| - \frac{1}{2}\right)^2 + \left(|y| - \frac{1}{2}\right)^2 \geq \frac{1}{2}\right\}.$$

Intuition suggests that the presence of "$\geq$" indicates that A is closed. Showing that A is closed is equivalent to showing that its complement

$$A^c = \left\{(x, y) \in \mathbb{R}^2 \setminus \{(0, 0)\} : \left(|x| - \frac{1}{2}\right)^2 + \left(|y| - \frac{1}{2}\right)^2 < \frac{1}{2}\right\} \cup \{(0, 0)\}$$

is open. Observe now that $O = (0, 0)$ is an interior point of A^c. Indeed, if $(x, y) \neq O$ and $(x, y) \in B(O, 1)$, then $x^2 + y^2 < 1$ and hence $|x| < 1$ and $|y| < 1$. This entails $x^2 < |x|$ and $y^2 < |y|$ so that $x^2 + y^2 < |x| + |y|$, namely $(x, y) \in A^c$. Thus O is an interior point of A^c. Next observe that

$$A^c = B_1 \cup B_2 \cup B_3 \cup B_4 \cup \{O\},$$

where B_1 is the open ball $B_1 = B(Q, \sqrt{2}/2)$ with $Q = (1/2, 1/2)$, namely

$$B_1 = \left\{(x, y) \in \mathbb{R}^2 : \left(x - \frac{1}{2}\right)^2 + \left(y - \frac{1}{2}\right)^2 < \frac{1}{2}\right\},$$

where B_2 is its symmetric with respect to the x-axis and B_3 and B_4 are the symmetric of B_1 and B_2 with respect to the y axis, respectively. Indeed,

- $(x, y) \mapsto (x, -y)$ maps the ball B_2 of center $(1/2, -1/2)$ and radius $\sqrt{2}/2$ onto B_1 and *vice versa*;
- $(x, y) \mapsto (-x, y)$ maps the ball B_3 of center $(-1/2, 1/2)$ and radius $\sqrt{2}/2$ onto B_1 and *vice versa*;

1.7 Guided Exercises

- $(x, y) \mapsto (-x, -y)$ maps the ball B_4 of center $(-1/2, -1/2)$ and radius $\sqrt{2}/2$ onto B_1 and *vice versa*.

If from symmetry considerations it follows that all of them are open, then their union is open. Since O is an interior point of A^c and $A^c = B_1 \cup B_2 \cup B_3 \cup B_4 \cup \{O\}$, it would follow that A^c is open, hence A is closed (Fig. 1.24).

All is left to be proven is then the general fact that that if $\Omega \subset \mathbb{R}^2$ is open, then the sets that are symmetric to Ω with respect to the axes are also open. Indeed, let Ω_1 denote the symmetric set with respect to the x-axis, namely $\Omega_1 = \{(x, y) \in \mathbb{R}^2 : (x, -y) \in \Omega\}$. Suppose that $P_1 = (x_1, y_1) \in \Omega_1$. Then $Q_1 = (x_1, -y_1) \in \Omega$, and since Ω is open there is $\delta > 0$ such that $B(Q_1, \delta) \subset \Omega$. Take $P = (x, y) \in B(P_1, \delta)$. Then

$$\|(x, -y) - (x_1, -y_1)\| = \|(x, y) - (x_1, y_1)\| < \delta$$

and hence $(x, -y) \in B(Q_1, \delta) \subset \Omega$ and $(x, y) \in \Omega_1$. Therefore $B(P_1, \delta) \subset \Omega_1$ and Ω_1 is open, as desired. Similarly one shows that the symmetric set with respect to the y-axis is open.

To conclude the exercise, observe that A is not bounded because the unbounded half line $L = \{(0, y) \in \mathbb{R}^2 : y \geq 2\}$ is contained in A since

$$f(0, y) = \frac{0 + y^2}{0 + |y|} = y \geq 2$$

on all points of L.

This exercise requires a little bit of thought. The main observation is actually of algebraic nature, and is the following simple fact:

$$x^2 + y^2 < |x| + |y| \iff \left(|x| - \frac{1}{2}\right)^2 + \left(|y| - \frac{1}{2}\right)^2 < \frac{1}{2}.$$

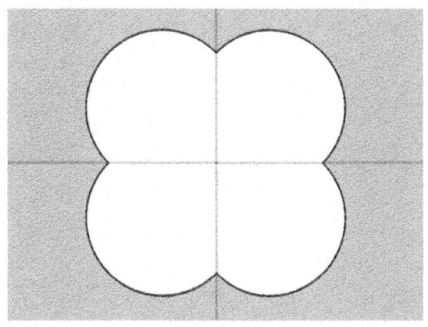
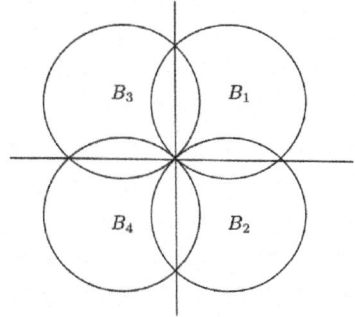

Fig. 1.24 Left: the superlevel set of $(x, y) \mapsto (x^2 + y^2)/(|x| + |y|)$, namely the points (x, y) for which $x^2 + y^2 \geq |x| + |y|$; right: the four circles of Exercise 1.9

This calls for considering the open ball B_1 where absolute values are not there and the inequality is simply

$$\left(x - \frac{1}{2}\right)^2 + \left(y - \frac{1}{2}\right)^2 < \frac{1}{2}$$

and then to shift B_1 around under the symmetries $(x, y) \mapsto (-x, y)$, $(x, y) \mapsto (x, -y)$ and $(x, y) \mapsto (-x, -y)$, which produce three other open balls. The union of these four balls together with O forms the set A^c.

1.10 Find a family $\{A_n\}_{n>0}$ of closed connected sets in $\mathbb{R}^2$ with $A_n \supset A_{n+1}$ for every positive integer n and such that $\bigcap_{n>0} A_n$ is not connected.

Answer. Define

$$A_n = \left\{(x, y) \in \mathbb{R}^2 : x > 0, \ y \geq \frac{1}{x}\right\} \cup \left\{(x, y) \in \mathbb{R}^2 : x \geq 0, \ y \leq \frac{1}{n}\right\} =: B \cup C_n.$$

Evidently $A_{n+1} \subset A_n$ because $C_{n+1} \subset C_n$. This follows at once from the fact that if $y \leq 1/(n+1)$ then *a fortiori* it is also $y \leq 1/n$. Both B and C_n are closed and hence A_n is closed, for every $n > 0$. Also, both B and C_n are connected and they intersect. Indeed, the point $(x, y) = (n + 1, y_n)$ where y_n is any value in the open interval $(1/(n+1), 1/n)$ is both in B and in C_n. Hence A_n is connected thanks to Proposition 1.3. However,

$$\bigcap_{n>0} A_n = \bigcap_{n>0} (B \cup C_n) = B \cup \bigcap_{n>0} C_n = B \cup D$$

where $D = \{(x, y) \in \mathbb{R}^2 : x \geq 0, \ y \leq 0\}$, because $y \leq 1/n$ for every positive integer n if and only if $y \leq 0$. The union $B \cup D$ is manifestly disconnected, because it is the union of two disjoint closed sets.

The main idea here is actually quite simple: one can keep cutting off larger and larger portions of a connected set until in the end one has indeed performed a full cut. This may be done in several different ways, of course. The proposed construction, depicted in Fig. 1.25, mimics the progressive effect of felling a tree by means of consecutive ax swings.

1.11 Draw the set $A = \{(x, y) \in \mathbb{R}^2 \setminus \{(0, 0)\} : (x^2 + y^2)^2 - 3y^2\sqrt{x^2 + y^2} \leq 2y^3\}$ after interpreting it in polar coordinates.

Answer. First of all, we observe that if F is the polar-coordinates map (1.18), then

$$F^{-1}(A) = \{(\theta, \rho) \in [0, 2\pi) \times (0, +\infty) : \rho^4 - 3\rho^3 \sin^2 \theta \leq 2\rho^3 \sin^3 \theta\}.$$

Hence, if $(\theta, \rho) \in F^{-1}(A)$, then, since $\rho > 0$,

$$\rho \leq \varphi(\theta) := 3\sin^2\theta + 2\sin^3\theta.$$

1.7 Guided Exercises

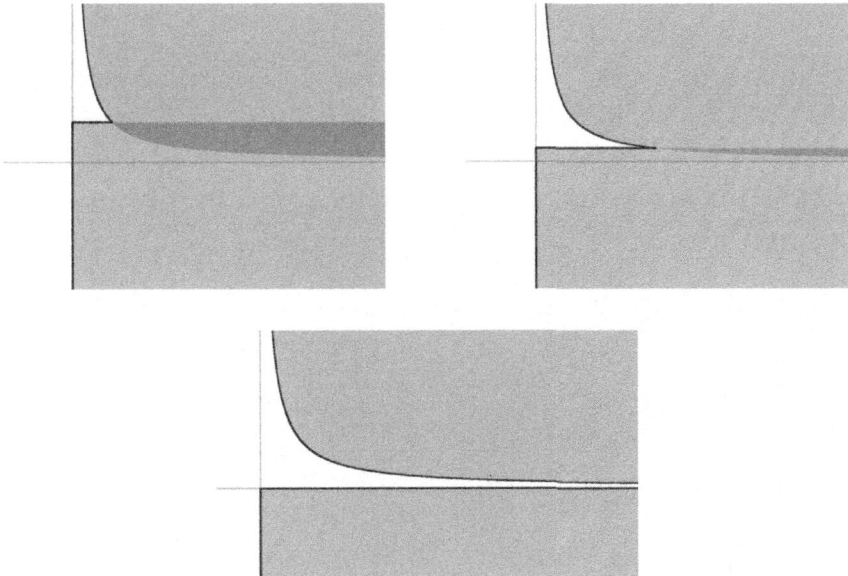

Fig. 1.25 Top: the sets A_1 and A_3 in the family described in Exercise 1.10; bottom: the intersection of all the A_n

It is then necessary to see what φ looks like. Now,

$$\varphi'(\theta) = 6\cos\theta\sin\theta + 6\sin^2\theta\cos\theta = 3\sin 2\theta(1 + \sin\theta)$$

is positive in $(0, \pi/2) \cup (\pi, 3\pi/2)$ so that φ is increasing in $[0, \pi/2]$ and in $[\pi, 3\pi/2]$, and decreasing in $[\pi/2, \pi]$ and in $[3\pi/2, 2\pi]$, see Fig. 1.26, left. The mapping φ restricted to $[0, \pi]$ is even with respect to $\pi/2$ and its restriction to $[\pi, 2\pi]$ is even with respect to $3\pi/2$. The set A is depicted in Fig. 1.26, right. To see this, notice that as θ increases from 0 to $\pi/2$ the value of ρ increases from 0 to 5, thereby spanning the first-quadrant quarter of A. As θ grows from $\pi/2$ to π the value of ρ decreases from 5 to 0 in a symmetric fashion, because of the symmetry of φ. Similar considerations apply to what happens when θ ranges in $[\pi, 3\pi/2]$ and then in $[3\pi/2, 2\pi]$.

There are essentially no difficulties in analyzing the polar-coordinates version of A, namely the set $F^{-1}(A)$ which is just the region between the θ-axis and the graph of φ, that is $0 < \rho \le \varphi(\theta)$ (because ρ is positive). Once this is achieved, the shape of A requires a little thought and is best understood by considering the curve

$$\theta \mapsto (\varphi(\theta)\cos\theta, \varphi(\theta)\sin\theta)$$

which surrounds A.

1.12 Draw the set $F(A)$ where F is the polar-coordinates map (1.18) and where A consists of those $(\theta, \rho) \in [0, 2\pi) \times (0, +\infty)$ which satisfy

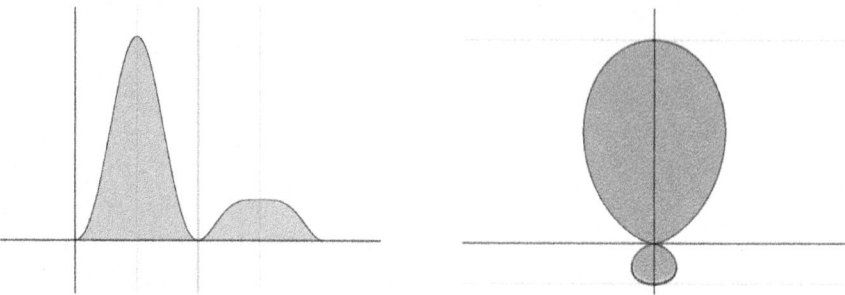

Fig. 1.26 Left: the set $F^{-1}(A)$ of Exercise 1.11 as ipograph of the mapping $\varphi(\theta) = 3\sin^2\theta + 2\sin^3\theta$; right: the set A

$$\begin{cases} \rho \le 1 \\ \rho^2 \le (\sqrt{3}+1)\Big(\rho\big[|\cos\theta| + |\sin\theta|\big] - 1\Big). \end{cases}$$

Establish if $F(A)$ is connected.

Answer. Clearly, $(x, y) \in F(A)$ if and only if

$$\begin{cases} 0 < x^2 + y^2 \le 1 \\ x^2 + y^2 \le (\sqrt{3}+1)\Big(\big[|x| + |y|\big] - 1\Big). \end{cases}$$

The second inequality is equivalent to

$$\Big(x^2 - |x|(\sqrt{3}+1)\Big) + \Big(y^2 - |y|(\sqrt{3}+1)\Big) \le -(\sqrt{3}+1)$$

and this, upon completing the squares, is equivalent to

$$\Big(|x| - \frac{\sqrt{3}+1}{2}\Big)^2 + \Big(|y| - \frac{\sqrt{3}+1}{2}\Big)^2 \le 2 + \sqrt{3} - (\sqrt{3}+1) = 1 \quad (1.22)$$

The set of points that corresponds to the second equation in the defining system for A is the set symmetric with respect to the coordinate axes that in the first quadrant is just the disk centered in $P_0 = \big((\sqrt{3}+1)/2, (\sqrt{3}+1)/2\big)$ and radius 1. A similar situation was discussed in Exercise 1.9 (Fig. 1.27).

Now, the distance from the center of any of these circles to the origin $O = (0, 0)$ is the same, and is equal to $d(P_0, O) = \sqrt{2 + \sqrt{3}}$ which is a number in $(1, 2)$. Therefore each of them meets the unit disk in the portion depicted in Fig. 1.27 that has the typical shape of the intersection of two overlapping disks. The set A is not connected because the two open half planes

1.7 Guided Exercises

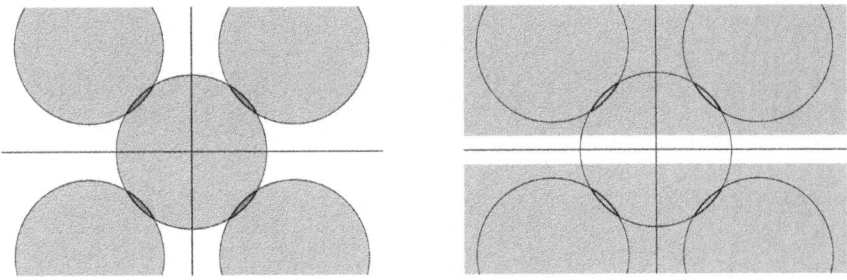

Fig. 1.27 Left: the set $F(A)$ of Exercise 1.12 is the intersection of the closed unit disk with four closed non intersecting disks, and is not connected; right: two half planes disconnecting $F(A)$

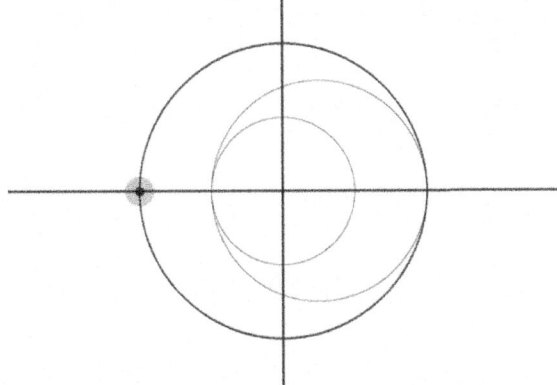

Fig. 1.28 The circles C_0, C_1 and C_2, the point P_0 ad a small open ball centered at P_0 in Exercise 1.13

$$U_1 = \left\{(x,y) \in \mathbb{R}^2 : y > \frac{y_0 - 1}{2} = \frac{\sqrt{3} - 1}{4}\right\}$$

$$U_2 = \left\{(x,y) \in \mathbb{R}^2 : y < -\frac{y_0 - 1}{2} = \frac{1 - \sqrt{3}}{4}\right\}$$

meet the conditions in Definition 1.12, where y_0 is the y-coordinate of $P_0 = (x_0, y_0)$.

As in Exercise 1.9, the main point is to achieve an expression in Cartesian coordinates that is a "symmetrized" version of the equation of a disk, namely Eq. (1.22), which therefore describes four disks. Upon computing the centers and radii, that are all equal to 1, it is clear that the four closed disks to not meet eachother but all of them meet the closed unit disk in a (closed) portion. Their union is manifestly disconnected: the horizontal open half planes at a distance from the x-axis which is half the distance of the four disks will disconnect the portions because they actually separate the upper two disks from the lower two ones.

1.13 Consider the sequences $(x_n)_{n\geq 0}$ and $(r_n)_{n\geq 0}$ defined by

$$\begin{cases} x_{2n} = 0 \\ x_{2n+1} = \frac{1}{2^{n+2}} \end{cases} \quad \text{and} \quad \begin{cases} r_{2n} = \frac{1}{2^n} \\ r_{2n+1} = \frac{3}{2^{n+2}} \end{cases}$$

respectively, and put

$$C_n = \{(x, y) \in \mathbb{R}^2 : (x - x_n)^2 + y^2 = r_n^2\}.$$

Is the set $C = \bigcup_{n\geq 0} C_n$ open? Is it closed? Is it connected?

Answer. Clearly, C is the union of countably many circles, more precisely, the circles centered at $(x_n, 0)$ with radius r_n as n ranges in $\mathbb{N}$. In order to show that C is not open it is enough to find a point $P_0 \in C$ such that in any open neighborhood of P_0 there are points not belonging to C. This can be proved by observing that there is indeed a leftmost point $P_0 \in C$. In order to formulate this precisely observe first that if $(x, y) \in C_n$, then of course $|x - x_n| \leq r_n$ so that $x_n - r_n \leq x \leq x_n + r_n$. Now,

$$x_{2n} - r_{2n} = -\frac{1}{2^n} \geq -1, \quad x_{2n+1} - r_{2n+1} = -\frac{1}{2^{n+1}} > -1,$$

so that in all cases if $(x, y) \in C_n$, then $x \geq x_n - r_n \geq -1$. It follows that if $(x, y) \in C$ then necessarily $(x, y) \in C_n$ for some integer $n \geq 0$ and hence $x \geq -1$. Now, the point $P_0 = (-1, 0) \in C_0$ because C_0 is the unit circle centered at the origin, so that $P_0 \in C$. Any open ball $B = B(P_0, \delta)$ contains points with x-cordinates strictly less than -1, which cannot belong to any of the circles C_n, hence *a fortiori* not in C.

Next, C is not closed either. Indeed, the point $P_n = (x_n + r_n, 0)$ lies on C_n and the sequence $(P_n)_{n\geq 0}$ converges to $(0, 0)$ because if $n = 2m$ or $n = 2m + 1$ then

$$x_n + r_n = \frac{1}{2^m} \to 0.$$

It follows that $O = (0, 0) \in \overline{C}$. However, $O \notin C$ because the origin does not belong to any of the circles simply because $x_n^2 \neq r_n^2$ for all $n \geq 0$. Therefore C is not closed because it is strictly contained in its closure.

The set C, however, is connected. This is best seen by observing that each circle C_n intersects the next one C_{n+1}. Indeed,

$$\left(\frac{1}{2^n}, 0\right) \in C_{2n} \cap C_{2n+1}, \quad \left(-\frac{1}{2^{n+1}}, 0\right) \in C_{2n+1} \cap C_{2n+2}.$$

Since each circle is connected, the conclusion follows directly from Proposition 1.3.

It is not easy to have a clear picture of C, but enough can be understood: the circles decrease rapidly in radius and sort of wiggle down to the origin, properly enclosing it. There is a "leftmost" point P_0 that lies on the unit circle. This immediately reveals that any point to its left is not in C and hence that any ball centered at P_0 contains points not in C. The shrinking of the circles calls for understanding what happens in the limit. Here it can be observed that the rightmost points of the circles actually form a sequence which converges to the origin, which is thus in the closure of C but clearly in none of the circles. Thus C is not closed. Finally, the wiggling is such that each circle touches the next, so they form a chain of connected sets. These facts are illustrated in Fig. 1.28.

1.14 Consider the function $f \colon \mathbb{C} \to \mathbb{C}$ given by $f(z) = \dfrac{1}{z^2 - 2z}$. Find its real and imaginary parts, specifying their domains.

Answer. Evidently, $\mathrm{Dom}(f) = \mathbb{C} \setminus \{0, 2\}$. Further, if $z = x + iy$, then

$$z^2 - 2z = (x+iy)^2 - 2(x+iy) = x^2 - y^2 - 2x + i(2xy - 2y).$$

Therefore

$$\begin{aligned}
f(z) &= \frac{1}{x^2 - y^2 - 2x + i(2xy - 2y)} \\
&= \frac{x^2 - y^2 - 2x - i(2xy - 2y)}{\bigl(x^2 - y^2 - 2x + i(2xy - 2y)\bigr)\bigl(x^2 - y^2 - 2x - i(2xy - 2y)\bigr)} \\
&= \frac{x^2 - y^2 - 2x - i(2xy - 2y)}{(x^2 - y^2 - 2x)^2 + (2xy - 2y)^2}.
\end{aligned}$$

Thus, if $f(z) = u(x, y) + i v(x, y)$, then

$$\mathrm{Dom}(u) = \mathrm{Dom}(v) = \mathbb{R}^2 \setminus \{(0, 0), (2, 0)\}$$

and

$$u(x, y) = \frac{x^2 - y^2 - 2x}{(x^2 - y^2 - 2x)^2 + (2xy - 2y)^2}, \qquad v(x, y) = \frac{-2xy + 2y}{(x^2 - y^2 - 2x)^2 + (2xy - 2y)^2}.$$

There is nothing to understand here, just do the algebra right.

1.15 Describe the subset of the complex plane $C = \left\{ \dfrac{-is}{1 - is} : s \in \mathbb{R} \right\}$.

Answer. Multiplying and dividing by the complex conjugate of the denominator gives

$$\frac{-is}{1 - is} \cdot \frac{1 + is}{1 + is} = \frac{s^2}{1 + s^2} - i \frac{s}{1 + s^2}.$$

Observe that for $s = 0$ we obtain $0 \in C$ and for $s = \pm 1$ the corresponding points are $(1/2, \mp 1/2)$. Furthermore

$$\lim_{s \to \pm \infty} \frac{s^2}{1 + s^2} = 1, \qquad \lim_{s \to \pm \infty} \frac{s}{1 + s^2} = 0,$$

Fig. 1.29 The locus contains the three black dots and in the limit it reaches 1, the red dot

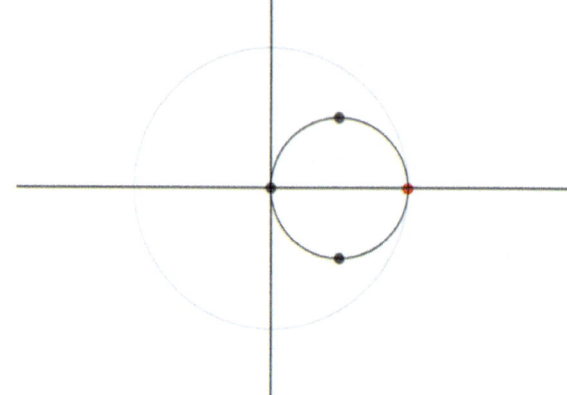

so that as the real parameter tends to $\pm\infty$ the corresponding point tends to $1 \in \mathbb{C}$. An educated guess is that C might represent a circle, centered at half-way (Fig. 1.29). Indeed:

$$\left(\frac{s^2}{1+s^2} - \frac{1}{2}\right)^2 + \left(-\frac{s}{1+s^2}\right)^2 = \left(\frac{2s^2 - (1+s^2)}{2(1+s^2)}\right)^2 + \frac{s^2}{(1+s^2)^2}$$
$$= \frac{1}{4}\left(\frac{1-s^2}{1+s^2}\right)^2 + \frac{1}{4}\left(\frac{2s}{1+s^2}\right)^2$$
$$= \frac{(1-s^2)^2 + 4s^2}{4(1+s^2)^2} = \frac{1}{4}.$$

All the points of the locus C lie on the circle $\mathscr{C}$ centered at $1/2 \in \mathbb{C}$ with radius $1/2$. The expression in the second line recalls, apart from the factor $1/4$, the identity

$$(\cos\theta, \sin\theta) = \left(\frac{1-s^2}{1+s^2}, \frac{2s}{1+s^2}\right)$$

which holds true for $s = \tan(\theta/2)$. This in fact shows that $C = \mathscr{C} \setminus \{1\}$.

Here the intuition comes from taking three points and a limit in the complex plane and the memory goes back to high-school: tracing curves point by point, plus a little bit of basic trigonometry.

1.8 Problems

1.16 Describe the domain of the following functions

$$f_1(x, y) = e^{\frac{1}{xy}} \qquad f_2(x, y) = \arctan \frac{1}{x^2 + y^2}$$

$$f_3(x, y) = \frac{1}{\sqrt{x + y - 1}} \qquad f_4(x, y) = \sqrt{\frac{9x^2 - 4y^2 - 36}{9 - x^2 - y^2}}$$

$$f_5(x, y) = \log \frac{2 - |x| - |y|}{(1 - x^2)(y^2 - 1)} \qquad f_6(x, y) = \sqrt{2x|xy|}.$$

1.17 Describe the level sets of the following functions

$$f_1(x, y) = e^{\frac{1}{xy}} \qquad f_2(x, y) = x^2 y$$

$$f_3(x, y) = \log(x^2 + 2y^2) \qquad f_4(x, y) = \arcsin \frac{x - y}{x + y}.$$

1.18 Find subsets $A, B \subset \mathbb{R}$ such that $A \cap \overline{B}, \overline{A} \cap B, \overline{A} \cap \overline{B}, \overline{A \cap B}$ are all different.

1.19 Find the closure of $E = \{1/n : n \in \mathbb{N} \setminus \{0\}\} \cup \{0\}$.

1.20 Establish whether $A = \{(x, y) \in \mathbb{R}^2 : 1 \leq x^2 + y^2 \leq 2x - 4y\}$ is open, closed, bounded, compact.

1.21 Describe the set $B = \{(x, y) \in \mathbb{R}^2 : e^x + x - 2 \leq 0, \ y \in [0, 1]\}$.

1.22 Establish whether $A = \{(x, y) \in \mathbb{R}^2 : 1 \leq x^2 + y^2 < 4x + 4y\}$ is open, closed, bounded, compact.

1.23 Show that $A = \{(x, y) \in \mathbb{R}^2 : 2|x| + 2|y| \leq x^2 + y^2\}$ is closed and unbounded.

1.24 Show that $A = \{(x, y) \in \mathbb{R}^2 : x^2 - y^2 \geq 9\}$ is not connected.

1.25 Decide which of the following sets is compact.

(a) $[0, 1)$ \qquad (b) $[0, +\infty)$

(c) $\mathbb{Q} \cap [0, 1]$ \qquad (d) $\{(x, y) \in \mathbb{R}^2 : x^2 + y^2 = 1\}$

(e) $\{(x, y) \in \mathbb{R}^2 : |x| + |y| = 1\}$ \qquad (f) $\{(x, y) \in \mathbb{R}^2 : x^2 + y^2 < 1\}$

(g) $\{(x, y) \in \mathbb{R}^2 : x \geq 1, \ 0 \leq y \leq 1/x\}$.

1.26 Decide which of the following sets is compact.

(a) $\left\{(x_1, \ldots, x_n) \in \mathbb{R}^n : \sum_{i=1}^n x_i^2 = 1\right\}$

(b) $\left\{(x_1, \ldots, x_n) \in \mathbb{R}^n : \sum_{i=1}^n x_i^2 \leq 1\right\}$

(c) $\{(x, y) \in \mathbb{R}^2 : y = \sin(1/x),\ x \in (0, 1]\}$

(d) $\{(x, y) \in \mathbb{R}^2 : y = 1/x,\ x \in (0, 1]\}$.

1.27 Find an example of a closed subset W of $\mathbb{R}^2$ with the property that its projection $p(W) = \{x \in \mathbb{R} : (x, y) \in W \text{ for some } y\}$ is not a closed subset of $\mathbb{R}$.

1.28 Establish whether $A = \{(x, y, z) \in \mathbb{R}^3 : x^2 + y^2 < z \leq 1\}$ is open, closed, bounded.

1.29 Let $A = \{(x, y) \in \mathbb{R}^2 : x^2 + y^2 < 1\}$ and $B = \{(x, y) \in \mathbb{R}^2 : (x - 2)^2 + y^2 \leq 1\}$. Establish which of A, B and $A \cup B$ is either connected or pathwise connected.

References

1. Manetti, M.: Topology, Unitext, 153, La Matematica per il 3+2. Springer (2023)
2. Lang, S.: Linear algebra. Undergraduate Texts in Mathematics, 3rd edn. Springer, New York (1989)

Chapter 2
Limits and Continuity

In Chap. 1 the basic topological notions for the Analysis in Euclidean space were introduced. The richness of the possible domains of functions of several variables has been illustrated and it should be clear that already in dimension $n = 2$ real valued functions may exhibit a variety of behaviours that functions of a single variable do not display. Graphs of functions of two variables may be visualized as surfaces, or, with a bit of informality, as less regular "landscapes". A very crucial notion for such functions, as it was in the case of a single variable, is the notion of continuity, which appeals to the visual picture that one can "walk along the graph' in order to reach a given point on it, or one cannot, the hyke is doable or not, the surface breaks or does not break at that point. A very efficient way to turn such suggestive images into quantitative notions, and in fact a very basic notion of Mathematics, is the notion of *limit*, which may be widely extended to yet more general setups.

2.1 Convergence Notions

The notion of limit for vector valued functions of several variables is an exact analogue of the notion given for real valued functions of a single variable, provided that one interprets the absolute values appearing in the latter with the appropriate norms for the former.

Definition 2.1 (*Finite limit at a point*) Suppose that P_0 is a limit point of $\Omega \subset \mathbb{R}^n$ (see Definition 1.9) and take a vector valued function $F \colon \Omega \to \mathbb{R}^m$. The vector $L \in \mathbb{R}^m$ is called the limit of F as P tends to P_0 if for any neighborhood $\mathscr{U}$ of L there exists a neighborhood $\mathscr{V}$ of P_0 such that the image under F of $\mathscr{V} \cap \Omega \setminus \{P_0\}$ is contained in $\mathscr{U}$. In this case one writes

$$\lim_{P \to P_0} F(P) = L$$

or $F(P) \to L$ as $P \to P_0$. Equivalently, the limit of F as $P \to P_0$ is L if and only if for every $\varepsilon > 0$ there exists $\delta > 0$ such that if $P \in \Omega$ and $0 < \|P - P_0\| < \delta$, then $\|F(P) - L\| < \varepsilon$. In this case one says that F converges to L or that it tends to L.

The scalar case, namely when $m = 1$, takes a slightly simpler form, whereby in fact the limit of $f: \Omega \to \mathbb{R}$ is $\ell \in \mathbb{R}$ if and only if for every $\varepsilon > 0$ there exists $\delta > 0$ such that if $P \in \Omega$ and $0 < \|P - P_0\| < \delta$, then $|f(P) - \ell| < \varepsilon$. As a matter of fact, the vector valued case can be reduced to the scalar case in the sense that if $F = (f_1, \ldots, f_n)$ and $L = (\ell_1, \ldots, \ell_n)$, then

$$\lim_{P \to P_0} F(P) = L \quad \Longleftrightarrow \quad \lim_{P \to P_0} f_j(P) = \ell_j, \quad 1 \leq j \leq n.$$

An easy case where it is possible to infer that $\lim_{P \to P_0} F(P) = 0$ is when

$$\|F(P)\| \leq C \|P - P_0\|$$

for some $C > 0$, because one selects $\delta = \varepsilon/C$. It is worthwhile observing that if $F: \Omega \subset \mathbb{R}^n \to \mathbb{R}^m$, then the norm appearing in the left hand side is in $\mathbb{R}^m$ whereas that in the right hand side is in $\mathbb{R}^n$. A similar observation applies to the norms appearing in Definition 2.1 above.

Diverging scalar valued functions at a limit point P_0 are defined similarly to the one variable case, namely

Definition 2.2 (*Diverging function at a point*) Suppose that P_0 is a limit point of $\Omega \subset \mathbb{R}^n$. The scalar valued function $f: \Omega \to \mathbb{R}$ is said to diverge to $+\infty$ (or to $-\infty$, respectively) if for any $K > 0$ there exists a neighborhood $\mathscr{V}$ of P_0 such that the image under f of $\mathscr{V} \cap \Omega \setminus \{P_0\}$ is contained in the interval $(K, +\infty)$ (or in the interval $(-\infty, -K)$, respectively). In this case one writes

$$\lim_{P \to P_0} f(P) = +\infty, \qquad \lim_{P \to P_0} f(P) = -\infty$$

or $f(P) \to \pm\infty$ as $P \to P_0$. Equivalently, $f(P) \to \pm\infty$ as $P \to P_0$ if and only if

$\forall K > 0 \ \exists \delta > 0 \ $ such that $P \in \Omega$ and $0 < \|P - P_0\| < \delta \implies f(P) > K$
$\forall K > 0 \ \exists \delta > 0 \ $ such that $P \in \Omega$ and $0 < \|P - P_0\| < \delta \implies f(P) < -K$.

There are of course many examples of diverging functions, for example the negative powers of the distance function, or logarithms of the distance function, both defined in the punctured space $\Omega = \mathbb{R}^n \setminus \{P_0\}$:

$$\lim_{P \to P_0} \frac{1}{\|P - P_0\|^\alpha} = +\infty$$

for every $\alpha > 0$, and $\lim_{P \to P_0} \log(\|P - P_0\|) = -\infty$.

2.1 Convergence Notions

The notion of "limit at infinity" for scalar valued functions is built on the idea that the sensible notion of punctured neighborhood of ∞ in $\mathbb{R}^n$ is any set $\mathscr{U}$ containing the complement of a ball $B(0, R)$. Hence any unbounded set Ω has ∞ as a limit point because if Ω is unbounded, then $\Omega \cap \mathscr{U} \neq \emptyset$ for any $\mathscr{U} \supset \mathbb{R}^n \setminus B(0, R)$. Indeed, any $P \in \Omega$ with $\|P\| \geq R$ (and such points P exist for otherwise Ω would be bounded) is such that $P \in \mathscr{U}$. A petty comment: a neighborhood of ∞ is a set of the form $\mathscr{U} \cup \{\infty\}$ where $\mathscr{U}$ is a punctured neighborhood of ∞.

Definition 2.3 (*Limits at infinity, scalar fields*) Suppose that $\Omega \subset \mathbb{R}^n$ is an unbounded set and take a scalar field $f \colon \Omega \to \mathbb{R}$. The real number ℓ is the limit at infinity of f if for any $\varepsilon > 0$ there exists a punctured neighborhood $\mathscr{U}$ of ∞ such that the image under f of $\mathscr{U} \cap \Omega$ is contained in the ball $B(\ell, \varepsilon)$. In this case one writes

$$\lim_{\|P\| \to +\infty} f(P) = \ell \quad \text{or} \quad \lim_{P \to \infty} f(P) = \ell.$$

The function is said to *diverge at infinity* to $+\infty$ (or to $-\infty$, respectively) if for any $K > 0$ there exists a punctured neighborhood $\mathscr{U}$ of ∞ such that the image under f of $\mathscr{U} \cap \Omega$ is contained in the interval $(K, +\infty)$ (or in the interval $(-\infty, -K)$, respectively). In these cases one writes:

$$\lim_{\|P\| \to +\infty} f(P) = +\infty, \qquad \lim_{\|P\| \to +\infty} f(P) = -\infty,$$

respectively, or, equivalently:

$$\lim_{P \to \infty} f(P) = +\infty, \qquad \lim_{P \to \infty} f(P) = -\infty.$$

The reader is invited to spell out the above definitions in all possible equivalent ways. It is worthwhile observing that the notion of limit at infinity requires the function to behave uniformly in all directions, and in fact, to have the same behaviour outside larger and larger balls over the whole domain.

Finally, it is possible to extend Definitions 2.2 and 2.3 to vector valued functions in rather obvious ways. Thus, if $F \colon \Omega \subset \mathbb{R}^n \to \mathbb{R}^m$, and if P_0 is a limit point of Ω, then the following is to be understood

$$\lim_{P \to P_0} F(P) = \infty \quad \Longleftrightarrow \quad \lim_{P \to P_0} \|F(P)\| = +\infty$$

and if Ω is unbounded

$$\lim_{\|P\| \to +\infty} F(P) = \infty \quad \Longleftrightarrow \quad \lim_{\|P\| \to +\infty} \|F(P)\| = +\infty.$$

As in the case of limits for functions of a single real variable, it is useful to adopt the notation $\overline{\mathbb{R}}$ used in [1] for the extended real line. Set-theoretically, this is the union $\mathbb{R} \cup \{\pm\infty\}$. When handling (possibly vector valued) functions of several variables, it is also very useful to introduce the so-called compactification of $\mathbb{R}^n$, which consists of the set

$$\overline{\mathbb{R}^n} = \mathbb{R}^n \cup \{\infty\}.$$

It is conceptually different from $\overline{\mathbb{R}}$ where one distinguishes two different directions. In the compactification, one adds the single point "∞". Using this idea, all the previous notions of limit can be subsumed in the following general definition.

Definition 2.4 (*Limit: general notion*) Suppose that $P_0 \in \overline{\mathbb{R}^n}$ is a limit point of $A \subset \mathbb{R}^n$. The function $F: A \to \mathbb{R}^m$ has the limit $L \in \overline{\mathbb{R}^m}$ if for every neighborhood $\mathscr{V}$ of L there exists a punctured neighborhood $\mathscr{U}$ of P_0 such that $F(\mathscr{U}) \subset \mathscr{V}$. In this case, one writes

$$\lim_{P \to P_0} F(P) = L$$

or also $F(P) \to L$ as $P \to P_0$.

The reader is urged to check that Definition 2.4 encompasses all the previous ones. In the case $m = 1$, one must distinguish between $L = +\infty$ and $L = -\infty$, which are both elements of $\overline{\mathbb{R}}$, and have very different neighborhoods.

For limits of all kinds the desired uniqueness result holds, in the sense that if the limit L, in the sense of Definition 2.4, exists, finite or infinite, then it is unique.

2.2 Results on Limits

Below is a list without too many comments of the basic results on limits, most of which are pretty straightforward versions of the corresponding results for single variable functions. In many circumstances, what matters are local properties, in the sense that a property holds *locally at* P_0 for the (possibly vector valued function) $F: A \to \mathbb{R}^m$ if P_0 is a limit point of A and if there exists a punctured neighborhood $\mathscr{U}$ of P_0 such that $F(P)$ enjoys the property for every $P \in A \cap \mathscr{U}$.

Theorem 2.1 (Sequential limits) *Suppose that $L \in \overline{\mathbb{R}^m}$, $P_0 \in \overline{\mathbb{R}^n}$, that P_0 is a limit point of $A \subset \mathbb{R}^n$ and take $F: A \to \mathbb{R}^m$. The following are equivalent:*

(i) $F(P) \to L$ as $P \to P_0$
(ii) $\lim_n F(P_n) = L$ for every sequence $(P_n)_{n \geq 0}$ of points in $A \setminus \{P_0\}$ with $P_n \to P_0$.

Theorem 2.2 (Permanence of sign) *Suppose that $\ell \in \overline{\mathbb{R}} \setminus \{0\}$, that $P_0 \in \overline{\mathbb{R}^n}$, is a limit point of $A \subset \mathbb{R}^n$ and take $f: A \to \mathbb{R}$. If $f(P) \to \ell$ as $P \to P_0$, then locally at P_0 the function f has the same sign of ℓ, where it is understood that the sign of $+\infty$ is positive and that the sign of $-\infty$ is negative.*

Theorem 2.3 (Squeeze theorem) *Suppose that $P_0 \in \overline{\mathbb{R}^n}$ is a limit point of $A \subset \mathbb{R}^n$ and suppose that the functions $f, g, h: A \to \mathbb{R}$ satisfy*

2.2 Results on Limits

(i) $f(P) \to \ell$ and $h(P) \to \ell$ as $P \to P_0$, where $\ell \in \overline{\mathbb{R}}$
(ii) $f(P) \le g(P) \le h(P)$ locally at P_0.

Then the limit $\lim_{P \to P_0} g(P)$ exists and $\lim_{P \to P_0} g(P) = \ell$.

Theorem 2.4 (Algebra of finite limits) *Suppose that $P_0 \in \overline{\mathbb{R}^n}$ is a limit point of $A \subset \overline{\mathbb{R}^n}$ and suppose that $F, G: A \to \mathbb{R}^m$ are such that $F(P) \to L \in \mathbb{R}^m$ and $g(P) \to M \in \mathbb{R}^m$ as $P \to P_0$. Then, as $P \to P_0$, the following holds:*

(i) $\|F(P)\| \to \|L\|$;
(ii) $F(P) + G(P) \to L + M$;
(iii) $F(P) \cdot G(P) \to L \cdot M$;
(iv) *for every* $\lambda \in \mathbb{R}$, $\lambda F(P) \to \lambda L$;
(v) *if* $m = 1$ *and* $M \ne 0$, *then* $F(P)/G(P) \to L/M$.

Notice that the assumption $M \ne 0$ in item (v) actually guarantees that the ratio $F(P)/G(P)$ makes sense in a punctured neighborhood of P_0, by permanence of sign.

Theorem 2.5 (Algebra of infinite limits) *Suppose that $P_0 \in \overline{\mathbb{R}^n}$ is a limit point of $A \subset \mathbb{R}^n$ and consider the scalar fields $f, g: A \to \mathbb{R}$. Then, as $P \to P_0$, the following holds:*

(ia) *if* $f(P) \to +\infty$ *and* $g(P) > m \in \mathbb{R}$ *locally at* P_0, *then* $f(P) + g(P) \to +\infty$;
(ib) *if* $f(P) \to -\infty$ *and* $g(P) < M \in \mathbb{R}$ *locally at* P_0, *then* $f(P) + g(P) \to -\infty$;
(iia) *if* $f(P) \to \pm\infty$ *and* $g(P) \to \ell > 0$ *or* $g(P) \to +\infty$, *then* $f(P)g(P) \to \pm\infty$;
(iib) *if* $f(P) \to \pm\infty$ *and* $g(P) \to \ell < 0$ *or* $g(P) \to -\infty$, *then* $f(P)g(P) \to \mp\infty$;
(iiia) *if* $f(P) \to \pm\infty$, *then* $1/f(P) \to 0$;
(iiib) *if* $f(P) \to 0$, *and* $f(P) > 0$ *locally at* P_0, *then* $1/f(P) \to +\infty$;
(iiic) *if* $f(P) \to 0$, *and* $f(P) < 0$ *locally at* P_0, *then* $1/f(P) \to -\infty$.

Theorem 2.6 (Limit of compositions) *Consider the possibly vector valued functions $F: A \subset \mathbb{R}^n \to \mathbb{R}^m$, $G: B \subset \mathbb{R}^m \to \mathbb{R}^d$ and take $P_0 \in \overline{\mathbb{R}^n}$, $Q_0 \in \overline{\mathbb{R}^m}$ limit points of A and B, respectively. Suppose that $F(A) \subset B$ and that*

$$\lim_{P \to P_o} F(P) = Q_0, \qquad \lim_{Q \to Q_0} G(Q) = L.$$

Suppose further that one of the following assumptions is satisfied:

(i) $Q_0 \in B$ *and* $G(Q_0) = L$
(ii) $Q_0 \notin B$
(iii) $F(P) \ne Q_0$ *locally at* P_0.

Then as $P \to P_0$ the limit of the composition $G \circ F$ exists and

$$\lim_{P \to P_0} G(F(P)) = \lim_{Q \to Q_0} G(Q) = L.$$

2.2.1 Uniformity and Polar Coordinates

Simple examples show that it is not possible, in general, to compute the limit of a function of, say, two variables by considering one coordinate at a time. Explicitly,

$$\lim_{x \to x_0} \left(\lim_{y \to y_0} f(x, y) \right) = \ell, \quad \lim_{y \to y_0} \left(\lim_{x \to x_0} f(x, y) \right) = \ell \;\not\Rightarrow\; \lim_{(x,y) \to (x_0, y_0)} f(x, y) = \ell.$$

For instance, at $P_0 = O = (0, 0)$, the function $f(x, y) = xy/(x^2 + y^2)$ has vasnishing iterated limits:

$$\lim_{x \to 0} \left(\lim_{y \to 0} \frac{xy}{x^2 + y^2} \right) = \lim_{y \to 0} \left(\lim_{x \to 0} \frac{xy}{x^2 + y^2} \right) = 0$$

because $f(x, 0) = 0$ for all $x \neq 0$ and similarly $f(0, y) = 0$ for all $y \neq 0$. However, the limit as $P \to O$ does not exist because along any line $y = mx$, when $x \neq 0$ the function is constant and equal to $f(x, mx) = m/(1 + m^2)$. It follows that in every punctured neighborhood of O the function attains every value in $[-1/2, 1/2]$ and this shows that the limit cannot exist.

Functions of two or more variables can display even more singular behaviours. As explained in detail in the guided Exercise 2.2, the limit along every straight line through the origin of the function $f(x, y) = xy^2/(x^2 + y^4)$ exists and is equal to 0, yet the limit as $P \to O$ does not exist. The reason is that the behaviour along the various lines is not uniform, in the precise sense described in Proposition 2.1 below.

In order to formulate the uniformity result alluded to above some notation is needed. Take a function $f: A \to \mathbb{R}$ and a limit point $P_0 = (x_0, y_0)$ of A. Upon using the polar coordinates (1.19), fix a positive ρ and let

$$\Theta_\rho = \{ \theta \in [0, 2\pi) : (x_0 + \rho \cos \theta, y_0 + \rho \sin \theta) \in A \}.$$

Thus, the set of points $\{(x_0 + \rho \cos \theta, y_0 + \rho \sin \theta) \in \mathbb{R}^2 : \theta \in \Theta_\rho\}$ is the intersection of A with the circle $C(P_0, \rho)$ centered at P_0 with radius ρ.

Proposition 2.1 *Let $P_0 = (x_0, y_0)$ be a limit point of A and take $f: A \to \mathbb{R}$ and $\ell \in \mathbb{R}$. The following are equivalent*

(i) $\lim\limits_{P \to P_0} f(P) = \ell$

(ii) $\lim\limits_{\rho \to 0} \sup\limits_{\theta \in \Theta_\rho} \left| f(x_0 + \rho \cos \theta, y_0 + \rho \sin \theta) - \ell \right| = 0.$

The meaning of item (ii) is that the (supremum) distance between the values of f on $C(P_0, \rho)$ and the real value ℓ tends to 0 as $\rho \to 0$. Thus, what really matters is a uniform behaviour along all directions, that is, a consistent behaviour along circles that shrink around the limit point. This principle can be generalized to diverging functions as described next:

2.3 Continuity

$$\lim_{P \to P_0} f(P) = +\infty \iff \lim_{\rho \to 0} \inf_{\theta \in \Theta_\rho} f(x_0 + \rho \cos\theta, y_0 + \rho \sin\theta) = +\infty$$

$$\lim_{P \to P_0} f(P) = -\infty \iff \lim_{\rho \to 0} \sup_{\theta \in \Theta_\rho} f(x_0 + \rho \cos\theta, y_0 + \rho \sin\theta) = -\infty$$

Furthermore, upon taking expanding circles, that is, by looking at what happens as $\rho \to +\infty$, limits at ∞ can also be treated:

$$\lim_{P \to \infty} f(P) = \ell \iff \lim_{\rho \to +\infty} \sup_{\theta \in \Theta_\rho} \left| f(x_0 + \rho \cos\theta, y_0 + \rho \sin\theta) - \ell \right| = 0$$

$$\lim_{P \to \infty} f(P) = +\infty \iff \lim_{\rho \to +\infty} \inf_{\theta \in \Theta_\rho} f(x_0 + \rho \cos\theta, y_0 + \rho \sin\theta) = +\infty$$

$$\lim_{P \to \infty} f(P) = -\infty \iff \lim_{\rho \to +\infty} \sup_{\theta \in \Theta_\rho} f(x_0 + \rho \cos\theta, y_0 + \rho \sin\theta) = -\infty.$$

The result of Proposition 2.1 clearly extends in dimension 3. If $f : A \subset \mathbb{R}^3 \to \mathbb{R}$ and $P_0 \in \overline{\mathbb{R}^3}$ is a limit point, for any $\rho > 0$ one takes the intersection of A with the sphere of radius ρ, and hence considers the set Ω_ρ consisting of the angles $(\theta, \varphi) \in [0, \pi] \times [0, 2\pi)$ for which

$$(x_0 + \rho \sin\theta \cos\varphi, y_0 + \rho \sin\theta \sin\varphi, z_0 + \rho \cos\theta) \in A.$$

Then the following are equivalent:

(i) $\lim_{P \to P_0} f(P) = \ell$

(ii) $\lim_{\rho \to 0} \sup_{(\theta,\varphi) \in \Omega_\rho} \left| f(x_0 + \rho \sin\theta \cos\varphi, y_0 + \rho \sin\theta \sin\varphi, z_0 + \rho \cos\theta) - \ell \right| = 0.$

The reader is invited to write out all other possibilities in dimension 3.

2.3 Continuity

In this section, A denotes a non empty subset of $\mathbb{R}^n$. Often A is assumed to be open but this is actually not always necessary. Whenever differentiability notions at a point $P_0 \in A$ are treated, like in Definition 3.1 or in Definition 3.2, it is enough to assume that A has non empty interior and that P_0 is an interior point.

Definition 2.5 (*Continuity of scalar functions*) Let $f : A \to \mathbb{R}$ and suppose that $P_0 \in A$ is a limit point of A. The function $f : A \to \mathbb{R}$ is said to be *continuous* at P_0 if

$$\lim_{P \to P_0} f(P) = f(P_0).$$

If P_0 is not a limit point of A, then f is said to be continuous at P_0. If B is a non empty subset of A and f is continuous at every point of B, then f is said to be continuous on B, and if f is continuous on A then f is simply said to be continuous.

By definition, a function is automatically continuous at the isolated points of A. As in the case of functions of a single variable, this fact is required in order to comply with the general definition of continuity, according to which a function is continuous if the inverse image of an open set (of $\mathbb{R}$) is open (in the relative topology of the domain as a subset of $\mathbb{R}^n$).

Continuity at the limit point P_0 can be reformulated by saying that for every $\varepsilon > 0$ there exists $\delta > 0$ such that whenever $P \in A$ satisfies $\|P - P_0\| < \delta$ it then follows that $|f(P) - f(P_0)| < \varepsilon$. This suggests how to extend the notion of continuity to vector valued functions, for if $F \colon A \to \mathbb{R}^m$, then one asks that for every $\varepsilon > 0$ there exists $\delta > 0$ such that whenever $P \in A$ satisfies $\|P - P_0\| < \delta$ it then follows that $\|F(P) - F(P_0)\| < \varepsilon$, where of course the former norm is that of $\mathbb{R}^n$ and the latter is that of $\mathbb{R}^m$. This is the correct notion. Alternatively, one may adopt the definition that follows, which turns out to be equivalent to the previous one.

Definition 2.6 (*Continuity of vector valued functions*) Let the vector valued function $F \colon A \to \mathbb{R}^m$ have components $(f_1, \ldots, f_m)$ and suppose that $P_0 \in A$ is a limit point of A. Then $F \colon A \to \mathbb{R}$ is said to be continuous at P_0 if every component f_j is continuous at P_0, $j = 1, \ldots, m$. If P_0 is not a limit point of A, then F is continuous at P_0. If B is a non empty subset of A and F is continuous at every point of B, then F is said to be continuous on B, and if F is continuous at A then F is simply said to be continuous.

In what follows, the results are formulated for scalar functions for simplicity. The reader is urged to check which of them can be extended to the case of vector valued functions and to write the corresponding statement.

Proposition 2.2 (Sequential continuity) *Suppose that $P_0 \in A$ is a limit point for A and take $f \colon A \to \mathbb{R}$. The following are equivalent:*

(i) *f is continuous at P_0;*
(ii) *for every sequence $(P_n)_{n \geq 0}$ of points of A such that $P_n \to P_0$ it holds that $\lim_n f(P_n) = f(P_0)$.*

Proposition 2.3 (Permanence of sign) *Suppose that $P_0 \in A$ is a limit point for A, take a continuous function $f \colon A \to \mathbb{R}$ and assume $f(P_0) \neq 0$. Then there exists a neighborhood of P_0 in the points of which f has the same sign as $f(P_0)$.*

The above Proposition is useful to establish whether a subset of $\mathbb{R}^n$ is open or not.

Corollary 2.1 *Suppose that $f \colon \mathbb{R}^n \to \mathbb{R}$ is continuous. Then for any $a \in \mathbb{R}$*

(i) *the sets $\{P \in \mathbb{R}^n : f(P) > a\}$ and $\{P \in \mathbb{R}^n : f(P) < a\}$ are open;*
(ii) *the sets $\{P \in \mathbb{R}^n : f(P) \geq a\}$ and $\{P \in \mathbb{R}^n : f(P) \leq a\}$ are closed.*

2.4 Topological Coda

Proposition 2.4 (Algebra of continuous functions) *Suppose that $f, g: A \to \mathbb{R}$ are continuous at $P_0 \in A$. Then also the functions $|f|$, $f + g$, fg, λf with $\lambda \in \mathbb{R}$ and f/g, if $g(P_0) \neq 0$, are continuous at P_0.*

From Proposition 2.4 it follows that linear combinations of continuous functions at a point are continuous at that point, so that the set

$$C(B) = \{f: B \to \mathbb{R} : f \text{ is continuous on } B\}$$

is a real vector space for any subset B of $\mathbb{R}^n$.

In the next proposition, the usual hypotheses under which the composition $g \circ f$ makes sense are tacitly assumed, as are those under which the various continuities are meaningful.

Proposition 2.5 (Continuity of compositions) *If f is continuous at P_0 and g is continuous at $f(P_0)$, then $g \circ f$ is continuous at P_0.*

Large families of elementary functions are continuous. All polynomials are continuous on $\mathbb{R}^n$ and every rational function is continuous (on its domain). For example, $f(x, y) = 1/xy$ is continuous (on its domain, which is $\mathbb{R}^2 \setminus \{\text{axes}\}$).

Proposition 2.5 holds true even in the general case in which one considers also vector valued functions, provided that all the indicated compositions make sense. For example consider the case where $I \subset \mathbb{R}$ is an open interval and $\gamma: I \to \mathbb{R}^3$ is a curve, that is, a continuous vector valued function $\gamma(t) = (x(t), y(t), z(t))$, representing the time evolution of a particle. If $d: \mathbb{R}^3 \to \mathbb{R}$ is the distance function from the origin, namely

$$d(x, y, z) = \sqrt{x^2 + y^2 + z^2},$$

then the composition $d \circ \gamma : I \to \mathbb{R}$ is a real valued function of a single variable that represents the distance from the origin of the particle at any given time $t \in I$. As the distance function is clearly continuous, $d \circ \gamma$ is continuous provided that each component $x(t)$, $y(t)$ and $z(t)$ is such.

2.4 Topological Coda

Many of the topological properties discussed in Chap. 1 can be analyzed very efficiently by means of continuous maps. The main fact on which this general statement rests is the very basic Theorem 2.7 below. Recall that given a (possibly vector valued) map $F: A \subset \mathbb{R}^n \to \mathbb{R}^m$ and a subset Ω of $\mathbb{R}^m$, the *inverse image* of Ω under F is by definition

$$F^{-1}(\Omega) = \{P \in A : F(P) \in \Omega\}.$$

Theorem 2.7 *Take a map* $F \colon A \subset \mathbb{R}^n \to \mathbb{R}^m$. *The following are equivalent:*

(i) *F is continuous on A;*
(ii) *the inverse image of every open set* $\Omega \subseteq \mathbb{R}^m$ *has the form* $F^{-1}(\Omega) = A \cap \mathcal{O}$, *where* $\mathcal{O} \subseteq \mathbb{R}^n$ *is an open set;*
(iii) *the inverse image of every closed set* $K \subseteq \mathbb{R}^m$ *has the form* $F^{-1}(K) = A \cap \mathcal{C}$ *where* $\mathcal{C} \subseteq \mathbb{R}^n$ *is a closed set.*

Because of the previous result, it is tempting to extend the notions of open and closed sets in $\mathbb{R}^n$ and introduce the notions of open and closed sets of a fixed subset $A \subseteq \mathbb{R}^n$ by declaring that $E \subseteq A$ is open in A if it is of the form $E = A \cap \mathcal{O}$ where $\mathcal{O} \subseteq \mathbb{R}^n$ is an open set in the usual sense. In this case E is said to be an open set in the relative topology of A. Similarly, one defines the closed subsets of A by taking intersections $A \cap \mathcal{C}$ with standard closed subsets $\mathcal{C}$, and calls them closed in the relative topology of A. With this understanding, Theorem 2.7 can be reformulated by saying that the following are equivalent:

(i) *F* is continuous on *A*;
(ii) the inverse image of every open set under *F* is an open set in the relative topology of *A*;
(iii) the inverse image of every closed set under *F* is a closed set in the relative topology of *A*.

Obviously, the above equivalence is true in particular when $A = \mathbb{R}^n$ and $m = 1$. It is now clear that if $f \colon \mathbb{R}^n \to \mathbb{R}$ is known to be continuous, then every set of the form $f^{-1}(\Omega)$ with Ω open in $\mathbb{R}$ is open in $\mathbb{R}^n$. Hence, if f is continuous, then:

- all sets of the form $f^{-1}((a,b))$, $f^{-1}((a,+\infty))$, $f^{-1}((-\infty,b))$ are open
- all sets of the form $f^{-1}([a,b])$, $f^{-1}([a,+\infty))$, $f^{-1}((-\infty,b])$ are closed.

This gives a quick way to establish when sets are closed or open.

The notions of relative open and closed sets to reinterpret Definition 1.12 by saying that a set $A \subset \mathbb{R}^n$ is disconnected if it is the union $A = V_1 \cup V_2$ of two nonempty and disjoint open sets V_1 and V_2 in the relative topology of A. If this does not happen, then A is said to be connected. Again, continuity may be of help in establishing connectedness issues, this time via direct images.

Proposition 2.6 *The image* $F(E)$ *of a connected set* $E \subseteq A \subseteq \mathbb{R}^n$ *under a continuous map* $F \colon A \to \mathbb{R}^m$ *is connected.*

It is worth observing that the 1-dimensional analogue of Proposition 2.6 is that the continuous image of an interval is an interval (Corollary 6.3 in [1]). Similarly continuous maps carry compact sets into compat sets, as stated next.

Proposition 2.7 *The image* $F(K)$ *of a compact set* $K \subseteq A \subseteq \mathbb{R}^n$ *under a continuous map* $F \colon A \to \mathbb{R}^m$ *is compact.*

The 1-dimensional analogue of Proposition 2.7 is that the continuous image of closed and bounded set is closed and bounded (Corollary 6.3 in [1] treats the case of intervals).

2.4.1 Curves

Curves are a pervasive object in Mathematics. Although there are several different notions of *curve*, the most widely used one owes its popularity to Mechanics, where it stands squarely at its very foundations because it models the trajectory in time of a massive point.

Definition 2.7 A curve is a continuous map $\gamma \colon I \to \mathbb{R}^n$ where $I \subseteq \mathbb{R}$ is an interval. When its components are to be emphasized, the curve is written as

$$\gamma(t) = (x_1(t), \ldots, x_n(t)).$$

In dimension 2 or 3 common notation is

$$\gamma(t) = (x(t), y(t)), \qquad \gamma(t) = (x(t), y(t), z(t)).$$

The *derivative* of a curve, provided that each of its components $x_j \colon I \to \mathbb{R}$ is of class C^1, is the curve $\dot\gamma \colon I \to \mathbb{R}^n$ given by

$$\dot\gamma(t) = (\dot x_1(t), \ldots, \dot x_n(t)) := \left(\frac{d}{dt}x_1(t), \ldots, \frac{d}{dt}x_n(t)\right) \tag{2.1}$$

often referred to as the *velocity* of the curve. The set $\gamma(I) \subset \mathbb{R}^n$ is called the *image* of the curve. A curve is called:

(i) a *path or an arc* if $I = [a, b]$;
(ii) *simple* if it is injective on the interior $\mathring{I}$;
(iii) *closed, or a loop*, if it is an arc and $\gamma(a) = \gamma(b)$;
(iv) a *Jordan curve*, if it is a simple loop.

The reader can visualize some of the above notions in Fig. 2.1. A few comments are in order. The variable of a curve is usually denoted by t, implicitly appealing to the idea of time. It is also very common to adopt Physicists' notation (2.1) for derivatives, which is spelled "γ-dot". It is of crucial importance to understand that a curve is a map, and not its image. Different curves may well have the same image as, for example, the arcs $\gamma_1, \gamma_2 \colon [0, 2\pi] \to \mathbb{R}^2$ given by

$$\gamma_1(t) = (\cos t, \sin t), \qquad \gamma_2(t) = (\cos 2t, \sin 2t).$$

They are different but have the same image, namely the unit circle, which is covered exactly once by γ_1 and twice by γ_2. Both are loops, γ_1 is simple but γ_2 is not because $\gamma_2(\pi/2) = \gamma_2(3\pi/2) = (-1, 0)$. Thus, the property of being simple cannot be detected from the image, though the presence of self-intersections reveals that the curve is not simple. Notice also that that γ_1 and γ_2 have different velocities.

Sometimes, curves, like points in Euclidean space, need to be thought of as column vectors. The image $\gamma(I)$ of a curve is necessarily a connected set because γ is

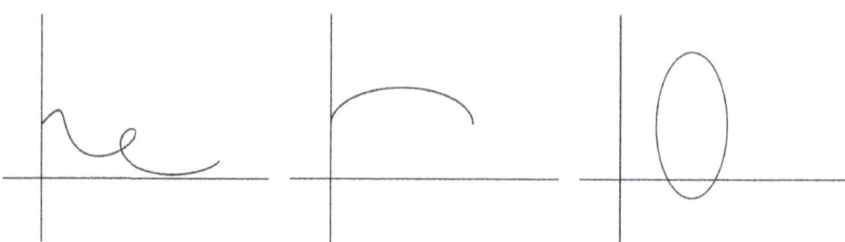

Fig. 2.1 Examples of images of curves. The curve on the left is neither simple (for it has a self-intersection) nor closed, the curve in the center is possibly simple but not closed, the curve on the right is closed and could be simple

Fig. 2.2 A connected set which is not path connected, namely the graph of $x \mapsto \sin(1/x)$ together with the limiting segment on the y-axis

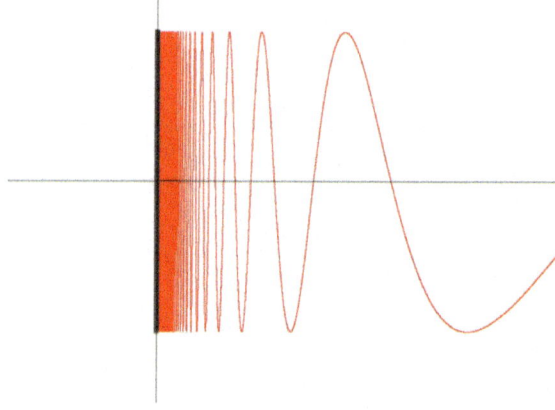

continuous and I is connected by definition. This crucial observation actually leads to the following important definition.

Definition 2.8 A subset $A \subseteq \mathbb{R}^n$ of Euclidean space is said to be *path connected, or arcwise connected* if for any pair of points $P, Q \in A$ there exists a path $\gamma : [a, b] \to A$ such that $\gamma(a) = P$ and $\gamma(b) = Q$.

Observe that a path connected set is necessarily connected, though the converse does not hold true. The planar set $A = \{(x, \sin(1/x)) \in \mathbb{R}^2 : x > 0\} \cup (\{0\} \times [-1, 1])$ turns out to be connected but not path connected. The details are left to the motivated reader. Likewise, convex sets and polygonally connected sets are all path connected, but the converse statements are false (Fig. 2.2).

2.5 Guided Exercises

Below is a list of basic guided exercises on limits and continuity. More guided exercises on continuity are to be found in the next chapter, where differentiability issues are considered.

2.1 Make an educated guess for the value of

$$\lim_{(x,y)\to(2,0)} \frac{x^3 - 6x^2 + 8x + xy^2 - 2y^2}{x - 2}$$

and then prove that the result is true by means of the definition.

Answer. Observe first that the domain of the rational function f at hand is manifestly $\mathbb{R}^2 \setminus \{(x, y) \in \mathbb{R}^2 : x = 2)\}$, so that $(2, 0)$ is a limit point of it. For $x \neq 2$ it is

$$f(x, 0) = \frac{x^3 - 6x^2 + 8x}{x - 2} = \frac{x(x - 2)(x - 4)}{x - 2} = x(x - 4)$$

and

$$\lim_{x \to 2} x(x - 4) = -4.$$

Therefore, the only possible value of the limit is -4. Furthermore,

$$|f(x, y) + 4| = \left| \frac{x^3 - 6x^2 + 8x + xy^2 - 2y^2}{x - 2} + 4 \right|$$

$$= \left| \frac{x^3 - 6x^2 + 8x + xy^2 - 2y^2 + 4x - 8}{x - 2} \right|$$

$$= \left| \frac{(x - 2)^3 + y^2(x - 2)}{x - 2} \right|$$

$$= \left| (x - 2)^2 + y^2 \right|.$$

Hence, for any fixed $\varepsilon > 0$ there exists $\delta = \sqrt{\varepsilon} > 0$ such that

$$0 < \|(x, y) - (2, 0)\| = \sqrt{(x - 2)^2 + y^2} < \delta = \sqrt{\varepsilon}$$
$$\implies |f(x, y) + 4| = (x - 2)^2 + y^2 < \varepsilon,$$

as desired.

This is a most basic example, which simply shows how to apply the definition.

2.2 Decide whether the limit $\lim_{(x,y)\to(0,0)} \dfrac{xy^2}{x^2 + y^4}$ exists or not, and, if yes, compute it.

Answer. The domain of the rational function f at hand is $\mathbb{R}^2 \setminus \{(0, 0)\}$, so that $(0, 0)$ is a limit point of it. On any (punctured) straight line of the form $y = mx$ it is

$$f(x, mx) = \frac{m^2 x^3}{x^2 + m^4 x^4} = \frac{m^2 x}{1 + m^4 x^2} \to 0$$

as $x \to 0$. Hence the limit, if existing, is necessarily equal to 0. However, on the parabola $x = y^2$, for $y \neq 0$, it is

$$f(y^2, y) = \frac{y^4}{y^4 + y^4} = \frac{1}{2}$$

hence the limit does not exist.

Here we see that walking along straight paths through the origin lead to zero, but the curve $x = y^2$ is fully contained in the level set $L_{1/2}$, so on this other path which approaches the origin f stays well above 0.

2.3 Compute, if existing, the limit as $(x, y) \to (0, 0)$ of $f(x, y) = \dfrac{xy}{\sqrt{x^2 + xy + y^2}}$.

Answer. Evidently,

$$\begin{aligned} \mathrm{Dom}(f) &= \{(x, y) \in \mathbb{R}^2 : x^2 + xy + y^2 > 0\} \\ &= \{(x, y) \in \mathbb{R}^2 : (x + \tfrac{1}{2}y)^2 + \tfrac{3}{4}y^2 > 0\} = \mathbb{R}^2 \setminus \{(0, 0)\}. \end{aligned}$$

Therefore the origin is a limit point of $\mathrm{Dom}(f)$. Next, on the punctured straight lines where $y = mx$ and $x \neq 0$ it is

$$f(x, mx) = \frac{mx^2}{\sqrt{x^2 + mx^2 + m^2 x^2}} = \frac{mx^2}{|x|\sqrt{1 + m + m^2}} = \frac{m|x|}{\sqrt{1 + m + m^2}},$$

and $f(x, mx) \to 0$ for $x \to 0$ for every $m \in \mathbb{R}$. Hence, if existing, the limit is necessarily 0. Using the polar coordinates

$$\lim_{(x,y)\to(0,0)} f(x, y) = 0 \quad \Longleftrightarrow \quad \lim_{\rho \to 0} \sup_{\theta \in [0, 2\pi]} |f(\rho \cos \theta, \rho \sin \theta)| = 0.$$

Now, for $\rho > 0$ and $\theta \in [0, 2\pi)$ it is

$$|f(\rho \cos \theta, \rho \sin \theta)| = \left| \frac{\rho^2 \cos \theta \sin \theta}{\rho \sqrt{\cos^2 \theta + \cos \theta \sin \theta + \sin^2 \theta}} \right| \leq \frac{\rho}{\sqrt{1 + \tfrac{1}{2} \sin 2\theta}} \leq \sqrt{2} \rho.$$

It follows that $\sup_{\theta \in [0, 2\pi]} |f(\rho \cos \theta, \rho \sin \theta)| \leq \sqrt{2} \rho$ and therefore

$$\lim_{\rho \to 0} \sup_{\theta \in [0, 2\pi)} |f(\rho \cos \theta, \rho \sin \theta)| = 0,$$

showing that the limit is indeed 0.

2.5 Guided Exercises

This is a classical instance of a polar coordinate uniformity argument. First one guesses what the limit is by handy restrictions, typically along straight lines, and then one tests if the candidate limit (0 in this case) is "reached" uniformly with respect to the angle of approach. This is done by very simple estimates that show that the distance of f from the limit is smaller than a function of the distance itself which tends to 0, indipendently of the direction.

2.4 Compute, if existing, the limit as $(x, y) \to (0, 1)$ of $f(x, y) = \dfrac{x^2(y-1)}{(y-1)^2 - x^2}$.

Answer. Clearly,

$$\text{Dom}(f) = \{(x, y) \in \mathbb{R}^2 : (y-1)^2 - x^2 \neq 0\}$$
$$= \{(x, y) \in \mathbb{R}^2 : (y - 1 - x)(y - 1 + x) \neq 0\}.$$

Now, the locus where $(y-1-x)(y-1+x) \neq 0$ is the complement, in the plane, of the two straight lines $y = 1 \pm x$, which intersect precisely at $P_0 = (0, 1)$. Hence P_0 is a limit point of $\text{Dom}(f)$. In order to guess what the possible limit is, observe that $f(x, 1) = 0$, so that that, if existing, the limit is 0. To check uniformity in polar coordinates, put

$$\Theta = [0, 2\pi) \setminus \left\{\frac{\pi}{4}, \frac{3\pi}{4}, \frac{5\pi}{4}, \frac{7\pi}{4}\right\}$$

for the set of directions that are sensible in this context. Now,

$$\sup_{\theta \in \Theta} |f(\rho \cos\theta, 1 + \rho \sin\theta)| = \sup_{\theta \in \Theta} \left|\frac{\rho^3 \cos^2\theta \sin\theta}{\rho^2(\sin^2\theta - \cos^2\theta)}\right|$$
$$= \sup_{\theta \in \Theta} \left|\frac{\rho \cos^2\theta \sin\theta}{\cos 2\theta}\right|$$
$$\geq \lim_{\theta \to \pi/4} \left|\frac{\rho \cos^2\theta \sin\theta}{\cos 2\theta}\right| = +\infty$$

for every $\rho > 0$. Hence the limit does not exist.

The natural guess for the candidate limit is of course 0, which is the value that f attains on the horizontal and vertical lines through the limit point. The suspect that the limit might not exist comes from the fact that two lines must be subtracted from the domain of f because the denominator vanishes on them. It is thus natural to look at what happens when approaching one of them along any circle of radius ρ centered at the limit point P_0 (see Fig. 2.3). In this case, the function diverges, thereby showing that the limit does not exist.

2.5 Find the domain of $f(x, y) = \dfrac{\sqrt{(x-1)^2 - y - 1}}{(x+y)^2}$ and compute, if existing, its limit as $(x, y) \to (1, -1)$.

Fig. 2.3 The domain of the function f in Exercise 2.4 is the complement of two lines intersecting at the limit point. Along the circles centered at it, f diverges when approaching the two lines

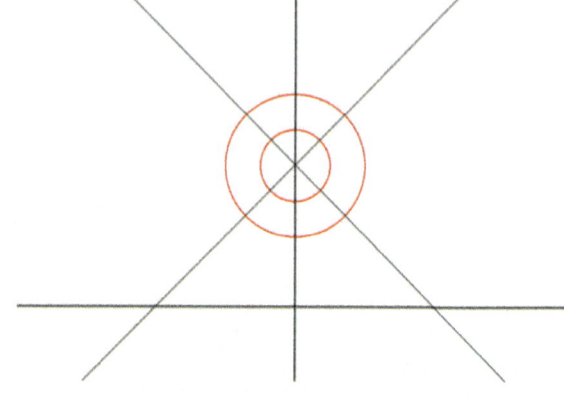

Fig. 2.4 The domain of the function f in Exercise 2.5 and the limit point

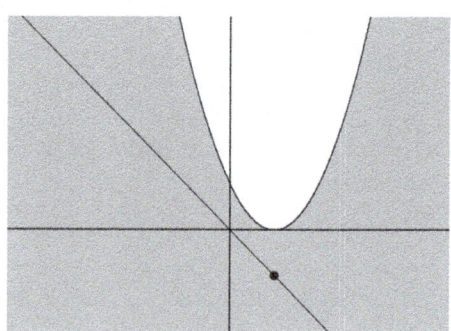

Answer. Clearly,

$$\text{Dom}(f) = \{(x, y) \in \mathbb{R}^2 : y \neq -x, \ y \leq (x-1)^2\}$$

and $P_0 = (1, -1)$ is therefore a limit point of $\text{Dom}(f)$ (see Fig. 2.4).

Consider first $f(1, y)$ for $y \leq 0$ and $y \neq -1$, namely

$$f(1, y) = \frac{\sqrt{-y} - 1}{(1+y)^2} = \frac{-(1+y)}{(1+y)^2(\sqrt{-y} + 1)} = \frac{-1}{(1+y)(\sqrt{-y} + 1)}.$$

Since

$$\lim_{y \to (-1)^{\pm}} \frac{-1}{(1+y)(\sqrt{-y} + 1)} = \mp\infty$$

the limit does not exist.

Here a simple restriction argument on a vertical line exhibits two different behaviours approching the limit point from different sides. The choice is suggested by the presence of y (and not y^2) inside the square root. In order to get rid of the square root a simple algebraic manipulation is enough.

2.5 Guided Exercises

2.6 Compute, if existing, the limit as $(x, y) \to (0, 0)$ of

$$f(x, y) = \frac{\arctan(xy \log(x^2 + y^2 + 1))}{(x^2 + y^2)^2}.$$

Answer. The domain of f is the punctured plane $\mathbb{R}^2 \setminus \{(0, 0)\}$ and hence $(0, 0)$ is a limit point thereof. For $x \neq 0$ it is $f(x, 0) = 0 \to 0$ as $x \to 0$. Hence the only candidate limit is 0. The same holds for $f(0, y)$. For $xy \neq 0$

$$f(x, y) = \frac{\arctan(xy \log(x^2 + y^2 + 1))}{xy \log(x^2 + y^2 + 1))} \frac{\log(x^2 + y^2 + 1)}{x^2 + y^2} \frac{xy}{x^2 + y^2},$$

and the first two factors both converge to 1 for $(x, y) \to (0, 0)$ (see the one-variable limits in [1]). The third factor, however, does not have a limit because, upon setting $g(x, y) = xy/(x^2 + y^2)$, it is $g(x, x) = 1/2$, while $g(x, 0) = g(0, y) = 0$. Thus, the required limit does not exist.

Just one observation: the product $\varphi \psi$ of two functions one of which, say φ, converges to a limit $\ell \neq 0$ and the other one, say ψ, does not have a limit, does not have a limit. Indeed, if it did tend to L (finite or not), then $\psi = (\varphi \psi)/\varphi$ would tend to L/ℓ, contrary to assumption.

2.7 Compute, if existing, the limit as $(x, y) \to (0, 0)$ of the restriction of

$$f(x, y) = \frac{x^2 + y^2}{2x}.$$

either to $A = \{(x, y) \in \mathbb{R}^2 : x^2 + y^2 - x < 0\}$ or to $B = \{(x, y) \in \mathbb{R}^2 : 0 < |y| < x\}$.

Answer. Observe first that f has a natural expression in polar coordinates, namely

$$f(\rho \cos \theta, \rho \sin \theta) = \frac{\rho}{\cos \theta},$$

which is well defined precisely when $x \neq 0$, that is, $\theta \in [-\pi, \pi) \setminus \{\pm \pi/2\}$ and $\rho > 0$. Now, upon fixing θ, the limit is clearly 0. Observe that in polar coordinates the set A corresponds to $A' = \{(\rho, \theta) : 0 < \rho < \cos \theta, -\frac{\pi}{2} < \theta < \frac{\pi}{2}\}$. However,

$$\sup_{\theta \in (-\frac{\pi}{2}, \frac{\pi}{2})} |f(\rho \cos \theta, \rho \sin \theta) - 0| = +\infty$$

so that the limit does not exist. Finally, in polar coordinates the set B corresponds $B' = \{(\rho, \theta) : 0 < \rho, -\frac{\pi}{4} < \theta < \frac{\pi}{4}\}$ and

$$\lim_{\rho \to 0} \sup_{\theta \in (-\frac{\pi}{4}, \frac{\pi}{4})} |f(\rho \cos \theta, \rho \sin \theta) - 0| = \lim_{\rho \to 0} \sqrt{2}\rho = 0.$$

Therefore the limit of the restriction of f to A does not exist and the limit of the restriction of f to B is 0.

Without using polar coordinates the argument needed to show that the restriction of f to A does not admit a limit would involve, for instance, considering circles of small radius tangent to the origin and contained in A, on which the function is constant and equal to the radius, which implies that there cannot be a limit. This amounts to lack of uniformity, a fact which is very clear in polar coordinates. Hence it is natural to use them. The case B does not display the same problem because B is a horizontal cone with aperture $\pi/2$ centered around the positive x-axis and hence θ is bounded away from $\pm\pi/4$.

2.8 Compute, if existing, the limit as $(x, y) \to (0, 0)$ of the function

$$f(x, y) = \frac{x^2 \log(1 + x^2 - 2y^2) + \sin(y^4)}{x^4 + y^4}$$

and of the restriction of f to $A = \{(x, y) \in \mathbb{R}^2 : |y| \le x^2\}$.

Answer. The domain of f is the set

$$B = \{(x, y) \ne (0, 0) : 2y^2 - x^2 < 1\},$$

which, for (x, y) sufficiently close to the origin (but different from the origin), contains the lines $\{x = y\}$ and $\{y = 0\}$. Since

$$f(x, x) = \frac{x^2 \log(1 - x^2) + \sin(x^4)}{2x^4} \to 0$$

for $x \to 0$ whereas

$$f(x, 0) = \frac{x^2 \log(1 + x^2)}{x^4} \to 1$$

for $x \to 0$, the limit as $(x, y) \to (0, 0)$ of f does not exist.

Consider now the restriction of f to A. It is not difficult to see that for small enough $r > 0$, it is

$$(B(O, r) \cap A) \setminus \{O\} \subseteq B,$$

that is, all points in a sufficiently small punctured neighborhood of the origin that belong to A are in the domain of f (see Fig. 2.5). Therefore the origin is an accumulation point of $A \cap B$. Observing that $x^2 - 2y^2 \to 0$ as $(x, y) \to (0, 0)$, it is possible to use the McLaurin expansion

$$\log(1 + t) = t + t\omega(t)$$

with $\omega(t) \to 0$ as $t \to 0$. Hence

$$x^2 \log(1 + x^2 - 2y^2) = x^2[(x^2 - 2y^2) + (x^2 - 2y^2)\omega(x^2 - 2y^2)]$$

2.5 Guided Exercises

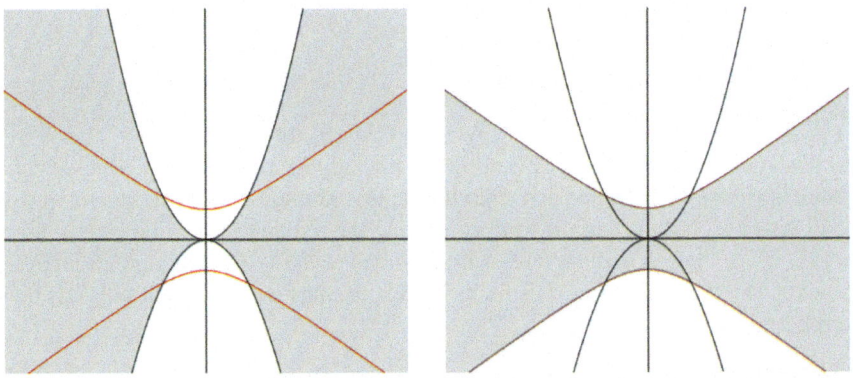

Fig. 2.5 On the left the set A (grey) and on the right the set B (grey) mentioned in Exercise 2.8; the loci $|y| = x^2$ and of $2y^2 - x^2 = 1$ are also depicted in black and in red, respectively

$$= x^4 - 2y^2 x^2 + x^2 (x^2 - 2y^2) \omega (x^2 - 2y^2)$$

Furthermore, $\sin y^4 \sim y^4$ as $y \to 0$ so that

$$f(x, y) = \frac{(x^2 - y^2)^2 + (x^4 - 2y^2 x^2) \omega (x^2 - 2y^2)}{x^4 + y^4}.$$

Now,

$$\frac{x^4 - 2y^2 x^2}{x^4 + y^4}$$

is bounded because $(x^2 - y^2)^2 \geq 0$ implies that $2y^2 x^2 \leq x^4 + y^4$ which, in turn, implies

$$\left| \frac{x^4 - 2y^2 x^2}{x^4 + y^4} \right| \leq \frac{x^4 + x^4 + y^4}{x^4 + y^4} \leq 2.$$

It follows that

$$\frac{(x^4 - 2y^2 x^2)}{x^4 + y^4} \omega (x^2 - 2y^2) \to 0.$$

Finally

$$\frac{(x^2 - y^2)^2}{x^4 + y^4} = 1 - \frac{2x^2 y^2}{x^4 + y^4}$$

and the assumption $|y| \leq x^2$ gives

$$\left| \frac{2x^2 y^2}{x^4 + y^4} \right| \leq \frac{2x^2 x^4}{x^4} \leq 2x^2 \to 0,$$

so that
$$\lim_{(x,y)\to(0,0)} f|_A(x,y) = 1.$$

This exercise requires the use of Taylor expansions for the numerator in order to achieve rational expressions. This allows to treat separately the main term, which is genuinely a rational function and tends to 1 under the assumption that $(x, y) \in A$ (for otherwise it wouldn't) and the term containing the remainder ω, which vanishes as $(x, y) \to (0, 0)$ even without the assumption that $(x, y) \in A$. Although the analysis is not trivial, the main idea of using Taylor expansions is natural and suggests how to proceed.

2.9 Compute, if existing, the limit as $(x, y) \to (0, 0)$ of the function $f(x, y) = y^x$.

Answer. The function $f(x, y) = y^x = e^{x \log y}$ is defined only for $y > 0$. Along the half-line $\{(x, x) : x > 0\}$ it is
$$\lim_{x \to 0^+} f(x, x) = \lim_{x \to 0^+} e^{x \log x} = 1$$
because $x \log x \to 0$. However, along the curve $\{(x, e^{-1/x}) : x > 0\}$, the points of which tend to $(0, 0)$ as $x \to 0^+$, it is
$$f(x, e^{-1/x}) = e^{x \log(e^{-1/x})} = e^{\frac{-x}{x} \log e} = e^{-1}.$$

Therefore the limit does not exist.

Here the suspect is $x \log y$. Indeed, along any power curve $x \mapsto y = x^n$ it tends to 0 as $x \to 0$ but it reveals quite clearly a more complex nature. Hence one is lead to consider level sets of the form $x \log y = c$ with $c \neq 0$, namely
$$x \log y = c \iff \log y = \frac{c}{x} \iff y = e^{\frac{c}{x}}.$$

Each of these level sets is a curve $y = y(x)$ for $x \neq 0$ that tends to the origin as $x \to 0^+$ if $c < 0$ and as $x \to 0^-$ if $c > 0$. Thus f takes on all possible values arbitrarily close to the origin.

2.10 Compute, if existing, the limit as $(x, y, z) \to (0, 0, 0)$ of
$$f(x, y, z) = \frac{\sin\left[(x^2 + \arctan y^2)z\right]}{x^2 + y^2 + z^2}.$$

Answer. Clearly, $(0, 0, 0)$ is a limit point of $\text{Dom}(f) = \mathbb{R}^3 \setminus \{(0, 0, 0)\}$. Now, for $z \neq 0$ it is $f(0, 0, z) = 0$ and hence the only possible limit is 0. Furthermore
$$\left|\frac{\sin\left[(x^2 + \arctan y^2)z\right]}{x^2 + y^2 + z^2}\right| \leq \frac{(x^2 + \arctan y^2)|z|}{x^2 + y^2 + z^2}$$

2.5 Guided Exercises

$$\leq \frac{(x^2+y^2)|z|}{x^2+y^2+z^2}$$
$$\leq |z| \leq \sqrt{x^2+y^2+z^2}.$$

Then the limit vanishes as it follows from the definition of limit, because $|f(P)| \leq \|P\|$.

This is a three dimensional example where all there is to do is to obtain the estimate $|f(P)| \leq \|P\|$, which is suggested by the presence of the norm-decreasing trigonometric functions $x \mapsto \sin x$ and $x \mapsto \arctan x$.

2.11 Compute, if existing, the limit as $(x, y, z) \to (0, 0, 0)$ of the function
$$f(x, y, z) = \frac{\sqrt{1-x^2}}{y^2+z^2}.$$

Answer. The function f is defined in the set Ω where the following inequalities are satisfied
$$\begin{cases} x^2 \leq 1 \\ y^2 + z^2 \neq 0, \end{cases}$$

namely the region between the planes $\{x = 1\}$ and $\{x = -1\}$ except the segment in the x-axis with $|x| \leq 1$ (where $y = z = 0$). Observe that if $z \neq 0$ then $(0, 0, z) \in \Omega$, and
$$\lim_{z \to 0} f(0, 0, z) = \lim_{z \to 0} \frac{1}{z^2} = +\infty.$$

It remains to be checked (e.g. using the definition) that the required limit is indeed $+\infty$. Now, if $\|(x, y, z)\| < \delta$ for δ small, say $\delta < \sqrt{3}/2$, then

$$\frac{\sqrt{1-x^2}}{y^2+z^2} \geq \frac{\sqrt{1-\delta^2}}{\delta^2} \geq \frac{1}{2\delta^2}.$$

This shows that for fixed $K > 0$, any small enough δ, say $\delta < \min\{\frac{\sqrt{3}}{2}, \frac{1}{\sqrt{2K}}\}$, is such that the points in Ω with $\|(x, y, z)\| < \delta$ satisfy

$$f(x, y, z) \geq \frac{1}{2\delta^2} > K,$$

as required.

The intuition is that the small denominator (smaller than $\|P\|$) and the numerator bounded below by a sufficiently small positive number ($\sqrt{1-x^2} \geq c$ if $x^2 \leq 1-c^2$, which is true if $\|P\|$ is small enough) cooperate to make the function big. A direct way of proving the result is to reason as above and get that for $\|P\| \to 0$ it holds

$$\frac{\sqrt{1-x^2}}{y^2+z^2} \geq \frac{\sqrt{1-\|P\|^2}}{\|P\|^2} \to +\infty.$$

2.12 Compute, if existing, the limit as $(x, y, z) \to (0, 0, 0)$ of the function

$$f(x, y, z) = \begin{cases} \dfrac{x\sqrt{|x|} + y^2}{|z|} & z^2 > x^2 + y^2 \\ \dfrac{x^3 + z^3}{x^2 + y^2} & z^2 \leq x^2 + y^2. \end{cases}$$

Answer. Inside the cone, that is, when $z^2 > x^2 + y^2$, one has $|z| \neq 0$. Outside the cone, that is when $z^2 \leq x^2 + y^2$, the indicated function is not defined if $x^2 + y^2 = 0$, in which case $z = 0$. Therefore the domain of f is $\mathbb{R}^3 \setminus \{O\}$ and the origin is an accumulation point. Observe that on the z-axis (except the origin of course) the function vanishes, so that the only candidate limit is 0.

Using the spherical coordinates (1.20), with $\rho > 0$, $\theta \in [0, \pi]$ and $\varphi \in [0, 2\pi)$, one must check uniformity, that is the three-dimensional version of Proposition 2.1, namely, writing $\Omega = [0, \pi] \times [0, 2\pi)$

$$\lim_{\rho \to 0} \sup_{(\theta, \varphi) \in \Omega} |F(\rho, \theta, \varphi)| = 0. \tag{2.2}$$

where $F(\rho, \theta, \varphi) = f(\rho \sin\theta \cos\varphi, \rho \sin\theta \sin\varphi, \rho \cos\theta)$. It is easy to see that the sets

$$A = \{(x, y, z) \in \mathbb{R}^3 : z^2 > x^2 + y^2\}, \quad B = \{(x, y, z) \in \mathbb{R}^3 : z^2 \leq x^2 + y^2\}$$

correspond to the values $\theta \in [0, \pi/4) \cup (3\pi/4, \pi]$, and $\theta \in [\pi/4, 3\pi/4]$, respectively. In the former case

$$|F(\rho, \theta, \varphi)| = \frac{\rho^{3/2} \sin\theta \cos\varphi \sqrt{|\sin\theta \cos\varphi|} + \rho^2 \sin^2\theta \sin^2\varphi}{\rho |\cos\theta|} \leq \frac{\sqrt{\rho} + \rho}{\sqrt{2}/2},$$

and in the latter

$$|F(\rho, \theta, \varphi)| = \frac{\rho^3 |\cos^3\varphi \sin^3\theta + \cos^3\theta|}{\rho^2 \sin^2\theta} \leq \frac{2\rho}{1/2}.$$

Therefore (2.2) holds true, which means that $f(x, y, z) \to 0$ as $(x, y, z) \to (0, 0, 0)$.

The geometry of the sets A and B must be understood first. A relevant observation is that the punctured z-axis lies inside A and on the punctured z-axis the function vanishes, so that 0 is the only possible limit value. Some of the geometry, furthermore, is also reflected in the function's symmetries. In the set B, for instance, the homogeneous polynomials appearing in the numerator and denominator allow, if written in spherical coordinates, to collect the appropriate powers of ρ and hence the function takes the form $F(\rho, \theta, \varphi) = \rho g(\theta)$, where g is a funcion of θ that turns out to be bounded in absolute value. In the set A this is not equally transparent but again the use of spherical coordinates yields the estimate $|F(\rho, \theta, \varphi)| \leq \sqrt{2}(\sqrt{\rho} + \rho)$.

2.5 Guided Exercises

For all practical purposes, the use of spherical coordinates in $\mathbb{R}^3$ is not dissimilar to that of polar coordinates in $\mathbb{R}^2$, in the sense that sometimes the combination of the geometry of the domain and the symmetries of the function can be exploited using these coordinates.

2.13 Compute, if existing, the limit as $\|P\| \to +\infty$ of $f(x, y) = \sqrt{\dfrac{x^2 + y^2}{1 + x^2 y^2}}$.

Answer. The function is defind in the whole plane. Now,

$$f(x, 0) = \sqrt{x^2} = |x| \to +\infty$$

for $x \to \pm\infty$. However,

$$f(x, x) = \sqrt{\dfrac{2x^2}{1 + x^4}} = \sqrt{\dfrac{2}{x^2 + \frac{1}{x^2}}} \to 0$$

for $x \to \pm\infty$. Hence the limit does not exist.

This is a simple limit because on different straight lines through the origin the function behaves differently at infinity. This is suggested by the fact that the rational function inside the square root is made of polynomials of different degrees in the numerator and denominator.

2.14 Compute, if existing, $\displaystyle\lim_{\|(x,y)\| \to +\infty} \dfrac{\sin(x+y)}{x^4 + y^4 + 1}$.

Answer. Clearly, the function at hand is defined in the whole plane, so the limit makes sense. Now, if $x^2 + y^2 \geq 1$, then

$$x^2 + y^2 \leq (x^2 + y^2)^2 = x^4 + y^4 + 2x^2 y^2 \leq 2(x^4 + y^4)$$

so that

$$x^4 + y^4 \geq \dfrac{1}{2}(x^2 + y^2).$$

Therefore

$$\left| \dfrac{\sin(x+y)}{x^4 + y^4 + 1} \right| \leq \dfrac{1}{x^4 + y^4 + 1} \leq \dfrac{2}{\|(x, y)\|^2}$$

whenever $\|(x, y)\| \geq 1$. Therefore the requested limit is 0.

The result is strongly suggested by the fourth powers appearing at the denominator and the bounded numerator. Indeed, outside large balls the fourth powers will dominate the second powers, which appear in the norm.

2.15 Compute, if existing, the limit as $\|(x, y)\| \to +\infty$ of $f(x, y) = \dfrac{e^{x^2 + |y|}}{x + y}$.

Answer. Notice that $\text{Dom}(f) = \{(x, y) \in \mathbb{R}^2 : x + y \neq 0\}$, which is an unbounded set. If $x \neq 0$, then

$$f(x, 0) = \frac{e^{x^2}}{x} \to \pm\infty$$

as $x \to \pm\infty$. Hence the limit at infinity does not exist.

Here the point is that the denominator has a different sign on the two half spaces that constitute the domain of f, whereas the numerator is positive, and in fact diverges at ∞ (show this!). Thus the behaviour of f outside large balls is not uniform.

2.16 Compute, if existing, the limit as $\|(x, y)\| \to +\infty$ of

$$f(x, y) = \frac{x^2 + y^2 - \log(x^2 + y + 1)}{|x| + |y| + 1}.$$

Answer. The domain of f is $D = \{(x, y) \in \mathbb{R}^2 : y > -1 - x^2\}$, so it contains the x-axis. Hence it is natural to observe first that

$$\lim_{x \to \pm\infty} \frac{x^2 - \log(x^2 + 1)}{|x| + 1} = +\infty.$$

In order to see if the limit is indeed $+\infty$, uniformity must be checked. In polar coordinates

$$f(\rho \cos\theta, \rho \sin\theta) = \frac{\rho^2 - \log(\rho^2 \cos^2\theta + \rho \sin\theta + 1)}{\rho(|\cos\theta| + |\sin\theta|) + 1}$$
$$\geq \frac{\rho^2 - \log(\rho^2 + \rho + 1)}{\rho(|\cos\theta| + |\sin\theta|) + 1}$$
$$\geq \frac{\rho^2 - \log(\rho^2 + \rho + 1)}{2\rho + 1},$$

and the right hand side is independent of θ. Therefore, from

$$\lim_{\rho \to +\infty} \frac{\rho^2 - \log(\rho^2 + \rho + 1)}{2\rho + 1} = +\infty$$

it follows that $f(x, y) \to +\infty$ as $\|(x, y)\| \to +\infty$.

The feeling is that the numerator is of order bigger than the denominator in terms of distance from the origin. This is almost exacly correct, in the sense that this function is actually larger in size than a function of ρ alone, for which this statement is completely correct. This observation implicitly entails a uniformity argument, which makes the conclusion legitimate.

2.17 Compute, if existing, the limit as $\|(x, y)\| \to +\infty$ of the function

$$f(x, y) = \frac{1 - 2x^4 - y^4}{3x^2 + 2xy + y^2}.$$

2.5 Guided Exercises

Answer. The domain of f is

$$\{(x, y) \in \mathbb{R}^2 : 3x^2 + 2xy + y^2 \neq 0\} = \{(x, y) \in \mathbb{R}^2 : (x + y)^2 + 2x^2 \neq 0\} = \mathbb{R}^2 \setminus \{(0, 0)\},$$

hence it is unbounded. In order to guess the possible limit, observe that the restriction to the y-axis diverges because

$$\lim_{y \to +\infty} f(0, y) = \lim_{y \to +\infty} \frac{1 - y^4}{y^2} = -\infty.$$

Using polar coordinates, it is immediate to see that

$$f(\rho \cos \theta, \rho \sin \theta) = \frac{1 - \rho^4(2 \cos^4 \theta + \sin^4 \theta)}{\rho^2(3 \cos^2 \theta + 2 \sin \theta \cos \theta + \sin^2 \theta)} = \frac{1 - \rho^4 g(\theta)}{\rho^2 h(\theta)},$$

where

$$g(\theta) = 2 \cos^4 \theta + \sin^4 \theta,$$
$$h(\theta) = 3 \cos^2 \theta + 2 \sin \theta \cos \theta + \sin^2 \theta = (\sin \theta + \cos \theta)^2 + 2 \cos^2 \theta \leq 6.$$

Since g is continuous and positive on $[0, 2\pi]$ it admits a positive minimum at θ_1. This means that for sufficiently large ρ

$$f(\rho \cos \theta, \rho \sin \theta) \leq \frac{1 - \rho^4 g(\theta_1)}{6\rho^2}.$$

Therefore

$$\lim_{\rho \to +\infty} \sup_{0 \leq \theta < 2\pi} f(\rho \cos \theta, \rho \sin \theta) \leq \lim_{\rho \to +\infty} \frac{1 - \rho^4 g(\theta_1)}{6\rho^2} = -\infty,$$

and hence the limit as $\|(x, y)\| \to +\infty$ of f is $-\infty$.

The main feature of this exercise is the use of the one-variable Weierstrass' theorem to obtain an upper bound for f without finding it explicitely.

2.18 Where is the function

$$f(x, y) = \begin{cases} \dfrac{\sqrt{e^{x^2+y^2}} - 1}{|x| + |y|} & (x, y) \neq (0, 0) \\ 1 & (x, y) = (0, 0) \end{cases}$$

continuous?

Answer. The domain of f is $\mathbb{R}^2$. Away from the origin f is the quotient of two continuous functions and the denominator never vanishes. Hence f is continuous on

$\mathbb{R}^2 \setminus \{(0, 0)\}$. As for the origin, observe that along each point of the axes, except of course the origin, f tends to 1:

$$\lim_{t \to 0} f(0, t) = \lim_{t \to 0} f(t, 0) = \lim_{t \to 0} \frac{\sqrt{e^{t^2} - 1}}{|t|} = \lim_{t \to 0} \sqrt{\frac{e^{t^2} - 1}{t^2}} = 1$$

hence the candidate limit is 1. However,

$$f(x, x) = \frac{\sqrt{e^{2x^2} - 1}}{2|x|} = \sqrt{\frac{e^{2x^2} - 1}{4x^2}} \to \frac{1}{\sqrt{2}}.$$

It follows that the limit does not exist. Hence f is not continuous at the origin.

The variables x and y play exactly the same role in f. It is thus natural to check what happens when either of them vanishes or when they are equal. In this case, the resulting function of a single variable exhibits different limits.

2.19 Establish for which values of $\alpha \in \mathbb{R}$ the function

$$f(x, y) = \frac{\left|e^{x+y-1} - 1\right|^\alpha}{(x + y - 1)^2},$$

admits a continuous extension at $(1, 0)$.

Answer. First of all $\text{Dom}(f) = \{(x, y) \in \mathbb{R}^2 : y \neq 1 - x\}$, the plane without a straight line passing through $(1, 0)$. Hence $(1, 0)$ is a limit point of $\text{Dom}(f)$ for every $\alpha \in \mathbb{R}$. On $\text{Dom}(f)$ the function may be expressed as

$$\frac{\left|e^{x+y-1} - 1\right|^\alpha}{|x+y-1|^\alpha} \frac{|x+y-1|^\alpha}{(x+y-1)^2} = \left|\frac{e^{x+y-1} - 1}{x+y-1}\right|^\alpha |x+y-1|^{\alpha-2}.$$

Now, it is easy to see that

$$\lim_{t \to 0} \frac{e^t - 1}{t} = 1 \quad \Longrightarrow \quad \lim_{(x,y) \to (1,0)} \frac{e^{x+y-1} - 1}{x+y-1} = 1. \qquad (2.3)$$

Indeed, from the definition of limit, for every $\varepsilon > 0$ there exists $\delta > 0$ such that if $0 < |t| < \delta$, then

$$\left|\frac{e^t - 1}{t} - 1\right| < \varepsilon.$$

Therefore, whenever $0 < |x + y - 1| < \delta$ it holds that

$$\left|\frac{e^{x+y-1} - 1}{x + y - 1} - 1\right| < \varepsilon.$$

Fig. 2.6 The strip around the straight line without the line itself is is a neighborhood of $(1, 0)$ intersected with $\text{Dom}(f)$ in Exercise 2.19

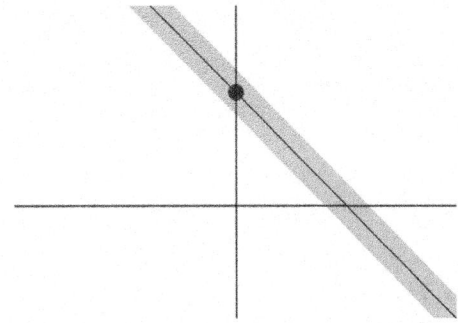

Now, the set

$$\mathscr{U} := \{(x, y) \in \mathbb{R}^2 : 0 < |x + y - 1| < \delta\}$$
$$= \{(x, y) \in \mathbb{R}^2 : -\delta < x + y - 1 < \delta, \ x + y - 1 \neq 0\}$$

is a neighborhood of $(1, 0)$ intersected with $\text{Dom}(f)$, see Fig. 2.6. This proves the claim (2.3). The very same argument shows that

$$\lim_{(x,y) \to (1,0)} |x + y - 1|^{\alpha - 2} = \begin{cases} 0 & \alpha > 2 \\ 1 & \alpha = 2 \\ +\infty & \alpha < 2. \end{cases}$$

It follows that

$$\lim_{(x,y) \to (1,0)} f(x, y) = \begin{cases} 0 & \alpha > 2 \\ 1 & \alpha = 2 \\ +\infty & \alpha < 2. \end{cases}$$

Therefore f admits a continuous extension at $(1, 0)$ if and only if $\alpha \geq 2$.

This exercise is meant to indicate that a suitable version of a change of variable formula for limits in higher dimensions should hold. In this case, one wants to use $x + y - 1$ as a new variable t. It can be shown directly that this is indeed possible because the set $\mathscr{U}$ where $|x + y - 1|$ is small and positive is a good set, namely it is a neighborhood of the limit point intersected with the domain of f, hence precisely of the form where it is required to show that the function is close to the limit. Furthermore, it should be noticed that the same argument would apply to show that the function admits a continuous extension for the same values of α at every point of the line $y = 1 - x$.

2.20 Find $\alpha \in \mathbb{R}$ such that the function

$$f(x,y) = \begin{cases} \dfrac{\sin(x^4 \sin y + x^2 y^2 \cos x)}{x^2 + y^2} & (x,y) \neq (0,0) \\ \alpha & (x,y) = (0,0) \end{cases}$$

is continuous in its domain.

Answer. Clearly the domain of f is $\mathbb{R}^2$ and f is continuous away from the origin since it is a ratio of compositions of continuous functions. In order to study the continuity at $(0,0)$ it is necessary to compute the limit as $(x,y) \to (0,0)$, if it exists. Now, since $f(x,0) = 0$ the only possible limit is 0. Further,

$$\left| \frac{\sin(x^4 \sin y + x^2 y^2 \cos x)}{x^2 + y^2} \right| \leq \frac{|x^4 \sin y + x^2 y^2 \cos x|}{x^2 + y^2}$$

$$\leq \frac{x^4 |\sin y| + x^2 y^2 |\cos x|}{x^2 + y^2}$$

$$\leq \frac{x^4 + x^2 y^2}{x^2 + y^2}$$

$$= x^2$$

$$\leq x^2 + y^2.$$

Then, since $|f(P)| \leq \|P\|^2 \leq \|P\|$ for small $\|P\| \neq 0$, the limit is 0. Therefore f is continuous at $(0,0)$, hence in its domain, if and only if $\alpha = 0$.

This is a standard and elementary estimate on the size of f, dictated by the norm decreasing function $x \mapsto \sin x$ and by the obvious boundedness of the trigonometric functions involved.

2.21 Determine for which $\alpha \in \mathbb{R}$ the function the limit as $(x,y) \to (1,0)$ of the function

$$f(x,y) = \begin{cases} \dfrac{\frac{1}{\sqrt{x-y}} - \frac{1}{\sqrt{e^{x+y}-1}} - y}{|y| + |x-1|} & (x,y) \neq (1,0) \\ \alpha & (x,y) = (1,0) \end{cases}$$

is continuous in its domain.

Answer. The domain of f is $\{(x,y) \in \mathbb{R}^2 : x > y\}$ and $(1,0)$ is an interior point for the domain. Observe that f is continuous in the open set $\text{Dom}(f) \setminus \{(1,0)\}$ where it is expressed as sum, ratio and composition of continuous functions. To establish continuity at $(1,0)$, observe that

$$f(x,y) = \frac{(1+(x-y-1))^{-1/2} - e^{\frac{1}{2}(1-x-y)} - y}{|y| + |x-1|}.$$

2.5 Guided Exercises

Since both $(x - y - 1)$ and $(1 - x - y)$ tend to 0 as $(x, y) \to (1, 0)$, it is natural to use the McLaurin expansions:

$$(1 + t)^{-1/2} = 1 - \frac{1}{2}t + \frac{3}{8}t^2 + t^2 \omega_1(t)$$

$$e^s = 1 + s + \frac{1}{2}s^2 + s^2 \omega_2(s),$$

where $\omega_1(t) \to 0$ and $\omega_2(s) \to 0$ at the origin, that is

$$(1 + (x - y - 1))^{-1/2} = 1 - \frac{1}{2}(x - y - 1) + \frac{3}{8}(x - y - 1)^2 + (x - y - 1)^2 \omega_1(x - y - 1)$$

$$e^{\frac{1}{2}(1-x-y)} = 1 + \frac{1}{2}(1 - x - y) + \frac{1}{8}(1 - x - y)^2 + \frac{1}{4}(1 - x - y)^2 \omega_2(\frac{1}{2}(1 - x - y)).$$

Therefore, substituting one finds that

$$f(x, y) = \frac{-y(x - 1) + \frac{1}{4}[(x - 1)^2 + y^2]}{|x - 1| + |y|}$$

$$+ \frac{(x - y - 1)^2}{|x - 1| + |y|} \omega_1(x - y - 1) - \frac{1}{4}\frac{(1 - x - y)^2}{|x - 1| + |y|} \omega_2(\frac{1}{2}(1 - x - y)).$$

By using polar coordinates centered at $(1, 0)$ the first summand becomes

$$\left| \frac{-\rho^2 \cos\theta \sin\theta + \frac{1}{4}\rho^2}{\rho(|\cos\theta| + |\sin\theta|)} \right| = \rho \left| \frac{\frac{1}{4} - \cos\theta \sin\theta}{|\cos\theta| + |\sin\theta|} \right| \leq \rho \frac{2}{m},$$

where m is the positive minimum of the continuous function $\theta \mapsto |\cos\theta| + |\sin\theta|$. The second summand, using again polar coordinates centered at $(1, 0)$, becomes

$$\frac{\rho^2 (\cos\theta - \sin\theta)^2}{\rho(|\cos\theta| + |\sin\theta|)} \leq \rho \frac{4}{m}$$

for the same m as above. The third summand is

$$\frac{1}{4} \frac{\rho^2 (\cos\theta + \sin\theta)^2}{\rho(|\cos\theta| + |\sin\theta|)} \leq \rho \frac{1}{m}.$$

All three summands tend to 0 uniformly in θ as $\rho \to 0$, *a fortiori* when multiplied by ω_1 and ω_2. Therefore the limit is 0, so that f is continuous at $(1, 0)$, hence in its domain, if and only if $\alpha = 0$

This exercise calls for a Taylor expansion. In this way the function can be written as the sum of a genuine rational function and of two remainder terms, each of which is easily handled in polar coordinates.

2.22 Compute, if existing, the continuous extension of the function

$$f(x, y) = \frac{1}{\sqrt{x^2 + y^2}} \int_0^{x^2 y} \frac{e^t - 1}{t} \, dt$$

at the origin.

Answer. Upon setting

$$h(t) = \begin{cases} \dfrac{e^t - 1}{t} & t \neq 0 \\ 1 & t = 0, \end{cases}$$

which is a continuous function on $\mathbb{R}$, the numerator of f becomes

$$\int_0^{x^2 y} h(t) \, dt,$$

a function well defined in $\mathbb{R}^2$ that vanishes on both the x-axis and the y-axis, as well as f. Therefore the only possible limit for f is 0. Now, in the neighborhood of $(0, 0)$ where $x^2 + y^2 \leq 1$, the function $(x, y) \mapsto |x^2 y|$ is bounded by 1 and so

$$\left| \int_0^{x^2 y} h(t) \, dt \right| \leq \int_0^{|x^2 y|} h(t) \, dt \leq x^2 |y| M$$

where M is an upper bound of h in $[0, 1]$. Therefore

$$|f(x, y)| \leq M \frac{x^2 |y|}{\sqrt{x^2 + y^2}}, \tag{2.4}$$

Using polar coordinates, the right hand side is

$$M \frac{\rho^3 \cos^2 \theta |\sin \theta|}{\rho} \leq M \rho^2.$$

Hence the limit is 0 and the continuous extension is

$$g(x, y) = \begin{cases} f(x, y) & (x, y) \neq (0, 0) \\ 0 & (x, y) = (0, 0). \end{cases}$$

It is quite clear that the main difficulty is that of understanding the behaviour of the integral when $(x, y) \to (0, 0)$. Clearly, its upper bound $x^2 y$ becomes small in size, so the idea is to estimate (roughly) the integral by the product of an upper bound for $|h|$ (which very well behaved, being continuous) and the size of the integration interval. This is precisely (2.4), which is easy to analyze in polar coordinates.

2.5 Guided Exercises

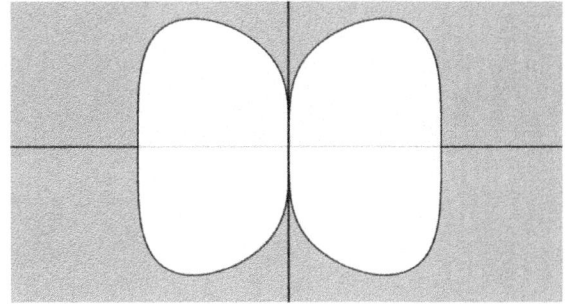

Fig. 2.7 In grey the set A in Exercise 2.23, which also includes the black boundary line, part of the x-axis and the whole y-axis

2.23 Consider $A = \left\{(x, y) \in \mathbb{R}^2 : |x| \leq x^4 + y^4\right\}$. Is A closed? Compact? Convex?

Answer. The function $f : \mathbb{R}^2 \to \mathbb{R}$ defined by $f(x, y) = x^4 + y^4 - |x|$ is continuous and $A = f^{-1}([0, +\infty))$. By Theorem 2.7 A is closed. The points $P_n = (0, n)$ are in A for any positive integer n. However, $\|P_n\| \to +\infty$, so that A is not bounded, hence not compact. Neither is A convex, because although $P = (0, 0) \in A$ and $Q = (2, 0) \in A$, the point $\frac{2}{3}P + \frac{1}{3}Q = (\frac{2}{3}, 0) \notin A$ (Fig. 2.7).

The inequality appearing in the definition of A contains a "$\leq$" sign, so the use of inverse image techniques is more or less obvious. The whole y-axis satisfies this inequality, and this takes care of the second question. Convexity requires a little thought. The point is that, on the x-axis, the inequality $|x| \leq x^4$ holds at the origin and for large values of x, namely $|x| \geq 1$ but not for small values, that is $0 < |x| < 1$.

2.24 Denote by A the domain of the function $f(x, y) = \sqrt{xy - x^3}$. Is A closed? Compact? Path connected?

Answer. First of all $(x, y) \in A$ if and only if $x(y - x^2) \geq 0$. Clearly, $A = g^{-1}([0, +\infty))$ where $g(x, y) = x(y - x^2)$ is obviously continuous, hence, by Theorem 2.7, A is closed. The points $P_n = (0, n)$ are in A for any positive integer n. However, $\|P_n\| \to +\infty$, so that A is not bounded, hence not compact. Finally, it is easily checked that $A = A_1 \cup A_2$ where

$$A_1 = \left\{(x, y) \in \mathbb{R}^2 : x \leq 0, \ y \leq x^2\right\} \quad A_2 = \left\{(x, y) \in \mathbb{R}^2 : x \geq 0, \ y \geq x^2\right\}.$$

Both A_1 and A_2 are path connected. Each point in A_1 may be joined with a vertictal segment to the x-axis remaining in A_1 (see Fig. 2.8). Similarly, each point in A_2 may be joined with a horizontal segment to the y-axis remaining in A_2.
Since the $O \in A_1 \cap A_2$, the set A is path connected.

The first two questions are dealt with exactly in the same way as in Exercise 2.23. To understand connectedness, one needs to have a precise picture of A, because the relevant inequality $x(y - x^2) \geq 0$ indicates that there are two sets to keep track of, in the sense that A is the union of two sets. The basic observation is that they touch at the origin.

Fig. 2.8 In grey the domain of the function in Exercise 2.24; here A_1 is the leftmost part and A_2 the rightmost one

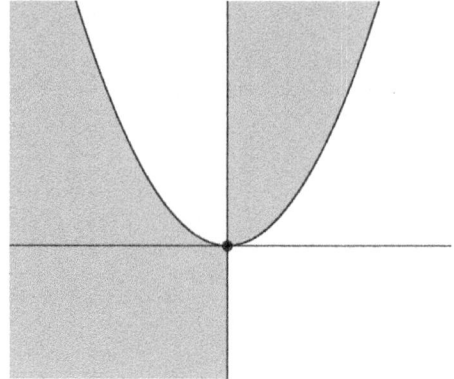

Fig. 2.9 In white the domain of the function in Exercise 2.25; notice that the line $x = y$ is not part of A as well as any of the points in the ellipse

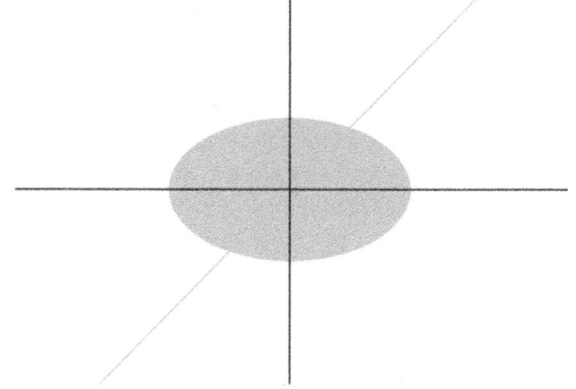

2.25 Denote by A the domain of the function

$$f(x, y) = \log(x^2 + 3y^2 - 1) - \log(|x - y|).$$

Is A open? Bounded? Path connected?

Answer. A is open because it is the intersection of the two open sets

$$A_1 = \left\{(x, y) \in \mathbb{R}^2 : x^2 + 3y^2 - 1 > 0\right\}, \quad A_2 = \left\{(x, y) \in \mathbb{R}^2 : x \neq y\right\}.$$

A is not bounded because the half line $\{(0, y) \in \mathbb{R}^2 : y \geq 1\}$ is in A (Fig. 2.9). Now, A_2 is not connected, being a disjoint union

$$\left\{(x, y) \in \mathbb{R}^2 : x < y\right\} \cup \left\{(x, y) \in \mathbb{R}^2 : x > y\right\}$$

of open sets, whose intersections with A_1 are both open. This shows that A is not connected.

Unlike what happens in Exercise 2.24, here A is the intersecion of two sets, because both logarithms must be well defined, and not the union. Both sets are open hence such will also be their intersection. Large values of y on the y-axis belong to both portions. As for path connectedness, the key intuition is that the line $x = y$ will cut the set in two disconnected pieces.

2.26 Denote by A the domain of the function

$$f(x, y) = \arcsin(1 - 4x^2 + 2y^2).$$

Is A closed? Bounded? Path connected?

Answer. Clearly

$$A = \{(x, y) \in \mathbb{R}^2 : -1 \leq 1 - 4x^2 + 2y^2 \leq 1\},$$

so that $A = A_1 \cap A_2$, where

$$A_1 = \{(x, y) \in \mathbb{R}^2 : 4x^2 - 2y^2 \leq 2\}, \qquad A_2 = \{(x, y) \in \mathbb{R}^2 : -4x^2 + 2y^2 \leq 0\},$$

which are both closed. Hence A is closed. Furthermore, the straight line with equation $y = \sqrt{2}x$ is contained in $A = A_1 \cap A_2$, so that A is not bounded. In order to show that A is connected it is enough to prove that both A_1 and A_2 are path connected because their intersection is not empty. To see that A_1 is path connected observe that if $(x, y) \in A_1$ then $(\lambda x, y) \in A_1$ for every $\lambda \in [0, 1]$ because

$$4x^2 - 2y^2 \leq 2 \implies 4\lambda^2 x^2 - 2y^2 \leq 4x^2 - 2y^2 \leq 2,$$

and the y-axis is contained in A_1. Hence every point in A_1 can be the joined to the origin with a polygonal path contained in A_1. To see that A_2 is path connected observe that if $(x, y) \in A_2$ then the curve $t \mapsto (tx, ty) \in A_2$ for every $t \in [0, 1]$ because

$$-4x^2 + 2y^2 \leq 0 \implies -4(tx)^2 + 2(ty)^2 = t^2(-4x^2 + 2y^2) \leq 0.$$

Therefore every point in A_2 can be the joined to the origin with a segment contained in A_2 (Fig. 2.10).

Here one must understand first the two (clearly closed) sets corresponding to the two obvious inequalities that must be satisfied. As the variables appear with square powers and opposit signs, regions related to the geometry of hyperbolae will certainly play a role. Both of them are path connceted and they meet at the origin.

2.27 Denote by A the domain of the function

$$f(x, y) = \sqrt{y \sin(\pi \cos x)}$$

Draw A. Is A closed? Path connected? Prove that the image of f is $[0, +\infty)$.

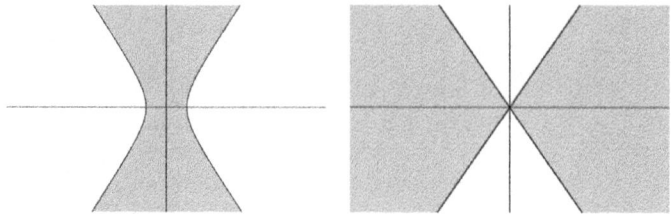

Fig. 2.10 On the left the set A_1 of Exercise 2.26 and on the right the set A_2, both in grey

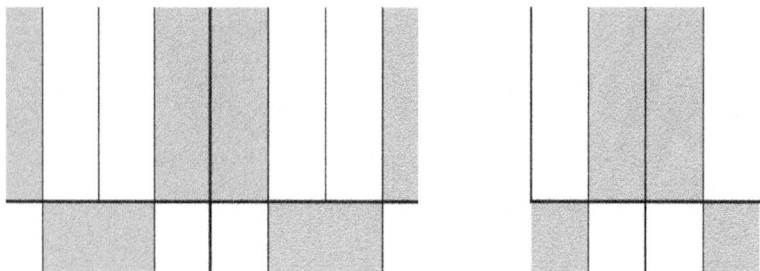

Fig. 2.11 In grey and black the set A of Exercise 2.27, and the part $-\pi \leq x < \pi$ on the right

Answer. Clearly, $A = \{(x, y) \in \mathbb{R}^2 : y \sin(\pi \cos x) \geq 0\}$, and $g(x, y) = y \sin(\pi \cos x)$ is a continuous function on $\mathbb{R}^2$. Hence A is certainly closed. In order to understand A, observe that if $(x, y) \in A$ then also $(x + 2k\pi, y) \in A$ for any integer $k \in \mathbb{Z}$. Hence it is enough to consider $x \in [-\pi, \pi)$. Now, observe that $g(-\pi, y) = g(0, y) = 0$ so that the lines $\{(-\pi, y) : y \in \mathbb{R}\}$ and $\{(0, y) : y \in \mathbb{R}\}$ are contained in A. Next, if $x \in [-\pi/2, \pi/2]$, then

$$0 \leq \cos x \leq 1 \implies 0 \leq \pi \cos x \leq \pi \implies 0 \leq \sin(\pi \cos x) \leq 1$$

so that for all $y \geq 0$ and $x \in [-\pi/2, \pi/2]$ it holds that $y \sin(\pi \cos x) \geq 0$. If instead $x \in (-\pi, -\pi/2) \cup (\pi/2, \pi)$, then

$$-1 < \cos x < 0 \implies -\pi < \pi \cos x < 0 \implies -1 < \sin(\pi \cos x) < 0,$$

so that for all $y \leq 0$ and $x \in (-\pi, -\pi/2) \cup (\pi/2, \pi)$ it holds that $y \sin(\pi \cos x) \geq 0$. Hence the set A is as in Fig. 2.11.

The set A is path connected because the x-axis is contained in A and every point in A may be joined by means of a vertical path to the x-axis remaining in A. Finally, being a square root, necessarily the image is contained in $[0, +\infty)$ and the image of $(\pi/3, y)$ with $y \geq 0$ is $\sqrt{y \sin(\pi/2)} = \sqrt{y}$ which fills $[0, +\infty)$ as y runs in $[0, +\infty)$.

Understanding the set in question requires a little attention. The presence of the cosine and sine functions reveals at once a certain periodicity in the x variable. This means that one can profitably focus on what happens for $x \in [-\pi, \pi)$. The second step is to realize that clearly the sign of $\cos x$ matters, and reveals the details. Path connectedness follows from the fact that the x-axis serves as a connecting line for all the various rectangles or rays whose union is A, because each of them is in fact either a single vertical ray or a union of vertical rays all of which are based on the x-axis. Indeed, they are all either of the form $\{(x_0, y) : y \geq 0\}$ or of the form $\{(x_0, y) : y \leq 0\}$ with x_0 fixed.

2.6 Problems

2.28 Compute if existing the following limits

(a) $\lim\limits_{(x,y)\to\infty} \dfrac{2x+3y}{x^2+y^2}$
(b) $\lim\limits_{(x,y)\to(0,0)} \dfrac{|x-y|}{\sqrt[3]{x^2}+\sqrt[3]{y^2}}$
(c) $\lim\limits_{(x,y)\to\infty} \dfrac{|x-y|}{\sqrt[3]{x^2}+\sqrt[3]{y^2}}$

(d) $\lim\limits_{(x,y)\to(0,0)} \dfrac{x^4+y^4}{x^2+y^2}$
(e) $\lim\limits_{(x,y)\to(0,0)} \dfrac{e^x-e^y}{|x|+|y|}$
(f) $\lim\limits_{(x,y)\to\infty} \dfrac{e^x-e^y}{|x|+|y|}$.

2.29 Consider the function

$$f(x,y) = \frac{\log(1+xy)}{\sqrt{x^2+y^2}}$$

and the set $A = \{(x, y) \in \mathbb{R}^2 : xy \geq 0\}$. Compute, if existing, the limits:

(a) $\lim\limits_{(x,y)\to(0,0)} f(x,y)$
(b) $\lim\limits_{\|(x,y)\|\to+\infty} f|_A(x,y)$.

2.30 Consider the function:

$$f(x,y) = \begin{cases} x\sin\frac{1}{y} + y\sin\frac{1}{x} & xy \neq 0 \\ 0 & xy = 0. \end{cases}$$

Compute, if existing, the limits:

(a) $\lim\limits_{x\to 0}\left[\lim\limits_{y\to 0} f(x,y)\right]$
(b) $\lim\limits_{y\to 0}\left[\lim\limits_{x\to 0} f(x,y)\right]$

(c) $\lim\limits_{(x,y)\to(0,0)} f(x,y)$
(d) $\lim\limits_{(x,y)\to\infty} f(x,y)$.

2.31 Compute, if existing, the following limits:

(a) $\lim_{(x,y,z)\to(0,0,0)} \dfrac{\sin x^2 - \sin y^2 + \sin z^2}{x^2 + y^2 + z^2}$

(b) $\lim_{(x,y)\to(0,0)} \dfrac{y \log(1 + y^3)}{x(x^2 + y^2)}$

(c) $\lim_{(x,y)\to(0,0)} \dfrac{\log(1 + x^2 + y^4)}{x^4 + y^2}$

(d) $\lim_{(x,y)\to(2,0)} \dfrac{\sin(x - 2)y}{x^2 - 4x + 4}$

(e) $\lim_{(x,y)\to\infty} e^{x^4+y^4} \sin\left(\dfrac{1}{x^2 + y^2}\right)$

(f) $\lim_{(x,y)\to(0,0)} \dfrac{x^3 - y^3}{x - y^2}$

(g) $\lim_{(x,y)\to(1,1)} \dfrac{x^3 - y^3}{x - y^2}$

(h) $\lim_{(x,y)\to(0,0)} \dfrac{x^2 \sin y - y^2 \cos x}{\sqrt[3]{x^2 + y^2}}$.

2.32 Compute, if existing and depending where indicated on the values of the parameter $\alpha > 0$, the following limits:

(a) $\lim_{(x,y)\to\infty} \dfrac{x^4 + y^4 - y^2 \sin(x^2 + y^2)}{x^2 + |y|^\alpha}$

(b) $\lim_{(x,y)\to\infty} \dfrac{\sqrt{x^2 + |y \sin y|}}{(y + 1)^2 e^y + \cos x + 2}$

(c) $\lim_{(x,y)\to(0,0)} \dfrac{x \arctan(x^2 - y) + y \sin(x - y^2)}{(x^2 + y^2)^\alpha}$

(d) $\lim_{(x,y)\to(0,0)} \dfrac{\sin x^3 + \sin y^3}{x^4 + y^4}$

(e) $\lim_{(x,y)\to(\frac{\pi}{4},\frac{\pi}{4})} \dfrac{\sin(x - y)^2 - \cos^2(x + y)}{(|x - \frac{\pi}{4}| + |y - \frac{\pi}{4}|)^\alpha}$

(f) $\lim_{(x,y)\to(0,0)} \dfrac{\int_0^{|x|+|y|} \frac{\sin t}{t} dt}{(|x| + |y|)^\alpha}$

(g) $\lim_{(x,y)\to(0,0)} \dfrac{e^{x^2} - \cos y}{\sqrt{x^2 + y^2}}$

(h) $\lim_{(x,y)\to(0,0)} \dfrac{\sqrt{x^2 + y^2}(e^x + y - 1)}{\sin(|x| + |y|)}$.

2.33 Compute, if existing, the following limit:

$$\lim_{(x,y,z)\to(0,0,1)} \dfrac{(x^2 - y^2)[\log(x + y + z) - \sin(y + x)]}{x^2 + y^2 + \tan^2(z - 1)}.$$

2.34 Consider $A = \{(x, y) \in \mathbb{R}^2 : |x| + |y| < 1\}$. Is A bounded, closed, open, convex, compact?

2.35 Consider $A = \{(x, y) \in \mathbb{R}^2 : x \geq x^2 + y^2 - 1\}$. Is A bounded? Is it connected?

2.36 Describe the topological features of the domain and the level sets of the function $f(x, y) = \arcsin(x^2 + y^2 - 2x - 4y)$.

2.37 Let $A_1 = \{(x, y) \in \mathbb{R}^2 : \max\{|x|, |y|\} \leq 1\}$ and $A_2 = \{(x, y) \in \mathbb{R}^2 : |x - 2| + |y| < 1\}$ and consider $A = A_1 \cup A_2$. Is A convex? Is it connected? Compact?

2.38 Consider $A = \left\{(x, y) \in \mathbb{R}^2 : \dfrac{x}{x^2 + y^2 + k^2} \geq 1\right\}$. For which $k \in \mathbb{R}$ is A closed?

2.39 Describe the topological features of the domain and the set of level $c = 0$ of the function $f(x, y) = \log(4x + 2y - x^2 - y^2) - \log 2y$.

2.40 Describe the topological features of the domain and the set of level $c = 0$ of the function $f(x, y) = \sqrt{2x + 2y - x^2 - y^2} - \sqrt{2x - 2y}$.

2.41 Consider the set $A = \{(x, y) \in \mathbb{R}^2 : -2 \leq y \leq |x| + 1, \ 2|x| \leq |y| + 1\}$ and describe its topological features.

2.42 Describe the topological features of the domain and the set of level through the point $(-3, 1)$ of the function $f(x, y) = \sqrt{\dfrac{2y + 3 - y^2}{y^2 - x}}$.

2.43 Consider the function

$$f(x, y) = \begin{cases} \dfrac{\sin(xy)}{\sqrt{x^2 + y^2}} & (x, y) \neq (0, 0) \\ a & (x, y) = (0, 0). \end{cases}$$

For which values of the parameter $a \in \mathbb{R}$ is the set $A_a = \{(x, y) \in \mathbb{R}^2 : f(x, y) < 1\}$ open? For which values is the set $B_a = \{(x, y) \in \mathbb{R}^2 : f(x, y) \leq 1\}$ closed?

Reference

1. Baronti, M., De Mari, F., van der Putten, R., Venturi, I.: Calculus Problems, Unitext, 101, La Matematica per il 3+2. Springer (2016)

Chapter 3
Differentiation

Perhaps the most powerful, yet elementary, tool in the analysis of functions of a single variable is the notion of derivative. It may be argued that it has both a very intuitive geometric flavour, the idea of line tangent to a curve at a point, and a just as natural interpretation as measure of growth. The many far-reaching applications of this one-variable concept strongly suggest that analogous concepts should be available for functions of several variables, and certainly indicate that such analogues are highly desirable. In the search of higher dimensional versions, however, one is immediately faced with the observation that while the idea of plane tangent to a surface at a point is a very direct, and possibly unique, way of extending the idea of line tangent to a curve, the idea of growth encounters basic ambiguities, for one has to first select a direction along which such growth is measured. The natural intuitive similarity between mathematical surfaces and physical terrain, the former being an abstraction of the latter, reveals indeed that when walking in mountaineous regions, upon selecting different directions one may find drastically different paths leading to the same point, and can often choose that with the desired degree of difficulty, namely of steepness. In one variable, on the contrary, there is essentially one and only one walking direction, hence a single possible slope. The analytic notions that arise are that of differentiability, which encodes the existence of the tangent (hyper)plane, and of directional derivative or partial derivative, which are weaker notions tailored to capture phenomena that are essentially one-dimensional.

3.1 Directional and Partial Derivatives

Definition 3.1 (*Directional and partial derivatives, gradient*) Let $A \subseteq \mathbb{R}^n$ be an open set and consider the scalar function $f : A \to \mathbb{R}$. Fix a point $P_0 \in A$ and a unit vector $Q \in \mathbb{R}^n$. If the limit

$$\lim_{t \to 0} \frac{f(P_0 + tQ) - f(P_0)}{t}$$

exists and is finite, that is, if it is a real number, then it is called the *directional derivative* of f at P_0 in the direction Q, and it is denoted

$$\frac{\partial f}{\partial Q}(P_0).$$

If $Q = e_j = (0, \ldots, 0, 1, 0, \ldots, 0)$ is one of the coordinate unit vectors, whereby the number 1 occurs at the jth position, then the corresponding directional derivative is called the jth *partial derivative* of f at P_0 and it is denoted

$$\frac{\partial f}{\partial x_j}(P_0) \quad \text{or} \quad f_{x_j}(P_0).$$

The vector the jth component of which is the jth partial derivative of f at P_0, provided that all of them exist, namely

$$\nabla f(P_0) = \left(\frac{\partial f}{\partial x_1}(P_0), \frac{\partial f}{\partial x_2}(P_0), \ldots, \frac{\partial f}{\partial x_n}(P_0) \right),$$

is called the *gradient* of f at P_0. Sometimes, for brevity, the partial derivatives of f are called the partials of f.

It is worth observing that, whenever existing at all points of the open set A, the gradient is a vector field, namely $\nabla f : A \subset \mathbb{R}^n \to \mathbb{R}^n$. For reasons that will become clear below, it is important to point out that the gradient of a function at a point should be thought of as a row vector. If $n = 2$ and if $f = f(x, y)$, unravelling the definition yields

$$\frac{\partial f}{\partial x}(x_0, y_0) = \lim_{t \to 0} \frac{f(x_0 + t, y_0) - f(x_0, y_0)}{t} \tag{3.1}$$

$$\frac{\partial f}{\partial y}(x_0, y_0) = \lim_{t \to 0} \frac{f(x_0, y_0 + t) - f(x_0, y_0)}{t}. \tag{3.2}$$

Thus, for functions of two variables

$$\nabla f(x_0, y_0) = \left(\frac{\partial f}{\partial x}(x_0, y_0), \frac{\partial f}{\partial y}(x_0, y_0) \right),$$

and similarly in the case $n = 3$.

A minute thought reveals that the directional derivative of f at the point P_0 in the direction Q is nothing but the usual derivative at the origin, in the sense of a single real variable, of the restriction of the function f to the line $\{P_0 + tQ : t \in \mathbb{R}\}$, namely the derivative of the one-variable function $t \mapsto f(P_0 + tQ)$ at the point $t = 0$ (see

Fig. 3.1). In particular, the partial derivatives, if existing, are the derivatives of f along the lines parallel to the coordinate axes passing through P_0. This observation clarifies also that the requirement that A should be open in Definition 3.1 can be relaxed by asking that P_0 is an interior point of A.

3.2 Differentiability

Unlike what happens for functions of a single variable, the existence of partial derivatives at a point, or even of all directional derivatives, does not imply that the graph of the function near that point is well approximated by the graph of a linear function. The natural geometric requirement in this latter sense, which has to do with the notion of tangent (hyper) plane, is subsumed in the definition that follows.

Definition 3.2 (*Differentiability, scalar functions*) Let $A \subseteq \mathbb{R}^n$ be an open set, $P_0 \in A$. The function $f: A \to \mathbb{R}$ is said to be *differentiable* at P_0, if there exists a linear map
$$L_{P_0}: \mathbb{R}^n \to \mathbb{R} \tag{3.3}$$
such that
$$\lim_{P \to P_0} \frac{f(P) - f(P_0) - L_{P_0}(P - P_0)}{\|P - P_0\|} = 0. \tag{3.4}$$

If f is differentiable at all points of A, then f is simply called differentiable. In this case, the map L_{P_0} is called the *differential* of f at P_0.

Fig. 3.1 The restriction of $f = f(x, y)$ to a line parallel to a coordinate axis

Again, the notion of differentiable function may be extended to vector valued functions $F: A \subset \mathbb{R}^n \to \mathbb{R}^m$ and it actually plays a major role in the analysis of smooth maps between domains of Euclidean spaces, particularly the study of vector fields, that is, the case when $n = m$. For vector valued functions, the existence of a linear map as in (3.3) is replaced by that of a linear map $L_{P_0}: \mathbb{R}^n \to \mathbb{R}^m$ for which it is required that the very same equality as in (3.4) is satisfied, where now the limit is taken in $\mathbb{R}^m$ and hence should hold componentwise. Thus, one can give the following equivalent definition.

Definition 3.3 (*Differentiability, vector valued functions*) Let $A \subseteq \mathbb{R}^n$ be open and $P_0 \in A$. The vector valued function $F: A \to \mathbb{R}^m$, with components $(f_1, \ldots, f_m)$, is differentiable at P_0 if each $f_j: A \to \mathbb{R}$ is differentiable at P_0, $j = 1, \ldots, m$. If F is differentiable at all points of A, then F is simply called differentiable.

Differentiability is a strong requirement, as clarified in the next result.

Theorem 3.1 *Let $A \subset \mathbb{R}^n$ and suppose that $f: A \to \mathbb{R}$ is differentiable at P_0. Then:*

(i) *f is continuous at P_0;*
(ii) *f has directional derivative in the direction Q at P_0 for every unit vector Q, hence in particular the gradient $\nabla f(P_0)$ exists; furthermore,*

$$\frac{\partial f}{\partial Q}(P_0) = \nabla f(P_0) \cdot Q \tag{3.5}$$

for every unit vector Q;
(iii) *the differential L_{P_0} in Definition 3.2 is given by $L_{P_0}(P) = \nabla f(P_0) \cdot P$.*

As alluded to above, the geometric meaning of differentiability is that if f is differentiable at the point $P_0 \in A$, then the hyperplane in $\mathbb{R}^{n+1}$ of equation

$$x_{n+1} = f(P_0) + \nabla f(P_0) \cdot (P - P_0)$$

is the *tangent plane* to the graph of f at P_0. If $n = 2$, it has the more explicit form

$$z = f(x_0, y_0) + \frac{\partial f}{\partial x}(x_0, y_0)(x - x_0) + \frac{\partial f}{\partial y}(x_0, y_0)(y - y_0). \tag{3.6}$$

It is important to stress that in order for this equation to represent the tangent plane to the graph of f at P_0 it is necessary that f is differentiable at P_0: the mere existence of the partial derivatives implies that Eq. (3.6) makes sense but it does not say that the plane with that equation is the tangent plane to the graph of f at P_0.

For example, the function

$$f(x, y) = \begin{cases} \dfrac{xy^2}{x^2 + y^2} & (x, y) \neq (0, 0) \\ 0 & (x, y) = (0, 0) \end{cases}$$

3.2 Differentiability

has directional derivatives in all directions Q at the origin $P_0 = (0, 0)$ but it is not differentiable at the origin. This latter statement may be seen from the fact that $\nabla f(0, 0) = (0, 0)$ because f vanishes along the axes, so that $\nabla f(P_0) \cdot Q = 0$ for every Q. Nonetheless, there exist unit vectors Q for which the directional derivative $\partial f / \partial Q$ does not vanish at the origin, for example $Q = (\sqrt{2}/2, \sqrt{2}/2)$. By item (ii) in Theorem 3.1 this implies that f is not differentiable at the origin. Clearly, (3.6) describes the plane $z = 0$, which is not the tangent plane to the graph of f at the origin, which does not exist (see Fig. 3.2).

The polynomial appearing in the right hand side of (3.6), namely

$$T_1 f(P) = f(P_0) + \nabla f(P_0) \cdot (P - P_0)$$

is called the *first order Taylor polynomial* of f at P_0. Observe that (3.4) says that if f is differentiable at P_0, then it admits the first order expansion

$$f(P) = T_1 f(P) + o(\|P - P_0\|).$$

One of the important results contained in Theorem 3.1, item (iii), is the fact that if f is differentiable at P_0, then the linear map L_{P_0} is given by the scalar product with the gradient. Thus, for a function that does have partial derivatives at the point P_0, its differentiability at P_0 is equivalent to the vanishing of

$$\lim_{P \to P_0} \frac{f(P) - f(P_0) - \nabla f(P_0) \cdot (P - P_0)}{\|P - P_0\|} \tag{3.7}$$

Item (iii) of Theorem 3.1 holds true with natural modifications for vector valued functions $F: A \to \mathbb{R}^m$. Indeed, if $F = (f_1, \ldots, f_m)$ is differentiable at P_0, then the linear map $L_{P_0}: \mathbb{R}^n \to \mathbb{R}^m$, which is again called the *differential* of f at P_0, is given by row-by-column multiplcation

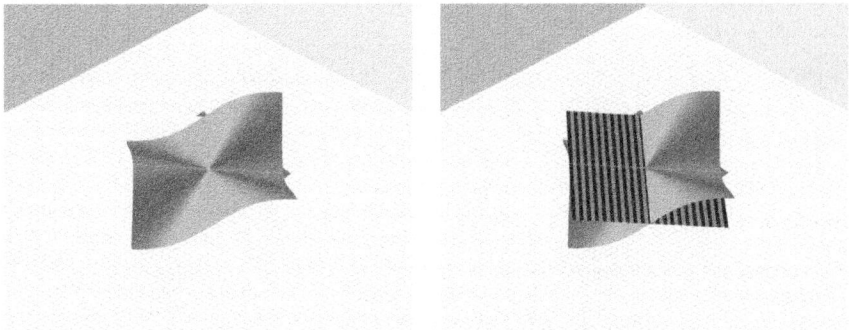

Fig. 3.2 The pinching of the graph of f accounts for its non differentiability

$$P \mapsto JF(P_0) \cdot P$$

where P is a column vector and $JF(P_0)$ is the *Jacobian matrix* of F at P_0, namely the $m \times n$ matrix the rows of which are the gradients of the components at P_0, namely:

$$JF(P_0) = \begin{bmatrix} \nabla f_1(P_0) \\ \nabla f_2(P_0) \\ \vdots \\ \nabla f_m(P_0) \end{bmatrix} = \begin{bmatrix} \frac{\partial f_1}{\partial x_1}(P_0) & \frac{\partial f_1}{\partial x_2}(P_0) & \cdots & \frac{\partial f_1}{\partial x_n}(P_0) \\ \frac{\partial f_2}{\partial x_1}(P_0) & \frac{\partial f_2}{\partial x_2}(P_0) & \cdots & \frac{\partial f_2}{\partial x_n}(P_0) \\ \vdots & \vdots & \cdots & \vdots \\ \frac{\partial f_m}{\partial x_1}(P_0) & \frac{\partial f_m}{\partial x_2}(P_0) & \cdots & \frac{\partial f_m}{\partial x_n}(P_0) \end{bmatrix}$$

Thus, when $m = n = 2$, the linear map $JF(P_0)$ associated to the vector field $F(x, y) = (f(x, y), g(x, y))$ at (x_0, y_0) is given by

$$JF(x_0, y_0)(x, y) = \begin{bmatrix} \frac{\partial f}{\partial x}(x_0, y_0) & \frac{\partial f}{\partial y}(x_0, y_0) \\ \frac{\partial g}{\partial x}(x_0, y_0) & \frac{\partial g}{\partial y}(x_0, y_0) \end{bmatrix} \begin{bmatrix} x \\ y \end{bmatrix}.$$

Evidently, when $m = 1$ the Jacobian matrix of the scalar function f is its gradient, namely $Jf = \nabla f$.

If $A \subset \mathbb{R}^n$ is an open set, then, in analogy with the case of functions of one variable, it is natural to introduce the vector space

$$C^1(A) = \left\{ f : A \to \mathbb{R} : \nabla f \text{ exists and is continuous on } A \right\}.$$

Recall that the continuity of the vector field $P \mapsto \nabla f(P)$ amounts to the existence and continuity of all the partial derivatives of f on A. Though the mere existence of partials is not enough to ensure differentiability, their continuity is, as stated in the next result.

Theorem 3.2 *If $f : A \to \mathbb{R}$ has partial derivatives on A that are continuous at P_0, then f is differentiable at P_0. If $f \in C^1(A)$, then f is differentiable on A.*

The next results are useful for arguing how differentiability properties behave under sums, products and compositions.

Theorem 3.3 *Suppose that $A \subset \mathbb{R}^n$ is open, $P_0 \in A$ and take $f, g : A \to \mathbb{R}$. Then:*

(i) *if f and g have partial derivatives at P_0, then so do $f + g$, fg and, if $g(P_0) \neq 0$, also f/g;*
(ii) *if f and g are differentiable at P_0, then such are $f + g$, fg and, if $g(P_0) \neq 0$, also f/g.*

Suppose further that $h : I \to \mathbb{R}$ is differentiable where $I \subset \mathbb{R}$ is an open set such that $f(A) \subset I$. If $f : A \to \mathbb{R}$ has partial derivatives at P_0, then so does $h \circ f$ and

$$\frac{\partial (h \circ f)}{\partial x_j}(P_0) = \frac{\partial f}{\partial x_j}(P_0) h'(f(P_0)). \tag{3.8}$$

Theorem 3.4 (Chain rule, I). *Suppose that:*

(i) *$I \subset \mathbb{R}$ is an open interval, $\gamma: I \to \mathbb{R}^n$ is a curve $\gamma(t) = (\gamma_1(t), \ldots, \gamma_n(t))$ each component of which is differentiable at t_0;*
(ii) *$A \subset \mathbb{R}^n$ is open and $f: A \to \mathbb{R}$ is differentiable at $\gamma(t_0)$*
(iii) *$\gamma(I) \subset A$.*

Then $f \circ \gamma$ is differentiable at t_0 and the following chain rule *holds*

$$(f \circ \gamma)'(t_0) = \nabla f(\gamma(t_0)) \cdot \gamma'(t_0) = \sum_{j=1}^{n} \frac{\partial f}{\partial x_j}(\gamma(t_0)) \gamma'_j(t_0). \tag{3.9}$$

Formula (3.9) is the simplest and perhaps most useful version of a much more general chain rule, which applies to compositions of vector valued functions. Here is the general statement.

Theorem 3.5 (Chain rule, II). *Suppose that:*

(i) *$B \subset \mathbb{R}^n$ is open and $F: B \to \mathbb{R}^d$ is differentiable at $P_0 \in B$;*
(ii) *$A \subset \mathbb{R}^d$ is open and $G: A \to \mathbb{R}^m$ is differentiable at $F(P_0) \in A$;*
(iii) *$F(B) \subset A$.*

Then $G \circ F: B \subset \mathbb{R}^n \to \mathbb{R}^m$ is differentiable at P_0 and the $m \times n$ Jacobian matrix of $G \circ F$ is the matrix product of the $m \times d$ Jacobian matrix of G at $F(P_0)$ with the $d \times n$ Jacobian matrix of F at P_0, that is:

$$J(G \circ F)(P_0) = JG(F(P_0)) \, JF(P_0). \tag{3.10}$$

Evidently, formula (3.9) is a particular instance of (3.10), whereby $m = n = 1$, so that the first Jacobian matrix in (3.9) is a row vector (hence a gradient) and the second is a column vector (hence a velocity vector).

3.3 Higher Order Derivatives and Taylor Expansions

As for functions of one variable, it is natural to introduce *higher order partial derivatives* of a function of several variables. Suppose that $A \subset \mathbb{R}^n$ is open and that $f: A \to \mathbb{R}$ has partial derivatives in A. Evidently, each of them is a funcion

$$\frac{\partial f}{\partial x_j}: A \to \mathbb{R}, \quad j = 1, \ldots, n$$

and it therefore makes sense to ask whether at some point in A, or even at all points of A, these admit, in turn, partial derivatives. In the affirmative case, say for example that the jth partial of f has kth partial derivative at P_0, one writes

$$\frac{\partial}{\partial x_k}\left(\frac{\partial f}{\partial x_j}\right)(P_0) = \frac{\partial^2 f}{\partial x_k \partial x_j}(P_0) = f_{x_k x_j}(P_0)$$

and calls this the second order partial derivative of f with respect to x_k and x_j at the point P_0. If $j = k$ it is customary to write

$$\frac{\partial^2 f}{\partial x_k^2}(P_0) = f_{x_k x_k}(P_0).$$

Clearly the procedure can be iterated and if $i_1, i_2, \ldots, i_\ell$ is a selection of possibly repeated indices, with $i_k \in \{1, 2, \ldots, n\}$, then the iterative procedure consists in asuming that the function has a partial derivative with respect to x_{i_ℓ} around P_0 which, in turn, has a partial derivative with respect to $x_{i_{\ell-1}}$ around P_0, and so on. Upon writing

$$I = (i_1, i_2, \ldots, i_\ell), \qquad |I| = i_1 + i_2 + \cdots + i_\ell,$$

the partial derivative of order $|I|$ of f at P_0 is then

$$\frac{\partial}{\partial x_{i_1}}\left(\frac{\partial}{\partial x_{i_2}}\left(\cdots\left(\frac{\partial}{\partial x_{i_{\ell-1}}}\left(\frac{\partial f}{\partial x_{i_\ell}}\right)\cdots\right)\right)\right)(P_0) = \frac{\partial^{|I|} f}{\partial x_{i_1} \partial x_{i_2} \ldots \partial x_{i_\ell}}(P_0),$$

also denoted

$$f_{x_{i_1} x_{i_2} \ldots x_{i_\ell}}(P_0).$$

In the case $n = 2$ there are four possible partial derivatives of order 2 of a function f of the variables (x, y), namely

$$\frac{\partial^2 f}{\partial x^2}, \quad \frac{\partial^2 f}{\partial x \partial y}, \quad \frac{\partial^2 f}{\partial y \partial x}, \quad \frac{\partial^2 f}{\partial y^2}$$

or, in different notation, f_{xx}, f_{xy}, f_{yx}, f_{yy}, respectively. In the current jargon the derivatives f_{xy} and f_{yx} are called the *mixed partials*. The four partial derivatives at the point P_0 are usually assembled in a square matrix, called the *Hessian matrix* of f at P_0, namely

$$Hf(P_0) = \begin{bmatrix} \frac{\partial^2 f}{\partial x^2}(P_0) & \frac{\partial^2 f}{\partial x \partial y}(P_0) \\ \frac{\partial^2 f}{\partial y \partial x}(P_0) & \frac{\partial^2 f}{\partial y^2}(P_0) \end{bmatrix}$$

For $n = 3$ there are nine partial derivatives of order 2 of a function f of the variables (x, y, z), and the Hessian matrix at P_0 is then, in different notation, the 3×3 matrix

3.3 Higher Order Derivatives and Taylor Expansions

$$Hf(P_0) = \begin{bmatrix} f_{xx}(P_0) & f_{xy}(P_0) & f_{xz}(P_0) \\ f_{yx}(P_0) & f_{yy}(P_0) & f_{yz}(P_0) \\ f_{zx}(P_0) & f_{zy}(P_0) & f_{zz}(P_0) \end{bmatrix}.$$

In general, the Hessian matrix at P_0 of $f : A \subset \mathbb{R}^n \to \mathbb{R}$ is the square $n \times n$ matrix

$$Hf(P_0) = \left[\frac{\partial^2 f}{\partial x_i \partial x_j}(P_0) \right]_{i,j}.$$

If $A \subset \mathbb{R}^n$ is an open set and if k is a positive integer, it is now possible to define the vector space $C^k(A)$ in analogy with the case of functions of one variable, namely

$$C^k(A) = \left\{ f : A \to \mathbb{R} : \text{ all partials of order } k \text{ of } f \text{ exist and are continuous on } A \right\}.$$

By looking at very simple functions of two variables, for examples monomials like $f(x, y) = x^p y^q$ with p and q positive integers, it often appears that the *a priori* different mixed partials f_{xy} and f_{yx} actually coincide wherever they exist. In the case just mentioned, it is clear that

$$\frac{\partial(x^p y^q)}{\partial x} = p x^{p-1} y^q, \qquad \frac{\partial(x^p y^q)}{\partial y} = q x^p y^{q-1},$$

so that, as anticipated, the equality

$$\frac{\partial^2(x^p y^q)}{\partial x \partial y} = pq x^{p-1} y^{q-1} = \frac{\partial^2(x^p y^q)}{\partial y \partial x}.$$

holds in $\mathbb{R}^2$. The next theorem addresses this issue in full generality.

Theorem 3.6 (Schwarz' theorem on equality of mixed partials) *Take $f : A \to \mathbb{R}$ where $A \subset \mathbb{R}^n$ is an open set. Suppose that for $i \neq j$ the mixed partial derivatives $f_{x_i x_j}$ and $f_{x_j x_i}$ exist in an open neighborhood of $P_0 \in A$ and are continuous at P_0. Then they are equal at P_0. Hence, if $f \in C^2(A)$, then for $i \neq j$*

$$\frac{\partial^2 f}{\partial x_i \partial x_j}(P) = \frac{\partial^2 f}{\partial x_j \partial x_i}(P)$$

at every point $P \in A$ and so the Hessian matrix is symmetric at every point $P \in A$.

The following famous example, due to Peano, shows that mixed partials may well exist but be different at a point. The function

$$f(x, y) = \begin{cases} xy \dfrac{x^2 - y^2}{x^2 + y^2} & (x, y) \neq (0, 0) \\ 0 & (x, y) = (0, 0) \end{cases}$$

is easily seen to have first order partial derivatives at all points in $\mathbb{R}^2$. Computing the limits of the appropriate difference quotients as in (3.1) and (3.2) it is also seen that the mixed partials at the origin exist and satisfy $f_{xy}(0, 0) = -1$, but $f_{yx}(0, 0) = 1$ (see Problem 3.42).

Theorem 3.7 (Second order Taylor expansion) *Let $A \subset \mathbb{R}^n$ be an open set, and take $P_0 \in A$. If $f \in C^2(A)$ then there exists a function $R: A \to \mathbb{R}$ such that*

$$f(P) = f(P_0) + \nabla f(P) \cdot (P - P_0) + \frac{1}{2} {}^t(P - P_0) H f(P_0)(P - P_0) + R(P) \tag{3.11}$$

for every $P \in A$, where the remainder function R satisfies

$$\lim_{P \to P_0} \frac{R(P)}{\|P - P_0\|^2} = 0,$$

that is, $R(P) = o(\|P - P_0\|^2)$.

Formula (3.11) can be written in explicit form using the coordinates $P = (x_1, \ldots, x_n)$ and $P_0 = (x_1^0, \ldots, x_n^0)$, namely

$$f(P) = f(P_0) + \sum_j \frac{\partial f}{\partial x_j}(P_0)(x_j - x_j^0)$$

$$+ \frac{1}{2} \sum_{i,j} \frac{\partial^2 f}{\partial x_i \partial x_j}(P_0)(x_i - x_i^0)(x_j - x_j^0) + o(\|P - P_0\|^2),$$

where all summation indices run in $\{1, 2, \ldots, n\}$. It is worth observing that under the assumption $f \in C^2(A)$, then for any vector $Y = (y_1, \ldots, y_n) \in \mathbb{R}^n$ the expression

$${}^t Y H f(P_0) Y = \sum_{i,j} \frac{\partial^2 f}{\partial x_i \partial x_j}(P_0) y_i y_j$$

is the *quadratic form* associated with the symmetric Hessian matrix of f at P_0 computed at Y, and represents the quadratic term in the expansion (3.11). In the case $n = 2$ the fully explicit form of the *second order Taylor polynomial* is thus

$$T_2 f(x, y) = f(P_0) + f_x(P_0)(x - x_0) + f_y(P_0)(y - y_0)$$
$$+ \frac{1}{2} f_{xx}(P_0)(x - x_0)^2 + f_{xy}(P_0)(x - x_0)(y - y_0) + \frac{1}{2} f_{yy}(P_0)(y - y_0)^2,$$

where of course $f(P_0) + f_x(P_0)(x - x_0) + f_y(P_0)(y - y_0)$ is the first order Taylor polynomial of f, namely $T_1 f(x, y)$.

3.4 Differentiation Under the Integral Sign

Many functions of several variables arise naturally when considering Riemann (or improper) integral functions in which the independent variables appear either in the bounds of integration or as a variable of the function that is integrated, whenever this depends on several variables only one of which is integrated. To simplify matters, consider first the case where the domain of the function is a simple rectangle R, namely a set of the form

$$R = [a, b] \times [c, d],$$

and take a continuous function $f: R \to \mathbb{R}$. For each triple (x, y, z) in the set

$$T := R \times [c, d] = [a, b] \times [c, d] \times [c, d]$$

it is thus meaningful to consider the integral $\int_y^z f(x, t)dt$, which is therefore a funcion of three variables (x, y, z). This integral is sometimes referred to as an integral depending on the *parameters* (x, y, z). It is often of interest to know how to take partial derivatives of $(x, y, z) \mapsto \int_y^z f(x, t)dt$ and, when this is actually possible, how to relate them to the partials of f. Statements relating such partial derivatives are thus known as results on *differentiation under the integral sign*. The next theorem contains the answer under the most favourable circumstances, namely when f is sufficiently regular, in the sense specified below.

Theorem 3.8 *Let $R = [a, b] \times [c, d]$ be a rectangle and put $T := R \times [c, d]$. Take a function $f: R \to \mathbb{R}$ and define $F: T \to \mathbb{R}$ by*

$$F(x, y, z) = \int_y^z f(x, t)dt.$$

(i) If f is continuous on R, then F is continuous on T and has continuous partial derivatives F_y and F_z given by the formulae:

$$\frac{\partial F}{\partial y}(x, y, z) = -f(x, y), \quad \frac{\partial F}{\partial z}(x, y, z) = f(x, z);$$

(ii) if f and $f_x(x, t)$ are continuous on R, then F has continuous partial derivative F_x in T given by the formula:

$$\frac{\partial F}{\partial x}(x, y, z) = \int_y^z \frac{\partial f}{\partial x}(x, t)dt,$$

and hence $F \in C^1(T)$;
(iii) if f and $f_x(x, t)$ are continuous on R and $\alpha, \beta: [a, b] \to [c, d]$ are differentiable functions, then the function $\Phi: [a, b] \to \mathbb{R}$ given by

$$\Phi(x) = \int_{\alpha(x)}^{\beta(x)} f(x,t)dt$$

is differentiable and

$$\Phi'(x) = \int_{\alpha(x)}^{\beta(x)} \frac{\partial f}{\partial x}(x,t)dt + f(x, \beta(x))\beta'(x) - f(x, \alpha(x))\alpha'(x) \quad (3.12)$$

for every $x \in [a, b]$.

The above statements can be extended *verbatim* to the case when $R = A \times [c, d]$ where $A \subset \mathbb{R}^n$ is a compact set with nonempty interior. The appropriate higher dimensional formulations are left as an exercise.

In the mathematical jargon it is stated that integration has a "smoothing effect" on functions, as first seen in the Fundamental Theorem of Calculus (see Theorem 10.8 in [1]) which asserts that the integral function of a continuous function is differentiable. Similar phenomena occur in higher dimensions. Suppose that $f: R \to \mathbb{R}$ has discountinuities distributed along the image of a curve. This means that there exists a continuous map $\gamma: [a, b] \to [c, d]$ such that $f: R \to \mathbb{R}$ is continuous on $R \setminus \Gamma$, where $\Gamma = \{(x, t) \in R : t = \gamma(x)\}$ is the graph of γ, see Fig. 3.3 for a picture.

Theorem 3.9 *Suppose that $f: R = [a, b] \times [c, d] \to \mathbb{R}$ has discountinuities distributed along the image of the continuous map $\gamma: [a, b] \to [c, d]$.*

(i) *If there exists a non negative and integrable function $g: [c, d] \to [0, +\infty)$ such that $|f(x, t)| \leq g(t)$ for all $(x, t) \in R$, then the function*

$$F(x, y, z) = \int_y^z f(x, t)dt.$$

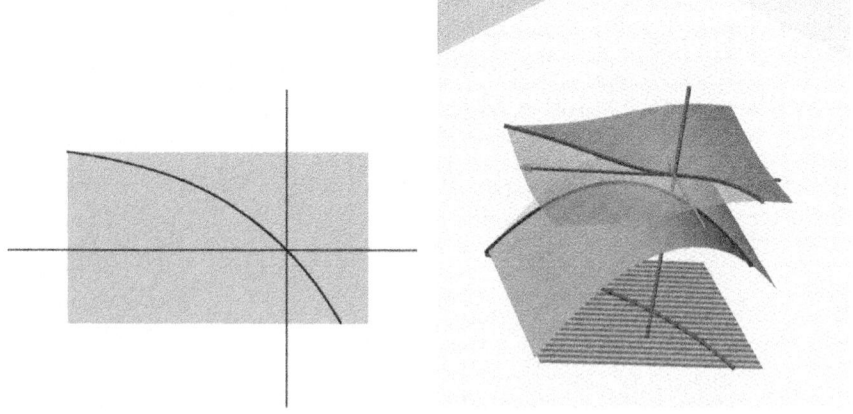

Fig. 3.3 On the left the image of the curve along which are located the discontinuities of the function whose graph, consisting of the two darker sufaces, is depicted on the right

3.4 Differentiation Under the Integral Sign

is continuous in $T = \mathbb{R} \times [c, d]$.

(ii) *If there exists a non negative and integrable function* $h: [c, d] \to [0, +\infty)$ *such that* $|f_x(x, t)| \leq h(t)$ *for all* $(x, t) \in R$, *then the partial derivative* F_x *is continuous* $T = \mathbb{R} \times [c, d]$ *and*

$$\frac{\partial F}{\partial x}(x, y, z) = \int_y^z \frac{\partial f}{\partial x}(x, t) dt.$$

General statements concerning differentiation under the integral sign are also available in the case where the parameter t runs in an unbounded interval I and functions defined on unbounded strips like $S = [a, b] \times I$. The integrals with respect to t must then be interpreted as improper integrals. Below the simple case $I = \mathbb{R}$ is treated.

Theorem 3.10 *Let* $S = [a, b] \times \mathbb{R}$ *and take a function* $f: S \to \mathbb{R}$.

(i) *If* f *is continuous on* S *and there exists a function* $g: \mathbb{R} \to [0, +\infty)$ *such that*

$$\int_{-\infty}^{+\infty} g(t) dt < +\infty, \qquad |f(x, t)| \leq g(t) \text{ for all } (x, t) \in S,$$

then the function

$$F(x) = \int_{-\infty}^{+\infty} f(x, t) dt$$

is continuous in $[a, b]$.

(ii) *If* f *is continuous on* S *and there exists a function* $h: \mathbb{R} \to [0, +\infty)$ *such that*

$$\int_{-\infty}^{+\infty} h(t) dt < +\infty, \qquad \left|\frac{\partial f}{\partial x}(x, t)\right| \leq h(t) \text{ for all } (x, t) \in S,$$

then $F \in C^1([a, b])$ *and*

$$F'(x) = \int_{-\infty}^{+\infty} \frac{\partial f}{\partial x}(x, t) dt.$$

As mentioned, conclusions similar to those of Theorem 3.10 hold if $\mathbb{R}$ is replaced by other unbounded intervals. The precise formulations are left as an exercise.

3.5 Guided Exercises

3.1 Consider the function

$$f(x, y) = \begin{cases} \dfrac{x|y|}{\sqrt{x^2 + y^2}} & (x, y) \neq (0, 0) \\ 0 & (x, y) = (0, 0). \end{cases}$$

Show that f is continuous at the origin, where it admits all directional derivatives but where it is not differentiable.

Answer. Using polar coordinates

$$|f(\rho \cos \theta, \rho \sin \theta)| = \left| \frac{\rho \cos \theta | \rho \sin \theta |}{\rho} \right| = |\rho \cos \theta | \sin \theta|| \leq \rho.$$

Therefore $f(x, y) \to 0$ for $(x, y) \to (0, 0)$, which shows that f is continuous at the origin.

Given any unit vector $Q = (\cos \theta, \sin \theta)$,

$$\frac{f(t \cos \theta, t \sin \theta) - f(0, 0)}{t} = \frac{t \cos \theta | t \sin \theta |}{t \sqrt{t^2}} = \cos \theta | \sin \theta |$$

hence

$$\frac{\partial f}{\partial Q}(0, 0) = \cos \theta | \sin \theta |$$

and, in particular, both partial derivatives vanish. Thus, f cannot be differentiable at $(0, 0)$ for otherwise the directional derivatives would all vanish as a consequence of

$$\frac{\partial f}{\partial Q}(0, 0) = \nabla f(0, 0) \cdot Q = 0.$$

This contradicts the previous formula (e.g. when $\theta = \pi/4$).

This simple problem draws the attention on the fact that a function whose gradient at a point vanishes cannot be differentiable at that point unless all the directional derivatives at that point vanish as well. This is due to formula (3.5). It thus gives an example of a function that has all directional derivatives at a point but no tangent plane to its graph at that point.

3.2 Consider the function

$$f(x, y) = \frac{|x|y^2}{\sqrt{x^2 + y^2}}.$$

(a) Show that f has a continuous extension g at the origin.
(b) Discuss the differentiability of g at the points $P_0 = (0, 0)$ and $P_1 = (0, 1)$.

3.5 Guided Exercises

Fig. 3.4 Exercise 3.2: sharp fold along the y-axis which flattens out at the origin

Answer. (a) Evidently, f is not defined at the origin. From $|y| \leq \sqrt{x^2 + y^2}$ it follows

$$|f(x, y)| = \frac{|x||y|^2}{\sqrt{x^2 + y^2}} \leq |x||y|,$$

so that $f(P) \to 0$ as $P \to (0, 0)$. Therefore

$$g(x, y) = \begin{cases} f(x, y) & (x, y) \neq (0, 0) \\ 0 & (x, y) = (0, 0) \end{cases}$$

is the desired extension.

(b) Since g vanishes on the axes, its gradient at the origin vanishes. Hence the limit in (3.7) is

$$\lim_{(x,y) \to (0,0)} \frac{g(x, y)}{\sqrt{x^2 + y^2}} = \lim_{(x,y) \to (0,0)} \frac{|x|y^2}{x^2 + y^2} = 0,$$

as it is easily established either by using again the inequality used in (a) or directly, by passing in polar coordinates. Therefore g is differentiable at P_0.

As for P_1, notice that

$$\frac{g(P_1 + t(1, 0)) - g(P_1)}{t} = \frac{g(t, 1) - g(0, 1)}{t} = \frac{|t|}{t\sqrt{1 + t^2}}$$

and this has no limit for $t \to 0$. Therefore g does not have the partial derivative at P_1 with respect to x and, *a fortiori* it is not differentiable at P_1.

This problem displays an interesting behaviour. The seemingly worrisome behaviour at the origin is actually no obstacle to the existence of the tangent plane to the graph of the (extended) function because the order with which the numerator vanishes is sufficiently high as compared to that of the denominator. However, problems do appear along the y-axis because of the presence of $|x|$ in the numerator. There is a sharp fold along the y-axis which flattens out exactly at the origin (see Fig. 3.4).

3.3 Consider the function

$$f(x, y) = \frac{2x + y}{\sqrt{4x^2 + y^2}}$$

(a) Draw the level sets of f at 0 and 1.
(b) Does f admit a continuous extension at the origin?
(c) Compute, if existing, the largest directional derivative of f at $(0, 1)$.

Answer. (a) The domain of f is $\mathrm{Dom}(f) = \mathbb{R}^2 \setminus \{(0, 0)\}$. Therefore

$$L_0 = \{(x, y) \in \mathrm{Dom}(f) : f(x, y) = 0\} = \{(x, y) \neq (0, 0) : y = -2x\}$$
$$L_1 = \{(x, y) \in \mathrm{Dom}(f) : f(x, y) = 1\} = \{(x, y) \neq (0, 0) : 2x+y = \sqrt{4x^2 + y^2}\}.$$

In order to describe L_1, observe that the equation $2x + y = \sqrt{4x^2 + y^2}$ for (x, y) in the domain of f is equivalent to the system

$$\begin{cases} 2x + y \geq 0 \\ (2x + y)^2 = 4x^2 + y^2 \\ (x, y) \neq (0, 0). \end{cases}$$

The first equation singles out the points above the graph of $x \mapsto -2x$ in the plane, the second is equivalent to $xy = 0$, which has the axes as solution locus, and the third rules out the origin. Hence L_1 consists of the two half open axes corresponding to positive values of x and y, respectively. The sets L_0 and L_1 are depicted in Fig. 3.5.

(b) Observe that on the x-axis it is $f(x, 0) = 2x/2|x|$, which has no limit as $x \to 0$. Therefore f has no continuous extension at the origin.

(c) Take the unit vector $Q = (\cos\theta, \sin\theta)$. Then, recalling the McLaurin expansion

$$\sqrt{1 + y} = 1 + \frac{1}{2}y + o(y),$$

Fig. 3.5 Exercise 3.3: level sets L_0 and L_1, respectively

3.5 Guided Exercises

the computation of the difference quotient at $(0, 1)$ in the direction θ is

$$\frac{f(t\cos\theta, 1+t\sin\theta) - f(0,1)}{t} = \frac{\frac{2t\cos\theta + 1 + t\sin\theta}{\sqrt{1+(2t\sin\theta + t^2\sin^2\theta + 4t^2\cos^2\theta)}} - 1}{t}$$

$$= \frac{2t\cos\theta + 1 + t\sin\theta - \sqrt{1 + (2t\sin\theta + t^2\sin^2\theta + 4t^2\cos^2\theta)}}{t\sqrt{1+(2t\sin\theta + t^2\sin^2\theta + 4t^2\cos^2\theta)}}$$

$$= \frac{2\cos\theta - 2t\cos^2\theta - t\sin^2\theta + o(t)}{\sqrt{1+(2t\sin\theta + t^2\sin^2\theta + 4t^2\cos^2\theta)}} \to 2\cos\theta$$

as $t \to 0$. The directional derivative corresponding to $(\cos\theta, \sin\theta)$ is thus $2\cos\theta$, which is therefore maximum for $\theta = 0$ and is equal to 2, namely $(\partial f/\partial x)(0, 1) = 2$.

Question (a) is meant to point out that level sets can be quite different one from the other and also that, in view of question (c), one partial derivative of f at $(0, 1)$, precisely $f_y(0, 1)$, vanishes because f is constantly equal to 1 for $x = 0$ and $y > 0$. Question (b) shows that along the x-axis f has a jump discontinuity at the origin because it reduces to the sign function. The more interesting question is (c), where the best approach is the direct computation of the difference quotients, because the computation of the gradient of f is quite tedious. It is certainly true that f is of class C^1 in $\mathbb{R}^2 \setminus \{(0,0)\}$, hence differentiable at $(0, 1)$, so that formula (3.5) holds true but the expression of the gradient, after some manipulation is

$$\nabla f(x,y) = \left(\frac{2y^2 - 4xy}{(4x^2+y^2)^{3/2}}, \frac{4x^2 - 2xy}{(4x^2+y^2)^{3/2}} \right).$$

Therefore $\nabla f(0, 1) = (2, 0)$, which yields the desired answer via formula (3.5). Interestingly, the answer is that the steepest direction at $(0, 1)$ is orthogonal to the flattest direction, for the maximal directional derivative is the x-partial, whereas, as seen in (a), the y-partial vanishes. The graph of f is locally like a wave: surfing it in one direction is a flat path, in the ortogonal direction it is very steep, see Fig. 3.6.

Fig. 3.6 Exercise 3.3: steep and flat paths

3.4 Consider the function

$$f(x, y) = \cos(\sqrt{x^2 + y^2}) \arctan x$$

(a) Where is f continuous?
(b) Where does f admit partial derivatives?
(c) Where is f differentiable?

Answer. (a) Being the product of two continuous functions, f is continuous on $\mathbb{R}^2$. Evidently, the first factor is the composition of $(x, y) \to \sqrt{x^2 + y^2}$, which is continuous, with the cosine function, obviously continuous.

(b) Since the cosine and arctangent functions are differentiable as well as the polynomial $(x, y) \mapsto x^2 + y^2$, by Theorem 3.3 f is differentiable at least for $(x, y) \neq 0$, and applying formula (3.8)

$$\frac{\partial f}{\partial x}(x, y) = -\sin(\sqrt{x^2 + y^2}) \frac{x}{\sqrt{x^2 + y^2}} \arctan x + \frac{\cos(\sqrt{x^2 + y^2})}{1 + x^2}$$

$$\frac{\partial f}{\partial y}(x, y) = -\sin(\sqrt{x^2 + y^2}) \frac{y}{\sqrt{x^2 + y^2}} \arctan x.$$

For $(x, y) = (0, 0)$, it is necessary to consider the difference quotients as in (3.1) and (3.2), namely

$$\frac{\partial f}{\partial x}(0, 0) = \lim_{h \to 0} \frac{f(h, 0) - f(0, 0)}{h} = \lim_{h \to 0} \frac{\cos(|h|) \arctan h}{h} = 1$$

$$\frac{\partial f}{\partial y}(0, 0) = \lim_{h \to 0} \frac{f(0, h) - f(0, 0)}{h} = \lim_{h \to 0} \frac{0 - 0}{h} = 0.$$

In conclusion, f admits partial derivatives at all points of $\mathbb{R}^2$.

(c) Away from the origin, both partials are manifestly continuous, hence f is differentiable. In order to establish differentiability at the origin, the limit (3.7) must be evaluated. Now,

$$\frac{f(x, y) - f(0, 0) - \nabla f(0, 0) \cdot (x, y)}{\sqrt{x^2 + y^2}} = \frac{\cos(\sqrt{x^2 + y^2}) \arctan x - x}{\sqrt{x^2 + y^2}}.$$

Upon adding and subtracting $\arctan x$ in the numerator, the computation of the limit as $(x, y) \to 0$ of the above expression reduces to that of (the sum of) the two limits

$$\lim_{(x,y) \to (0,0)} \frac{(\cos(\sqrt{x^2 + y^2}) - 1)}{\sqrt{x^2 + y^2}} \arctan x, \quad \lim_{(x,y) \to (0,0)} \frac{\arctan x - x}{\sqrt{x^2 + y^2}}$$

because these do exist, as it is seen next. The first vanishes because the first factor tends to $-1/2$ and the second to 0. As for the second,

$$\frac{\arctan x - x}{\sqrt{x^2+y^2}} = \frac{\arctan x - x}{x} \cdot \frac{x}{\sqrt{x^2+y^2}}$$

and while the second factor is bounded by 1, the first tends to 0, as one sees, for example, using the third order expansion $\arctan t = t - t^3/3 + o(t^3)$. Therefore, f is differentiable also at the origin.

This exercise is quite standard. All questions require to look up the definitions and use just a little bit of common sense, namely, that of treating separately the points at which some substantial difficulty might arise, due to the very definition of f, from those where no obstacles are to be expected. Only the computation of the limit (3.7) requires a little attention, but again nothing more than standard techniques are necessary.

3.5 Consider the function

$$f(x,y) = \begin{cases} \dfrac{x^4 \sqrt[3]{y^2}}{x^6+y^2} & (x,y) \neq (0,0) \\ 0 & (x,y) = (0,0). \end{cases}$$

(a) Where is f continuous?
(b) Does f admit partial derivatives at $(0,0)$? And at $(1,0)$?
(c) Establish for which unit vectors Q the directional derivative $(\partial f/\partial Q)(0,0)$ exists

Answer. (a) Away from the origin f is the ratio of continuous functions, hence it is continuous. As for the origin, we observe that along the axes f vanishes, and in fact for every $m \in \mathbb{R}$

$$f(x, mx) = \frac{x^2 \sqrt[3]{m^2 x^2}}{x^4 + m^2} \to 0$$

as $x \to 0$. This of course is not enough to ensure that $f(P) \to 0$ as $P \to 0$. Indeed, along the cubic $y = x^3$ it is

$$f(x, x^3) = \frac{x^4 \sqrt[3]{x^6}}{x^6 + x^6} = \frac{1}{2},$$

so that $\lim_{P \to 0} f(P)$ does not exist, and, *a fortiori*, f is not continuous at the origin.

(b) Since f vanishes along the axes, the partial derivatives at the origin exist and are both equal to 0. As for $(1,0)$,

$$\frac{\partial f}{\partial x}(1,0) = \lim_{h \to 0} \frac{f(1+h, 0) - f(1,0)}{h} = \lim_{h \to 0} \frac{0-0}{h} = 0,$$

so $f_x(1,0) = 0$. However,

$$\lim_{h \to 0^\pm} \frac{f(1, h) - f(1, 0)}{h} = \lim_{h \to 0^\pm} \frac{\sqrt[3]{\frac{h^2}{1+h^2}} - 0}{h} = \lim_{h \to 0^\pm} \frac{1}{\sqrt[3]{h}(1 + h^2)} = \pm\infty,$$

so that $f_y(1, 0)$ does not exist.

(c) Take $Q = (\cos\theta, \sin\theta)$. The existence (and vanishing) of partial derivatives at $(0, 0)$ has already been established, and also follows from the following general computation:

$$\lim_{t \to 0} \frac{f(t\cos\theta, t\sin\theta) - f(0, 0)}{t} = \lim_{t \to 0} \frac{\frac{t^4 \cos^4\theta \sqrt[3]{t^2 \sin^2\theta}}{t^6 \cos^6\theta + t^2 \sin^2\theta} - 0}{t}$$

$$= \lim_{t \to 0} t^{5/3} \frac{\cos^4\theta \sqrt[3]{\sin^2\theta}}{t^4 \cos^6\theta + \sin^2\theta} = 0.$$

Therefore, $(\partial f/\partial Q)(0, 0) = 0$ for every unit vector Q.

This exercise is standard. In (a) the task is to fiddle with powers and realize that there are level curves (corresponding to non zero levels) that approach the origin, to wit, along the cubic $y = \alpha^3 x^3$ (where α is a fixed real number), and away from the origin, the function is constantly equal to $\alpha^2/(1 + \alpha^6)$. As for (b), the fractional power of y at the numerator indicates that the restriction of f to the line $x = 1$ (or to any line x=constant) is a function of y which is not differentiable at the origin. Hence the partial at $(1, 0)$ with respect to y does not exist. At the origin, however, the restriction of f to any line is differentiable with derivative equal to 0, hence the answer to question (c) is positive for all directions.

3.6 Consider the function

$$f(x, y) = \frac{e^{xy} - 1}{\sqrt{x^2 + y^2}}$$

(a) Show that f admits a continuous extension g at the origin.
(b) Is g differentiable at the origin?

Answer. (a) Evidently, f is defined in $\mathbb{R}^2 \setminus \{(0, 0)\}$. Along the punctured axes, f vanishes. Outside the axes it is $xy \neq 0$ and from

$$\frac{e^{xy} - 1}{\sqrt{x^2 + y^2}} = \frac{e^{xy} - 1}{xy} \cdot \frac{xy}{\sqrt{x^2 + y^2}},$$

it follows that continuity at the origin can be reduced to that of the second factor, because the first factor tends to 1 as $(x, y) \to (0, 0)$. Using polar coordinates, the absolute value of the second factor is

$$\left| \frac{\rho^2 \cos\theta \sin\theta}{\rho} \right| = |\rho \cos\theta \sin\theta| \leq \rho \to 0,$$

whence the continuity at the origin of

3.5 Guided Exercises

$$g(x, y) = \begin{cases} f(x, y) & (x, y) \neq (0, 0) \\ 0 & (x, y) = (0, 0). \end{cases}$$

(b) Since g vanishes along the coordinate axes, its partials at the origin vanish and $\nabla g(0, 0) = (0, 0)$. This entails that the limit (3.7) for g becomes

$$\lim_{(x,y) \to (0,0)} \frac{e^{xy} - 1}{\sqrt{x^2 + y^2}\sqrt{x^2 + y^2}} = \lim_{(x,y) \to (0,0)} \frac{e^{xy} - 1}{x^2 + y^2}.$$

However, along the line $y = x$

$$\lim_{x \to 0} \frac{e^{x^2} - 1}{2x^2} = \frac{1}{2}$$

so that the previous limit cannot be 0. Hence g is not differentiable at the origin.

This simple exercise calls for the observation that locally around the origin this function behaves very much like

$$h(x, y) = \frac{xy}{\sqrt{x^2 + y^2}},$$

as shown below in Fig. 3.7. The continuous extension of h at the origin is almost visibly non-differentiable because of the corner at $y = x$ displayed by $h(x, x) = |x|/\sqrt{2}$.

3.7 Consider the function

$$f(x, y) = \begin{cases} \dfrac{x^3 \sin(xy + \sqrt{x^2 + y^2})}{x^2 + y^2} & (x, y) \neq (0, 0) \\ 0 & (x, y) = (0, 0). \end{cases}$$

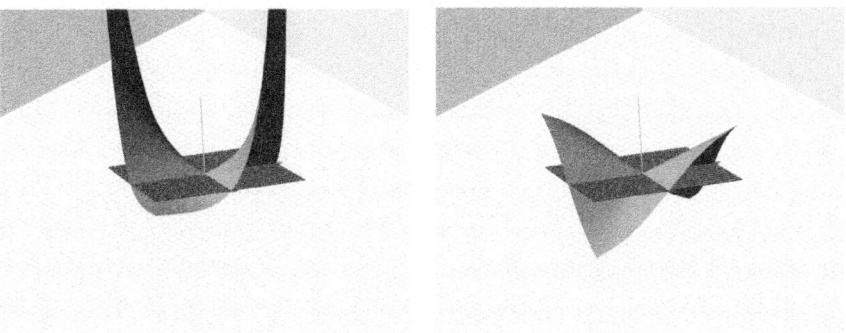

Fig. 3.7 Exercise 3.6: local graphs of f and h, respectively

(a) Is f continuous at the origin?
(b) Is f differentiable at the origin?
(c) Compute, if existing, the directional derivative of f at $(1, 0)$ with respect to the unit vector Q that has the same direction as the vector $(6, -8)$.

Answer. (a) Along the y-axis the function vanishes. Whenever $x \neq 0$ the estimate

$$|f(x, y)| \le \frac{|x^3|}{x^2 + y^2} \le \frac{|x^3|}{x^2} = |x|$$

holds. Therefore $f(x, y) \to 0$ as $(x, y) \to (0, 0)$ and f is continuous at the origin.

(b) Since

$$\lim_{x \to 0} \frac{f(x, 0) - f(0, 0)}{x} = \lim_{x \to 0} \frac{x^3 \sin(|x|)}{x^3} = 0$$

$$\lim_{y \to 0} \frac{f(0, y) - f(0, 0)}{x} = 0,$$

the gradient of f at the origin is $(0, 0)$ and the limit (3.7) becomes

$$\lim_{(x,y) \to (0,0)} \frac{x^3 \sin(xy + \sqrt{x^2 + y^2})}{(x^2 + y^2)\sqrt{x^2 + y^2}}.$$

In polar coordinates, the above expression is

$$\frac{\rho^3 \cos^3 \theta \sin(\rho + \rho^2 \cos \theta \sin \theta)}{\rho^3} = \cos^3 \theta \sin(\rho + \rho^2 \cos \theta \sin \theta).$$

Now,

$$|\cos^3 \theta \sin(\rho + \rho^2 \cos \theta \sin \theta)| \le |\rho + \rho^2 \cos \theta \sin \theta|$$
$$= \rho|1 + \rho \cos \theta \sin \theta|$$
$$\le \rho(1 + \rho) \to 0$$

as $\rho \to 0$. It follows that

$$\lim_{(x,y) \to (0,0)} \frac{x^3 \sin(xy + \sqrt{x^2 + y^2})}{(x^2 + y^2)\sqrt{x^2 + y^2}} = 0$$

and hence f is differentiable at the origin.

(c) Clearly, f is of class C^1, hence differentiable, on $\mathbb{R}^2 \setminus \{(0, 0)\}$ and in particular at $(1, 0)$. Therefore,

$$\frac{\partial f}{\partial Q}(1, 0) = \nabla f(1, 0) \cdot Q,$$

where
$$Q = \frac{(6, -8)}{\|(6, -8)\|} = \frac{(6, -8)}{\sqrt{36 + 64}} = (\frac{3}{5}, -\frac{4}{5}).$$

It remains to compute the partials of f at $(1, 0)$. For $x > 0$
$$f(x, 0) = x \sin(|x|) = x \sin x =: \varphi(x),$$

hence $f_x(1, 0) = \varphi'(1) = \sin 1 + \cos 1$. Further,
$$f(1, y) = \frac{\sin(y + \sqrt{1 + y^2})}{1 + y^2} =: \psi(y),$$

and
$$\psi'(y) = \frac{\cos(y + \sqrt{1 + y^2})(1 + \frac{y}{\sqrt{1 + y^2}})(1 + y^2) - 2y \sin(y + \sqrt{1 + y^2})}{(1 + y^2)^2},$$

so that $f_y(1, 0) = \psi'(0) = \cos 1$. In conclusion
$$\frac{\partial f}{\partial Q}(1, 0) = \nabla f(1, 0) \cdot Q = (\sin 1 + \cos 1, \cos 1) \cdot (\frac{3}{5}, -\frac{4}{5}) = \frac{3}{5} \sin 1 - \frac{1}{5} \cos 1.$$

This exercise deserves two observations. The first is that wherever f is differentiable, then formula (3.5) holds, so that the computation splits in finding the unit vector associated with $(6, -8)$, which is essentially trivial, and then in computing the gradient. The second observation is that a sometimes quicker way to compute the partials of a function, say f, at (x_0, y_0) is to compute $\varphi'(x_0)$ and $\psi'(y_0)$, where
$$\varphi(x) = f(x, y_0), \qquad \psi(y) = f(x_0, y).$$

Upon substituting directly the given value x_0, or y_0, the function of two variables often becomes a simple function of a single variable, thereby reducing the calculation of a partial to that of an ordinary derivative. The other standard way is to compute the derivative with respect to one variable, say x, by treating the other, say y, as a parameter, and then to substitute (x, y) with (x_0, y_0) at the end. As usual, which way to go depends both on taste and on the specific situation at hand.

3.8 Consider the function
$$g(x, y) = \begin{cases} f(x, y) & (x, y) \in \text{Dom}(f) \\ 0 & (x, y) \notin \text{Dom}(f), \end{cases}$$

where $f(x, y) = x\sqrt{x^2 - \cos^2 y} - x^2$.

(a) Draw Dom(f).
(b) Is g differentiable at $P_0 = (0, \pi/2)$?

Answer. (a) Evidently, the domain of f is the set

$$\{(x, y) \in \mathbb{R}^2 : x^2 \geq \cos^2 y\} = \{(x, y) \in \mathbb{R}^2 : |x| \geq |\cos y|\},$$

which is depicted in Fig. 3.8.

(b) Quite clearly, g vanishes along the y axis because none of the points $(0, y)$ is in the domain of f unless $\cos y = 0$ in which case $f(0, y) = 0$ (see again Fig. 3.8). This implies that $g_y(0, \pi/2) = 0$. Furthermore,

$$\lim_{t \to 0} \frac{g(t, \frac{\pi}{2}) - g(0, \frac{\pi}{2})}{t} = \lim_{t \to 0} \frac{t\sqrt{t^2} - t^2}{t} = 0$$

so that also $g_x(0, \pi/2) = 0$. Thus, the limit in (3.7) is 0 if $(x, y) \notin \text{Dom}(f)$, whereas if $(x, y) \in \text{Dom}(f)$ it is

$$\lim_{(x,y) \to (0, \frac{\pi}{2})} \frac{g(x, y) - g(0, \frac{\pi}{2})}{\sqrt{x^2 + (y - \frac{\pi}{2})^2}} = \lim_{(x,y) \to (0, \frac{\pi}{2})} \frac{x\left[\sqrt{x^2 - \cos^2 y} - x\right]}{\sqrt{x^2 + (y - \frac{\pi}{2})^2}}.$$

Finally,

$$\left|\frac{x}{\sqrt{x^2 + (y - \frac{\pi}{2})^2}}\right| \leq \frac{|x|}{|x|} = 1$$

and $\sqrt{x^2 - \cos^2 y} - x \to 0$ as $(x, y) \to (0, \pi/2)$ whenever $(x, y) \in \text{Dom}(f)$. Therefore the limit (3.7) exists and is equal to 0, which implies that g is differentiable at P_0.

This exercise displays a slight difficulty of geometric nature. It is important to understand that the point P_0 sits at the corner between the two regions in which the

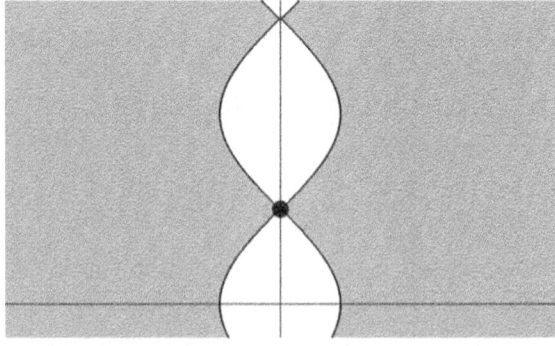

Fig. 3.8 Exercise 3.8: in grey the domain of f, including the boundary; the big dot is P_0

3.5 Guided Exercises

function g is defined by different formulae. This gives more or less immediately that $\nabla g(P_0) = 0$. The analysis of the limit (3.7) which is involved in the notion of differentiability is simpler than it appears at first sight. Indeed, in the "white" region, outside $\mathrm{Dom}(f)$, the function vanishes and so does the ratio that appears in (3.7). In the "grey" region, inside $\mathrm{Dom}(f)$, the ratio in (3.7) is actually the product between a function bounded by 1 and $f(x,y)/x$, which tends to 0 as $(x,y) \to (0, \pi/2)$.

3.9 Consider the function $f(x,y) = \log\left(\sqrt{y^2 - x} - 2y - 1\right)$.

(a) Draw $\mathrm{Dom}(f)$.

(b) Find, if existing, all real numbers a, b, c such that

$$\lim_{(x,y) \to (0,-1)} \frac{f(x,y) + ax + by + c}{\sqrt{x^2 + (y+1)^2}} = 0.$$

(c) Write, if meaningful, the equation of the tangent plane to the graph of f in the point $(-4, 0, f(-4, 0))$.

Answer. (a) The domain of f is the set

$$\mathrm{Dom}(f) = \left\{(x,y) \in \mathbb{R}^2 : y^2 \geq x, \ \sqrt{y^2 - x} > 2y + 1\right\}$$

$$= \left\{(x,y) \in \mathbb{R}^2 : y^2 \geq x, \ y < -\frac{1}{2}\right\}$$

$$\cup \left\{(x,y) \in \mathbb{R}^2 : y^2 \geq x, \ y \geq -\frac{1}{2}, \ y^2 - x > (2y+1)^2\right\}$$

$$= \left\{(x,y) \in \mathbb{R}^2 : y^2 \geq x, \ y < -\frac{1}{2}\right\}$$

$$\cup \left\{(x,y) \in \mathbb{R}^2 : y^2 \geq x, \ y \geq -\frac{1}{2}, \ x < -3y^2 - 4y - 1\right\}$$

and it is drawn in Fig. 3.9. The two sets forming $\mathrm{Dom}(f)$ are analyzed in Fig. 3.10.

(b) The request is that of finding a polinomial P of degree at most one such that

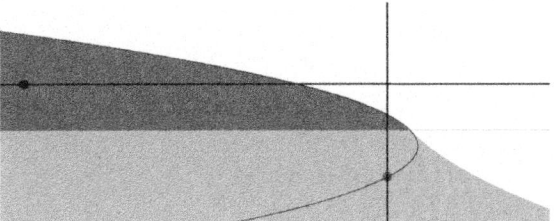

Fig. 3.9 The domain of f in Exercise 3.9 as the union of two sets; the two dots are, from left to right, the points $(-4, 0)$ and $(0, -1)$ that are relevant to questions (c) and (b), respectively

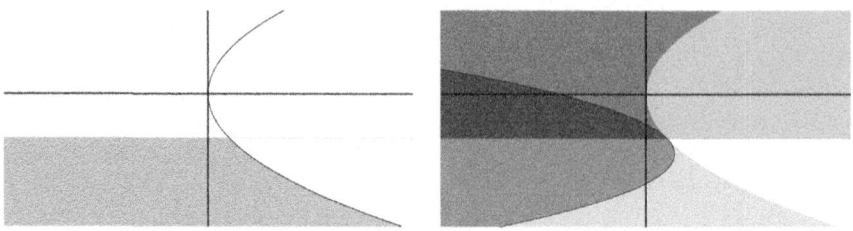

Fig. 3.10 Exercise 3.9: the first set given by two inequalities, the second (darker grey) by three; their union is the domain of f

$$\lim_{(x,y)\to(0,-1)} \frac{f(x,y) - P(x,y)}{\sqrt{x^2 + (y+1)^2}} = 0.$$

This is manifestly equivalent to proving that f is differentiable at $(0, -1)$. In such case, P is the first order Taylor polynomial of f at $(0, -1)$ and it is given by:

$$P(x, y) = f(0, -1) + \nabla f(0, -1) \cdot (x, y + 1).$$

Observe that $(0, -1)$ is an interior point of the domain of $(x, y) \mapsto \sqrt{y^2 - x}$, which is thus of class C^1 in a neighborhood of $(0, -1)$. Hence $(x, y) \mapsto \sqrt{y^2 - x} - 2y - 1$ is also of class C^1 in a neighborhood of $(0, -1)$, and strictly positive. It follows that f of class C^1 in a neighborhood of $(0, -1)$ and hence differentiable at $(0, -1)$. Now, a direct computation gives that in a neighborhood of $(0, -1)$

$$\frac{\partial f}{\partial x}(x, y) = -\frac{1}{2\sqrt{y^2 - x}\left(\sqrt{y^2 - x} - 2y - 1\right)}$$

$$\frac{\partial f}{\partial y}(x, y) = \frac{1}{\sqrt{y^2 - x} - 2y - 1}\left(\frac{y}{\sqrt{y^2 - x}} - 2\right),$$

so that $\nabla f(0, -1) = (-1/4, -2)$. Since $f(0, -1) = \log 2$, the first order Taylor polynomial of f at $(0, -1)$ is

$$P(x, y) = \log 2 - 2 - \frac{1}{4}x - 2y$$

and the answer is $a = 1/4$, $b = 2$ and $c = 2 - \log 2$.

(c) As before, it is immediate to see that f is of class C^1 in a neighborhood of $(-4, 0)$, so that f is actually differentiable at $(-4, 0)$. Therefore the required tangent plane does exist and its equation is

$$z = f(-4, 0) + \nabla f(-4, 0) \cdot (x + 4, y) = -\frac{1}{4}x - 2y - 1,$$

because $f(-4, 0) = 0$ and the previous computations give $\nabla f(-4, 0) = (-1/4, -2)$.

Question (a) is useful in order to have a clear picture of the domain of f, in view of (b) and (c). It requires to solve a simple inequality of the form $\sqrt{\varphi(x, y)} > \psi(x, y)$, which splits naturally into the systems

$$\begin{cases} \varphi(x, y) \geq 0 \\ \psi(x, y) < 0, \end{cases} \qquad \begin{cases} \varphi(x, y) \geq 0 \\ \psi(x, y) \geq 0 \\ \varphi(x, y) > \psi^2(x, y). \end{cases}$$

Each gives rise to a subset of $\mathbb{R}^2$ and their union is $\text{Dom}(f)$. The main point of question (b) is to understand that the issue is of course differentiability at $(0, -1)$. The point $(0, -1)$ is well inside the domain of f (first system), where f is the composition of functions of class C^1, hence differentiable. Once this is realized, then everything reduces to a standard computation of partial derivatives and then of the linearization at $(0, -1)$. Question (c) is essentially analogous, though formulated differently and at the point $(-4, 0)$ that is in turn well inside the other natural region (second system) that forms the domain of f.

3.10 Consider the function

$$f(x, y) = \begin{cases} \dfrac{x^3 + y^3}{\sqrt{x^2 + y^4}} & (x, y) \neq (0, 0) \\ \alpha & (x, y) = (0, 0). \end{cases}$$

(a) Determine all values of $\alpha \in \mathbb{R}$, if any, for which f is continuous at $(0, 0)$.
(b) Determine all values of $\alpha \in \mathbb{R}$, if any, for which f is differentiable at $(0, 0)$.

Answer. (a) Notice that if $|x| \leq 1$, then $x^2 \geq x^4$ and hence whenever $(x, y) \neq (0, 0)$ and $|x| \leq 1$

$$|f(x, y)| \leq \frac{|x^3 + y^3|}{\sqrt{x^4 + y^4}} =: h(x, y).$$

In polar coordinates,

$$h(\rho \cos \theta, \rho \sin \theta) = \frac{\rho^3 |\cos^3 \theta + \sin^3 \theta|}{\rho^2 \sqrt{\cos^4 \theta + \sin^4 \theta}} = \frac{|\cos^3 \theta + \sin^3 \theta|}{\sqrt{\cos^4 \theta + \sin^4 \theta}} \rho.$$

The denominator is continuous as a function of $\theta \in [0, 2\pi)$ and hence, by Weierstrass' theorem, it attains a positive minimum m at some $\bar{\theta} \in [0, 2\pi)$. Therefore

$$h(\rho \cos \theta, \rho \sin \theta) \leq \frac{2}{m} \rho,$$

so that

$$\lim_{\rho \to 0} \sup_{\theta \in [0, 2\pi]} |f(\rho\cos\theta, \rho\sin\theta)| \leq \lim_{\rho \to 0} \sup_{\theta \in [0, 2\pi]} h(\rho\cos\theta, \rho\sin\theta) \leq \lim_{\rho \to 0} \frac{2}{m}\rho = 0.$$

This proves that $f(x, y) \to 0$ as $(x, y) \to (0, 0)$. Therefore f is continuous precisely for $\alpha = 0$.

(b) As continuity is a necessary for differentiability, necessarily $\alpha = 0$. Now,

$$\lim_{t \to 0} \frac{f(t, 0) - f(0, 0)}{t} = \lim_{t \to 0} \frac{t^3}{|t|t} = 0$$

$$\lim_{t \to 0} \frac{f(0, t) - f(0, 0)}{t} = \lim_{t \to 0} \frac{t^3}{t^3} = 1$$

shows that $\nabla f(0, 0) = (0, 1)$. Thus,

$$\frac{f(x, y) - f(0, 0) - \nabla f(0, 0) \cdot (x, y)}{\sqrt{x^2 + y^2}} = \frac{\frac{x^3 + y^3}{\sqrt{x^2 + y^4}} - (0, 1) \cdot (x, y)}{\sqrt{x^2 + y^2}}$$

$$= \frac{x^3 + y^3 - y\sqrt{x^2 + y^4}}{\sqrt{x^2 + y^4}\sqrt{x^2 + y^2}}.$$

This latter expression evaluated at the points (x, x), with $x \geq 0$, becomes

$$\frac{2x^3 - x^2\sqrt{1 + x^2}}{x^2\sqrt{2}\sqrt{1 + x^2}} = \frac{2x - \sqrt{1 + x^2}}{\sqrt{2}\sqrt{1 + x^2}}$$

which tends to $-\sqrt{2}/2$ as $x \to 0$. It follows that even if the limit (3.7) existed (and it actually does not, as explained below) it would not be equal to 0, so that f is not differentiable at $(0, 0)$.

The first part of this exercise is standard, and it reduces to checking that the only possible value of $f(0, 0)$ is 0 in order for f to be differentiable at the origin. Thus, there is no way that by changing α the function becomes differentiable at the origin. The fact that the limit (3.7) does not exist can be argued by a polar coordinate argument. Indeed, it becomes

$$\frac{\rho(\cos^3\theta + \sin^3\theta) - \sin\theta\sqrt{\cos^2\theta + \rho^2\sin^4\theta}}{\sqrt{\cos^2\theta + \rho^2\sin^4\theta}}$$

which tends to $-\sin\theta$ whenever $\theta \notin \{\pi/2, 3\pi/2\}$, or to 0 if $\theta \in \{\pi/2, 3\pi/2\}$, hence the limit depends on the direction θ.

3.11 Consider the function $f(x, y) = \pi^{(x+y)} - (x + y)^\pi$.

3.5 Guided Exercises

(a) Describe the domain of f and determine at which of its points f is continuous, where f admits partial derivatives, where it is differentiable and where it is of class C^1.
(b) Write the second order Taylor polynomial of f at the point $(\pi/2, \pi/2)$.
(c) Does f admit a continuous extension g at the points of the line $x + y = 0$?

Answer. (a) By definition, $(x+y)^\pi = e^{\pi \log(x+y)}$ hence the domain of f is

$$\mathrm{Dom}(f) = \{(x, y) \in \mathbb{R}^2 : x + y > 0\}$$

and is depicted in Fig. 3.11.
Both $t \to \pi^t$ and $t \to t^\pi$ are continuous and differentiable functions for $t > 0$, and so f is continuous on $\mathrm{Dom}(f)$, where it admits partial derivatives and is differentiable because of Theorem 3.5, since such is $(x, y) \mapsto x + y$. Due to symmetry considerations, they coincide and are given by

$$f_x(x, y) = f_y(x, y) = \pi^{(x+y)} \log \pi - \pi(x+y)^{\pi-1}.$$

In particular, they are manifestly continuous so that $f \in C^1(\mathrm{Dom}(f))$.

(b) Clearly, the partials computed above are both differentiable, so that the second order partial derivaties of f do exist. Symmetry considerations, together with the above calculations, give

$$f_{xy}(x, y) = f_{yx}(x, y) = f_{xx}(x, y) = f_{yy}(x, y) = \pi^{(x+y)}(\log \pi)^2 - \pi(\pi - 1)(x+y)^{\pi-2}.$$

It follows that at $P_0 = (\pi/2, \pi/2)$

$$f(P_0) = \pi^\pi - \pi^\pi = 0$$
$$f_x(P_0) = f_y(P_0) = \pi^\pi(\log \pi - 1)$$
$$f_{xy}(P_0) = f_{yx}(P_0) = f_{xx}(P_0) = f_{yy}(P_0) = \pi^\pi(\log \pi)^2 - \pi^\pi + \pi^{\pi-1}.$$

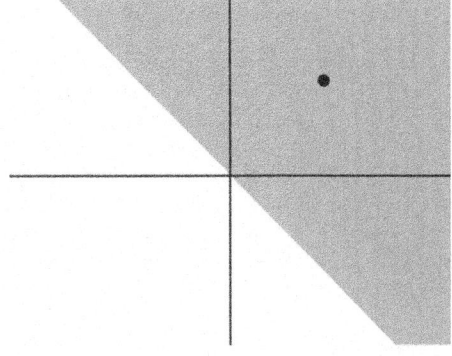

Fig. 3.11 Exercise 3.11: the domain of f and the point $(\pi/2, \pi/2)$

Therefore, the second order Taylor polynomial of f at P_0 is

$$T_2(x, y) = \pi^\pi (\log \pi - 1)\left[\left(x - \frac{\pi}{2}\right) + \left(y - \frac{\pi}{2}\right)\right]$$
$$+ \frac{1}{2}\left(\pi^\pi (\log \pi)^2 - \pi^\pi + \pi^{\pi-1}\right)\left[\left(x - \frac{\pi}{2}\right)^2 + 2\left(x - \frac{\pi}{2}\right)\left(y - \frac{\pi}{2}\right) + \left(y - \frac{\pi}{2}\right)^2\right].$$

(c) Take any point (x_0, y_0) satisfying $x_0 + y_0 = 0$. Since

$$\lim_{(x,y) \to (x_0, y_0)} \pi^{(x+y)} - e^{\pi \log(x+y)} = 1 - 0 = 1,$$

a continuous extension g of f is

$$g(x, y) = \begin{cases} \pi^{(x+y)} - (x+y)^\pi & x + y > 0 \\ 1 & x + y = 0. \end{cases}$$

This is almost an exercise on a function of a single variable, because indeed if $\varphi(t) = \pi^t - t^\pi$, which is defined for $t > 0$, then

$$f(x, y) = \varphi(x + y).$$

This is the guiding idea. Thus there is only one first derivative and one second derivative. The rest is essentially an application of rules, except for (c) which again rests on the observation that $\varphi(t) \to 1$ as $t \to 0^+$.

3.12 Consider the function

$$f(x, y) = \begin{cases} \dfrac{2x^2 y + xy^2}{x^2 + y^2} & (x, y) \neq (0, 0) \\ 0 & (x, y) = (0, 0). \end{cases}$$

(a) Compute, whenever existing, $(\partial f / \partial Q)(0, 0)$, where Q is a unit vector.
(b) Is f differentiable at $(0, 0)$?
(c) Is ∇f continuous at $(0, 0)$?

Answer. (a) Let $Q = (\cos \theta, \sin \theta)$. Then

$$\lim_{t \to 0} \frac{f(t \cos \theta, t \sin \theta) - f(0, 0)}{t} = \lim_{t \to 0} \frac{2t^3 \cos^2 \theta \sin \theta + t^3 \cos \theta \sin^2 \theta}{t^3}$$
$$= 2 \cos^2 \theta \sin \theta + \cos \theta \sin^2 \theta$$

shows that all directional derivatives exist and

$$\frac{\partial f}{\partial Q}(0, 0) = 2 \cos^2 \theta \sin \theta + \cos \theta \sin^2 \theta.$$

(b) From the above expression for directional derivatives it follows, upon choosing either $\theta = 0$ or $\theta = \pi/2$, that

$$\frac{\partial f}{\partial x}(0,0) = \frac{\partial f}{\partial y}(0,0) = 0.$$

Consequently, f cannot be differentiable at $(0,0)$ because if it were differentiable then for every unit vector Q it would be

$$\frac{\partial f}{\partial Q}(0,0) = \nabla f(0,0) \cdot Q = 0$$

because $\nabla f(0,0) = (0,0)$. But this contradicts the above computation that shows, for example, that for $Q_0 = (\sqrt{2}/2, \sqrt{2}/2)$ it is

$$\frac{\partial f}{\partial Q_0}(0,0) = \frac{3}{4}\sqrt{2}.$$

(c) Since away from the origin f is a rational function, its partial derivatives certainly exist in $\mathbb{R}^2 \setminus \{(0,0)\}$, and, as shown above, they also exist at the origin. If ∇f were continuous at the origin, then f would be differentiable at the origin, and this is not true as established in (b).

This simple exercise is almost of theoretical nature: any function that has all possible directional derivatives at a point P_0 and vanishing gradient at P_0, can only be differentiable at P_0 if all the directional derivatives at P_0 vanish. Informally speaking, this is because the vanishing gradient at P_0 implies that the tangent plane, if existing, would have to be horizontal, whereas a nonvanishing $(\partial f/\partial Q)(P_0)$ means that walking along the direction Q a steep path is to be found, see Fig. 3.12. By the same token, if $\nabla f(P_0) = (0,0)$, then ∇f cannot be continuous at P_0 if there exists a direction Q along which the corresponding directional derivative does not vanish.

3.13 Consider the function $f(x, y) = \max\{(y - x)(y - x^3), 0\}$.

(a) Where is f continuous?
(b) Where does f admit partial derivatives?
(c) Is f differentiable at $(0,0)$?
(d) Compute, if existing, the $\lim_{P \to \infty} f(P)$.

Answer. (a) First of all, it is necessary to understand the domain Ω on which $f(x, y)$ is positive, namely where $(y - x)(y - x^3) > 0$. This is given by the points (x, y) where $y - x$ and $y - x^3$ have the same sign. In Fig. 3.13 and in Fig. 3.14 this is clarified.

It is clear that f is continuous in Ω and also in $\mathbb{R}^2 \setminus \Omega$, where it vanishes; in particular, f is continuous on $\partial\Omega$, that is, along the curves $y = x$ and $y = x^3$ which belong to $\mathbb{R}^2 \setminus \Omega$ because the polynomial $(y - x)(y - x^3)$ is continuous on $\mathbb{R}^2$. Hence f is continuous everywhere in $\mathbb{R}^2$.

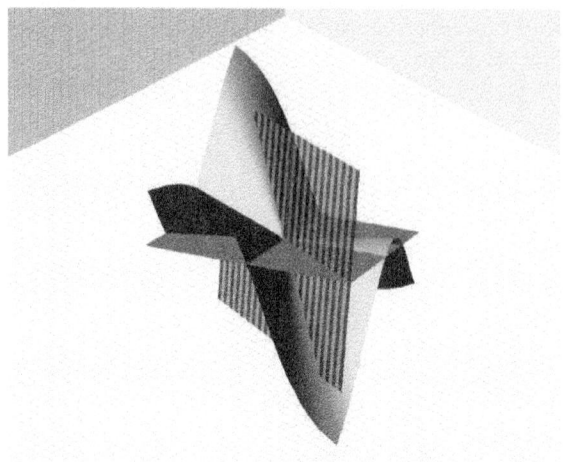

Fig. 3.12 Exercise 3.12: steep path for $\theta = \pi/4$

Fig. 3.13 Exercise 3.13: in grey the domains $y - x > 0$ and $y - x^3 > 0$, respectively

Fig. 3.14 Exercise 3.13: in grey the domain Ω on which $(y - x)(y - x^3) > 0$, and the three points at which the two lines $y = x$ and $y = x^3$ intersect: $(-1, -1)$, $(0, 0)$ and $(1, 1)$

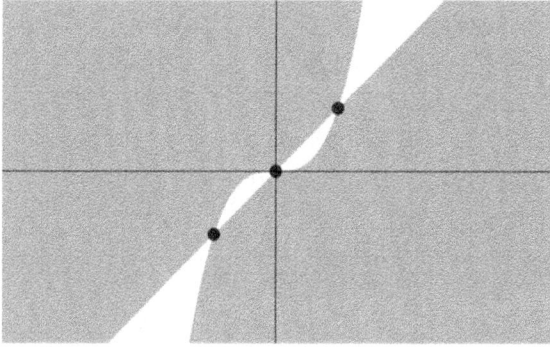

(b) On $\mathbb{R}^2 \setminus \overline{\Omega}$, that is, where $(y - x)(y - x^3) < 0$ both partials vanish, whereas on Ω

$$\frac{\partial f}{\partial x}(x, y) = -(y - x^3) - (y - x)3x^2 = 4x^3 - 3x^2 y - y$$

3.5 Guided Exercises

$$\frac{\partial f}{\partial y}(x, y) = (y - x^3) + (y - x) = 2y - x - x^3.$$

Since the coordinate axes belong to Ω and since $(y - x)(y - x^3)$ vanishes at the origin, we deduce that the preceding formulae hold also at $(0, 0)$. Analogous arguments hold for the points $(-1, -1)$ and $(1, 1)$. Now we take $(x, y) \in \partial\Omega \setminus \{(0, 0), (1, 1), (-1, -1)\}$. Along the line $y = x$ the function admits partial derivatives provided that

$$\begin{cases} 4x^3 - 3x^2y - y = 0 \\ 2y - x - x^3 = 0 \\ y = x \end{cases} \implies \begin{cases} 4x^3 - 3x^2 x - x = 0 \\ 2x - x - x^3 = 0 \end{cases} \implies x(x^2 - 1) = 0,$$

namely for $x = 0, \pm 1$. Similarly, along the curve $y = x^3$ the function admits partial derivatives provided that

$$\begin{cases} 4x^3 - 3x^2y - y = 0 \\ 2y - x - x^3 = 0 \\ y = x^3 \end{cases} \implies \begin{cases} 4x^3 - 3x^5 - x^3 = 0 \\ 2x^3 - x - x^3 = 0 \end{cases} \implies x(x^2 - 1) = 0,$$

again for $x = 0, \pm 1$, which correspond to the points $(-1, -1)$, $(0, 0)$ and $(1, 1)$. Therefore, on the boundary $\partial\Omega$, the function admits partial derivatives only at the points $(-1, -1)$, $(0, 0)$ and $(1, 1)$.

(c) As $f(0, 0) = 0$ and $\nabla f(0, 0) = (0, 0)$, the limit in (3.7) is

$$\lim_{(x,y) \to (0,0)} \frac{f(x, y)}{\sqrt{x^2 + y^2}}.$$

If $(x, y) \in \mathbb{R}^2 \setminus \Omega$ this is clearly 0 because f vanishes. Now, on Ω

$$\frac{f(x, y)}{\sqrt{x^2 + y^2}} = \frac{(y - x)(y - x^3)}{\sqrt{x^2 + y^2}} = \frac{y^2 - xy - yx^3 + x^4}{\sqrt{x^2 + y^2}}$$

so that, using polar coordinates, the following estimate holds

$$\left| \frac{f(\rho \cos\theta, \rho \sin\theta)}{\rho} \right| = \frac{|\rho^2 \sin^2\theta - \rho^2 \cos\theta \sin\theta - \rho^4 \sin\theta \cos^3\theta + \rho^4 \cos^4\theta|}{\rho}$$

$$= \rho \left| \sin^2\theta - \cos\theta \sin\theta - \rho^2(\sin\theta \cos^3\theta + \cos^4\theta) \right|$$

$$\leq \rho \left(\sin^2\theta + |\cos\theta \sin\theta| + \rho^2 |\sin\theta \cos^3\theta| + \rho^2 \cos^4\theta \right)$$

$$\leq \rho(2 + 2\rho^2).$$

This clearly implies that when $(x, y) \in \Omega$

$$\lim_{(x,y) \to (0,0)} \frac{f(x, y)}{\sqrt{x^2 + y^2}} = 0$$

and hence that f is differentiable at the origin.

(d) On the line $x = y$ the function vanishes, whereas for $y = -x$

$$f(x, -x) = 2x(x + x^3) = 2x^2 + 2x^4,$$

which diverges to $+\infty$ as $x \to \pm\infty$. It follows that $\lim_{P \to \infty} f(P)$ does not exist.

Part (a) could actually be treated by observing that the minimum of two continuous functions is continuous because

$$\max(f(P), g(P)) = \frac{1}{2}\Big(f(P) + g(P) + |f(P) - g(P)|\Big).$$

The function of this exercise is a polynomial that is cut below the $z = 0$ plane and therefore calls for a careful description of the zero-level curve of the polynomial, the cutting edge. This is the union of two curves, one is a straight line and the other is a cubic, and they meet at three distinct points. The sharp adjustment of the graph of a polynomial to that of a flat plane generates problems along the cutting edge, as it is displayed by the fact that along the edge the gradient exists only at the three special points. This behaviour can be directly guessed by looking at Fig. 3.14 because of all the points along the grey-white border it is only at the three special points that the vertical and horizontal lines centered at the point safely remain inside the grey region, so that the restriction of f does not sharply cross from a non zero polynomial to a constant. At the origin f behaves nicely because the polynomial contains only terms of second and fourth degree and thus flattens at the origin enough for a tangent plane to exist. The answer to (d) is almost obvious because the white region extends to infinity and so does the grey region, too, but f is flat on the former and polynomial on the latter.

3.14 Consider the functions $f, g : \mathbb{R}^2 \to \mathbb{R}$ defined by

$$f(x, y) = \begin{cases} 1 & xy \in \mathbb{Q} \\ 0 & xy \notin \mathbb{Q} \end{cases}, \qquad g(x, y) = xf(x, y).$$

(a) Are f and g continuous at $(0, 0)$?
(b) Do f and g admit partial derivatives at $(0, 0)$?

Answer. (a) The function f is not continuous at $(0, 0)$ because if

$$R_n = \left(\frac{1}{n}, \frac{1}{n}\right), \qquad S_n = \left(\frac{\sqrt{2}}{n}, \frac{1}{n}\right)$$

3.5 Guided Exercises

then $R_n \to (0, 0)$ and $S_n \to (0, 0)$, but $f(R_n) = 1$ and $f(S_n) = 0$ for all n, so that $\lim_{P \to (0,0)} f(P)$ does not exist. The function g is continuous at $(0, 0)$ because

$$|g(x, y)| \leq |x|$$

implies that $\lim_{P \to (0,0)} g(P) = 0 = g(0, 0)$.

(b) Since $f(x, 0) = f(0, y) = 1$ for every $x, y \in \mathbb{R}$, the functions admits partials at the origin, and both vanish. Finally, since $g(x, 0) = xf(x, 0) = x$, and $g(0, y) = 0$ it is $\nabla g(0, 0) = (1, 0)$.

This exercise needs a little ingenuity, because it is impossible to appeal to any reasonable visualization. In (a) the issue is to find sequences $P_n = (x_n, y_n)$ converging to the origin and for which $x_n y_n$ is either rational or irrational, on which, by definition, the function f attains the value 1 or 0, respectively. The presence of the factor x squeezes g to 0 near the origin. In (b) the answer is trivial for f once it is realized what happens along the coordinate axes.

3.15 Is the function $f : \mathbb{R} \times (-\pi/2, \pi/2) \to \mathbb{R}$ defined by

$$f(x, y) = \begin{cases} \dfrac{\sin x}{x} + \dfrac{y}{\sin y} & xy \neq 0 \\ 2 & xy = 0 \end{cases}$$

differentiable at $(0, 0)$?

Answer. Along the axes f is constantly equal to 2, including the origin. Thus both partials of f vanish at the origin. The limit in (3.7) becomes then

$$\lim_{(x,y) \to (0,0)} \frac{f(x, y) - f(0, 0)}{\sqrt{x^2 + y^2}} = \lim_{(x,y) \to (0,0)} \frac{\frac{\sin x}{x} + \frac{y}{\sin y} - 2}{\sqrt{x^2 + y^2}}$$

$$= \lim_{(x,y) \to (0,0)} \frac{\frac{\sin x - x}{x} + \frac{y - \sin y}{\sin y}}{\sqrt{x^2 + y^2}}.$$

From the McLaurin expansion $\sin t = t + t^2 \omega(t)$, with $\omega(t) \to 0$ as $t \to 0$, it follows

$$\lim_{(x,y) \to (0,0)} \frac{f(x, y) - f(0, 0)}{\sqrt{x^2 + y^2}} = \lim_{(x,y) \to (0,0)} \frac{\frac{x^2 \omega(x)}{x} - \frac{y^2 \omega(y)}{\sin y}}{\sqrt{x^2 + y^2}}$$

$$= \lim_{(x,y) \to (0,0)} \left(\frac{x}{\sqrt{x^2 + y^2}} \omega(x) - \frac{y}{\sin y} \frac{y}{\sqrt{x^2 + y^2}} \omega(y) \right) = 0,$$

where the conclusion is drawn from the inequalities

$$\left| \frac{x}{\sqrt{x^2 + y^2}} \right| \leq 1, \quad \left| \frac{y}{\sqrt{x^2 + y^2}} \right| \leq 1.$$

Therefore f is differentiable at $(0, 0)$.

The function at hand could be written as $f(x, y) = \text{sinc}(x) + 1/\text{sinc}(y)$, the sum of two functions of x and y separately, each smooth near $0 \in \mathbb{R}$, where sinc is the cardinal sine

$$\text{sinc}(x) = \begin{cases} \dfrac{\sin x}{x} & x \neq 0 \\ 1 & x = 0. \end{cases}$$

This fact translates, in the end, into the splitting of the limit in (3.7) as the sum of two limits that are essentially one-variable limits. The reader is urged to check whether, in general, the sum $\varphi(x) + \psi(y)$ of two functions of class $C^1(I)$, where $I \subset \mathbb{R}$ is, say, an open interval, is differentiable on $I \times I$. And what about the product?

3.16 Consider the function

$$f(x, y) = \begin{cases} 1 & x^3 < y < 4x^3 \\ 0 & y \leq x^3 \text{ or } y \geq 4x^3. \end{cases}$$

(a) Is f continuous at the origin?
(b) Establish if f admits partial derivatives at the points $(0, 0)$ and $(1, 1)$.
(c) For which unit vectors Q does f admit the directional derivative $(\partial f / \partial Q)(0, 0)$?

Answer. (a) Clearly, as depicted in Fig. 3.15, $f(0, y) = 0$ for every $y \in \mathbb{R}$ and for every fixed $m \in \mathbb{R}$ $f(x, mx) = 0$ for x small enough (see (c) below). This of course does not imply continuity at $(0, 0)$, and indeed restricting f along the cubic $y = 2x^3$ for $x > 0$ gives $f(x, 2x^3) = 1$, so that f is not continuous at $(0, 0)$.

(b) As already observed, f vanishes along the y axis, but actually also along the x axis (for $y = 0 < x^3$ when $x > 0$ and $4x^3 < 0 = y$ for $x < 0$). Hence the partials at the origin exist and are equal to 0. The point $(1, 1)$ sits at the boundary between Ω, the open region where f attains the value 1 (in white in Fig. 3.16) and $\mathbb{R}^2 \setminus \Omega$.

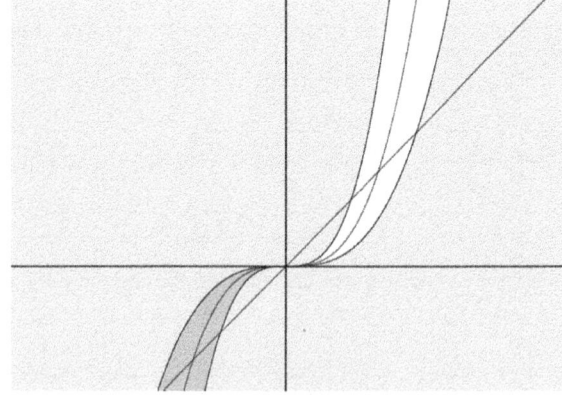

Fig. 3.15 Exercise 3.16: in white the region Ω where $f(x, y) = 1$; the regions where either $y \geq 4x^3$ or $y \leq x^3$ cover the grey part $\mathbb{R}^2 \setminus \Omega$ and intersect in the darker region in the third quadrant; the lines $y = mx$ and $y = 2x^3$ are also drawn

Fig. 3.16 Exercise 3.16: at (1, 1) the vertical and horizontal lines are transversal to $\partial\Omega$

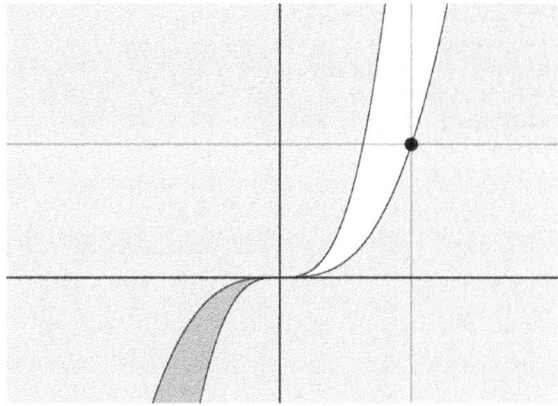

Actually, $(1, 1) \in \mathbb{R}^2 \setminus \Omega$ and hence $f(1, 1) = 0$. Also, the inequality $x^3 < y < 4x^3$ holds at the points $(1 + t, 1)$ with small enough $t < 0$ and at the points $(1, 1 + t)$ with small enough $t > 0$, so that at these points $f(x, y) = 1$, whence

$$\lim_{t \to 0^-} \frac{f(1+t, 1) - f(1, 1)}{t} = \lim_{t \to 0^-} \frac{1 - 0}{t} = -\infty$$

$$\lim_{t \to 0^+} \frac{f(1, 1+t) - f(1, 1)}{t} = \lim_{t \to 0^+} \frac{1 - 0}{t} = +\infty.$$

Therefore f does not admit partial derivatives at $(1, 1)$.

(c) Fix a unit vector $Q = (\cos\theta, \sin\theta)$ and consider the limit

$$\lim_{t \to 0} \frac{f(t\cos\theta, t\sin\theta) - f(0, 0)}{t} = \lim_{t \to 0} \frac{f(t\cos\theta, t\sin\theta)}{t}.$$

For $\theta \in [\pi, 2\pi)$ it is $f(t\cos\theta, t\sin\theta) = 0$, whence the existence and vanishing of the corresponding directional derivative.

Next, fix $\theta \in (0, \pi/2)$. There exists $t(\theta) > 0$ such that for $|t| < t(\theta)$ the point $(x, y) = (t\cos\theta, t\sin\theta)$ is such that $f(t\cos\theta, t\sin\theta) = 0$. Indeed, consider the functions $y = (\tan\theta)x$ and $y = 4x^3$ for $x > 0$. It is enough to show that they meet, as displayed in Fig. 3.17. Indeed, in the considered range for θ it is $\tan\theta > 0$ and

$$4x^3 = (\tan\theta)x \implies 4x^2 = \tan\theta \implies x = \frac{1}{2}\sqrt{\tan\theta}.$$

Thus, the two graphs certainly meet at $P = (\sqrt{\tan\theta}/2, \tan\theta\sqrt{\tan\theta}/2) \in \partial\Omega$ and

$$t(\theta) = \overline{OP} = \sqrt{\frac{\tan\theta}{4} + \frac{\tan^3\theta}{4}} = \frac{\sqrt{\tan\theta}}{2}\sqrt{1 + \tan^2\theta} = \frac{\sqrt{\tan\theta}}{2|\cos\theta|}$$

Fig. 3.17 Exercise 3.16: The point $P = (t(\theta)\cos\theta, t(\theta)\sin\theta) \in \partial\Omega$ where the graphs of the straight line $y = (\tan\theta)x$ and that of the cubic $y = 4x^3$ meet (here $0 < \theta < \pi/2$)

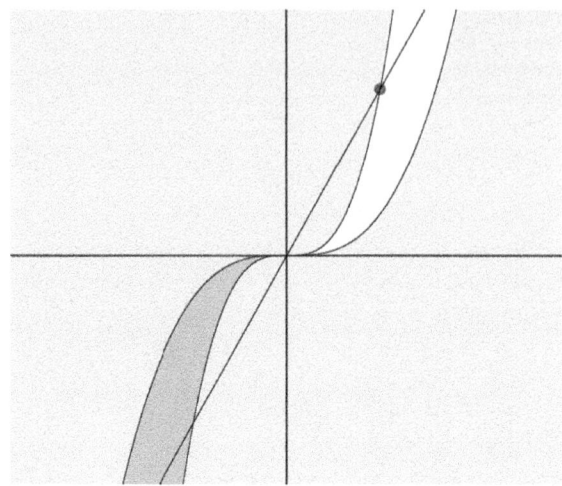

It follows that
$$\frac{\partial f}{\partial Q}(0,0) = \lim_{t \to 0} \frac{f(t\cos\theta, t\sin\theta)}{t} = 0.$$

This exercise has a geometric flavour. Question (a) is very easy, for it is enough to realize that the function f is the characteristic function of Ω and it is clear that one can approach the origin either on paths that are entirely outside Ω or entirely inside Ω. Question (b) can again be tackled by a geometric observation: along both the coordinate axes f vanishes because they are contained in $\mathbb{R}^2 \setminus \Omega$, whereas both the horizontal and the vertical lines passing through $(1, 1) \in \partial\Omega$ cross the boundary transversally at that point, hence the function exhibits a jump discontinuity when restricted to them. Question (c) is again clear from a geometric standpoint. First of all only unit vectors in the first quadrant may be problematic because the lines produced from them do cross Ω. This, however, does not happen close enough to the origin, so that the directional derivatives always do exist and vanish.

3.17 Consider the function $\min\{x^2 - y, y^2 - x\}$.

(a) Where is f continuous?
(b) Where does f admit partial derivatives?
(c) Is f differentiable at $(-1/2, -1/2)$?
(d) Compute, if existing, $\lim_{P \to \infty} f(P)$.

Answer. (a) It is important to understand the nature of the two regions
$$A := \{(x, y) \in \mathbb{R}^2 : x^2 - y < y^2 - x\}$$
$$B := \{(x, y) \in \mathbb{R}^2 : x^2 - y > y^2 - x\}$$

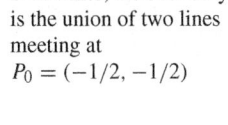

Fig. 3.18 Exercise 3.17: Region A in grey and region B in white; the boundary C is the union of two lines meeting at $P_0 = (-1/2, -1/2)$

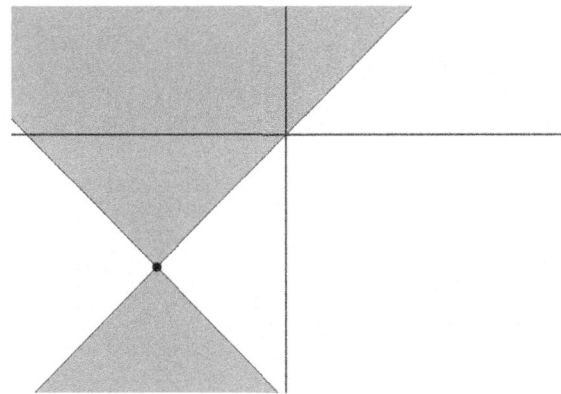

with common boundary $C := \partial A = \partial B = \{(x, y) \in \mathbb{R}^2 : x^2 - y = y^2 - x\}$. They are depicted in Fig. 3.18. With this notation, it is possible to write

$$f(x, y) = \begin{cases} x^2 - y & (x, y) \in A \\ y^2 - x & (x, y) \in B \\ x^2 - y = y^2 - x & (x, y) \in C. \end{cases}$$

Since both polynomials $x^2 - y$ and $y^2 - x$ are obviously continuous functions, the function f is continuous because it is the minimum of two continuous functions:

$$\min(f(P), g(P)) = \frac{1}{2}\Big(f(P) + g(P) - |f(P) - g(P)|\Big).$$

(b) On $A \cup B$ clearly f admits partial derivatives because it is the restriction of polynomials. Precisely, on A

$$\frac{\partial f}{\partial x} = 2x, \qquad \frac{\partial f}{\partial y} = -1$$

and on B

$$\frac{\partial f}{\partial x} = -1, \qquad \frac{\partial f}{\partial y} = 2y.$$

Along $C \setminus \{P_0\}$, the existence of partials is equivalent to the system of equalities

$$\begin{cases} 2x = -1 \\ -1 = 2y \end{cases}$$

which is not satisfied if $P \neq P_0$. Therefore the gradient does not exist in $C \setminus \{P_0\}$. As for P_0 both coordinate axes passing through P_0 belong each to one of the two regions, and hence they exist.

(c) Notice that $f(P_0) = 3/4$ and, from the previous calculation

$$\frac{\partial f}{\partial x}(P_0) = \frac{\partial f}{\partial y}(P_0) = -1$$

Therefore, whenever $(x - 1/2, y - 1/2) \in A$ the limit (3.7) becomes

$$\lim_{(x,y) \to (0,0)} \frac{f(-\frac{1}{2} + x, -\frac{1}{2} + y) - f(-\frac{1}{2}, -\frac{1}{2}) - \nabla f(-\frac{1}{2}, -\frac{1}{2})) \cdot (x, y)}{\sqrt{x^2 + y^2}}$$

$$= \lim_{(x,y) \to (0,0)} \frac{(-\frac{1}{2} + x)^2 - (y - \frac{1}{2}) - \frac{3}{4} + x + y}{\sqrt{x^2 + y^2}}$$

$$= \lim_{(x,y) \to (0,0)} \frac{x^2}{\sqrt{x^2 + y^2}} = 0,$$

because in polar coordinates

$$\left|\frac{\rho^2 \cos^2 \theta}{\rho}\right| \leq \rho$$

which tends to 0 independently of θ. Similarly, for $(x - 1/2, y - 1/2) \in B$ the limit (3.7) is

$$\lim_{(x,y) \to (0,0)} \frac{(y - \frac{1}{2})^2 - (x - \frac{1}{2}) - \frac{3}{4} + x + y}{\sqrt{x^2 + y^2}} = \lim_{(x,y) \to (0,0)} \frac{y^2}{\sqrt{x^2 + y^2}} = 0,$$

for the same reason as above. Finally, $(x - 1/2, y - 1/2) \in C$ if either $y = x$ or if $y = -x - 1$ and the limit (3.7) in both cases is:

$$\lim_{x \to 0} \frac{x^2}{\sqrt{2x^2}} = 0.$$

Therefore f is differentiable at $P_0 = (-1/2, -1/2)$.

(d) Manifestly the limit does not exist. Indeed, all the points $(x, y) = (-1/2, y)$ are in A and $f(-1/2, y) = (1/4) - y$ As $y \to \pm\infty$ evidently $\|(-1/2, y)\| \to +\infty$. However,

$$\lim_{y \to \pm\infty} f(-1/2, y) = \lim_{y \to \pm\infty} \frac{1}{4} - y = \mp\infty.$$

Therefore the limit does not exist.

3.5 Guided Exercises

A basic understanding of the definition of f is tantamount to understanding the nature of the set A, or, equivalently, B, and guides the approach to all issues. Continuity, actually, needs not be argued with formulae or limits because the very definition of f implies it. Indeed, f is obtained by choosing different polynomial expressions on two open sets, and their common value along the boundary C. The existence of partials on A and B is trivial. As for C, geometric intuition suggests that it is only at the vertex P_0 that the vertical and horizontal lines remain inside either A (the vertical line) or B (the horizontal line), thereby giving rise to smooth restrictions of f. At all other points of C the translated coordinate axes both cross from A to B and the corresponding restrictions of f exhibit non differentiable points at the crossing. Finally, differentiability at the vertex P_0 miracolously happens because the tangent planes to the graphs of the two distinct polynolmials coincide at P_0, fact that is to be guessed because their gradients coincide at P_0. Finally, limits at ∞ of polynomials are, in general, not to be expected to exist.

3.18 Consider the vector valued functions $F \colon \mathbb{R}^2 \to \mathbb{R}^3$ and $g \colon \mathbb{R}^3 \to \mathbb{R}^2$ given by

$$F(x, y) = (x^2, y^2, e^{xy}), \qquad G(x, y, z) = (xyz, x + y + z).$$

(a) Determine the explicit form of $G \circ F$ and of $F \circ G$
(b) Find $J(G \circ F)$ and $J(F \circ G)$.

Answer. (a) Computing

$$G \circ F(x, y) = G((F(x, y)) = G(x^2, y^2, e^{xy}) = (x^2 y^2 e^{xy}, x^2 + y^2 + e^{xy})$$
$$F \circ G(x, y, z) = F(G(x, y, z)) = F(xyz, x + y + z) = \big((xyz)^2, (x + y + z)^2, e^{xyz(x+y+z)}\big).$$
eUnALT

(b) Computing separately

$$JF(x, y) = \begin{bmatrix} 2x & 0 \\ 0 & 2y \\ ye^{xy} & xe^{xy} \end{bmatrix}, \qquad JG(x, y, z) = \begin{bmatrix} yz & xz & xy \\ 1 & 1 & 1 \end{bmatrix}.$$

Writing $F(x, y) = (t(x, y), u(x, y), v(x, y))$ or, for short, $F(x, y) = (t, u, v)$, where evidently $t = x^2$, $u = y^2$ and $v = e^{xy}$, and using formula (3.10),

$$J(G \circ F)(x, y) = JG(F(x, y)) \cdot JF(x, y)$$
$$= JG(t, u, v) \cdot JF(x, y)$$
$$= \begin{bmatrix} uv & tv & tu \\ 1 & 1 & 1 \end{bmatrix} \begin{bmatrix} 2x & 0 \\ 0 & 2y \\ ye^{xy} & xe^{xy} \end{bmatrix}$$
$$= \begin{bmatrix} 2xuv + ye^{xy}tu & 2ytv + xe^{xy}tu \\ 2x + ye^{xy} & 2y + xe^{xy} \end{bmatrix}$$

$$= \begin{bmatrix} (2xy^2 + x^2y^3)e^{xy} & (2x^2y + x^3y^2)e^{xy} \\ 2x + ye^{xy} & 2y + xe^{xy} \end{bmatrix}.$$

Similarly, writing now $G(x, y, z) = (U(x, y, z), V(x, y, z))$, or, for short, $G(x, y, z) = (U, V)$, where evidently $U(x, y, z) = xyz$ and $V(x, y, z) = x + y + z$, and using formula (3.10),

$$\begin{aligned} J(F \circ G)(x, y, z) &= JF(G(x, y, z)) \cdot JG(x, y, z) \\ &= JF(U, V) \cdot JG(x, y, z) \\ &= \begin{bmatrix} 2U & 0 \\ 0 & 2V \\ Ve^{UV} & Ue^{UV} \end{bmatrix} \begin{bmatrix} yz & xz & xy \\ 1 & 1 & 1 \end{bmatrix}. \end{aligned}$$

Computing the product and substituting the explicit form of U, V gives

$$\begin{bmatrix} 2xy^2z^2 & 2x^2yz^2 & 2x^2y^2z \\ 2(x+y+z) & 2(x+y+z) & 2(x+y+z) \\ e^{xyz(x+y+z)}yz(2x+y+z) & e^{xyz(x+y+z)}xz(x+2y+z) & e^{xyz(x+y+z)}xy(x+y+2z) \end{bmatrix}$$

for the required Jacobian $J(F \circ G)(x, y, z)$.

This exercise is a simple application of the chain rule for vector valued functions. The use of auxiliary symbols such as t, u, v or U, V may be handy, but is not strictly necessary.

3.19 Consider the vector valued function $F(x, y) = (xy - x^2, ye^y + g(x))$ where g is a real valued function of class $C^1(\mathbb{R})$.

(a) Can F be the gradient of a function $V : \mathbb{R}^2 \to \mathbb{R}$ if $g(x) = \arctan(h(x))$ with $h \in C^1(\mathbb{R})$ satisfying $h(0) = 1$?
(b) Find a g of class $C^1(\mathbb{R})$ and a $V : \mathbb{R}^2 \to \mathbb{R}$ such that ∇V is equal to the vector valued $F : \mathbb{R}^2 \to \mathbb{R}^2$ that corresponds to g.
(c) Compute $\nabla(G \circ F)$ when $G(x, y) = x + y$ and $g(x) = \sin x$.

Answer. (a) Suppose that a function $V : \mathbb{R}^2 \to \mathbb{R}$ with partial derivatives in $\mathbb{R}^2$ exists with $\nabla V(x, y) = F(x, y)$ if $g(x) = \arctan(h(x))$. Since $F \in C^1(\mathbb{R}^2)$ then $V \in C^2(\mathbb{R}^2)$ and the equalities

$$\begin{cases} V_x(x, y) = xy - x^2 \\ V_y(x, y) = ye^y + \arctan(h(x)) \end{cases} \implies \begin{cases} V_{yx}(x, y) = x \\ V_{xy}(x, y) = \dfrac{h'(x)}{1 + h^2(x)} \end{cases}$$

imply, by Schwarz' theorem, that for all $x \in \mathbb{R}$ it is

$$\frac{h'(x)}{1 + h^2(x)} = x$$

3.5 Guided Exercises

with $h(0) = 1$. Integrating this equality, and taking into account that $h(0) = 1$, yields

$$\int_0^x \frac{h'(t)}{1+h^2(t)} \, dt = \int_0^x t \, dt \implies \arctan(h(x)) - \frac{\pi}{4} = \frac{x^2}{2} \implies \arctan(h(x)) = \frac{x^2}{2} + \frac{\pi}{4}.$$

However, the image of the arctangent function is the interval $(-\pi/2, \pi/2)$ and this forces $x^2 < \pi/2$, so that h cannot be defined on the whole real line. The answer is therefore no.

(b) Computing as before

$$\begin{cases} V_x(x,y) = xy - x^2 \\ V_y(x,y) = ye^y + g(x) \end{cases} \implies \begin{cases} V_{yx}(x,y) = x \\ V_{xy}(x,y) = g'(x). \end{cases}$$

Schwarz' theorem gives this time $g'(x) = x$ everywhere, whence $g(x) = x^2/2 + c$, for some constant $c \in \mathbb{R}$. Hence, for any function F of the given form there may exist a function V which satisfies $\nabla V = F$ only if $g(x) = x^2/2 + c$, for some $c \in \mathbb{R}$. The simplest choice is of course $c = 0$. Then $F(x,y) = (xy - x^2, ye^y + x^2/2)$ and the gradient of V is therefore given by

$$\begin{cases} V_x(x,y) = xy - x^2 \\ V_y(x,y) = ye^y + \frac{x^2}{2}. \end{cases}$$

Any primitive of $x \mapsto V_x(x,y)$ is of the form

$$V(x,y) = \int (ty - t^2) \, dt = \frac{x^2}{2} y - \frac{x^3}{3} + C(y)$$

where C is any differentiable function of y. Taking the partial derivative of such a V with respect to y together with the above computation of V_y yields

$$V_y(x,y) = \frac{x^2}{2} + C'(y) = ye^y + \frac{x^2}{2},$$

whence $C'(y) = ye^y$. Choosing then

$$C(y) = \int_0^y te^t \, dt = ye^y - e^y + 1$$

gives $V(x,y) = \frac{x^2}{2} y - \frac{x^3}{3} + ye^y - e^y + 1$. It is immediate do check then indeed $\nabla V = F$, where $g(x) = x^2/2$ and hence $F(x,y) = (xy - x^2, ye^y + x^2/2)$.

(c) Computing:

$$\nabla G(x,y) = JG(x,y) = \begin{bmatrix} 1 & 1 \end{bmatrix}, \qquad JF(x,y) = \begin{bmatrix} y - 2x & x \\ \cos x & e^y(1+y) \end{bmatrix}.$$

By the chain rule for vector valued functions, keeping in mind that J_G is constant,

$$\nabla(G \circ F)(x, y) = JG \cdot JF(x, y)$$
$$= [1, 1] \begin{bmatrix} y - 2x & x \\ \cos x & e^y(1 + y) \end{bmatrix}$$
$$= [y - 2x + \cos x \quad x + e^y(1 + y)].$$

This exercise requires a natural and important observation, a direct consequence of the fact that Schwarz' theorem states that $V_{xy} = V_{yx}$ for any C^2 function V. Indeed, if $F = (F_1, F_2)$ is of class C^1 and $\nabla V = F$ holds in some domain, then V is of class C^2 and

$$\frac{\partial F_1}{\partial y} = V_{yx} = V_{xy} = \frac{\partial F_2}{\partial x}$$

must be true in the same domain. This is truly the *Leitmotiv* of the exercise.

In question (a), the equality above translates into a differential equation for h that is easily treated by separating variables and shows that h cannot be defined everywhere as is instead requested. With this example in mind, one is warned that not every possible function g can work, and that some attention must be paid in question (b). The differential equation for g determines the whole class of vector fields F for which the problem possibly makes sense, namely those F that correspond to a function g in the set $\{x^2/2 + c : c \in \mathbb{R}\}$. Since it is required to find just one pair (g, V), one tries with $c = 0$. Thus F is fixed and so are the two partials of the function V to be found. Here the second intuitive argument comes about: one passes from V_x to V_y by the following chain of operations:

$$V_x \xrightarrow{\int \cdot dx} V \xrightarrow{\frac{\partial}{\partial y}} V_y.$$

Since V_x and V_y are known, these operations must lead to conditions on the "constant of integration" which appears after the first conceptual operation, that of finding the primitive functions of V_x.

Question (c) is just a computation, stretching muscles after hard work.

3.20 Consider the vector valued function

$$F(s, t) = ((s - t) \sin s, (s - t) \cos s, \arcsin(s - t)).$$

(a) Determine $\text{Dom}(F)$.
(b) Write an equation in cartesian coordinates for the image $\text{Im}(F)$ of F.
(c) Verify that F is differentiable at $P_0 = (0, 0)$ using the definition.
(d) Compute the differential of F at $P_1 = (\pi/2, \pi/4)$.

3.5 Guided Exercises

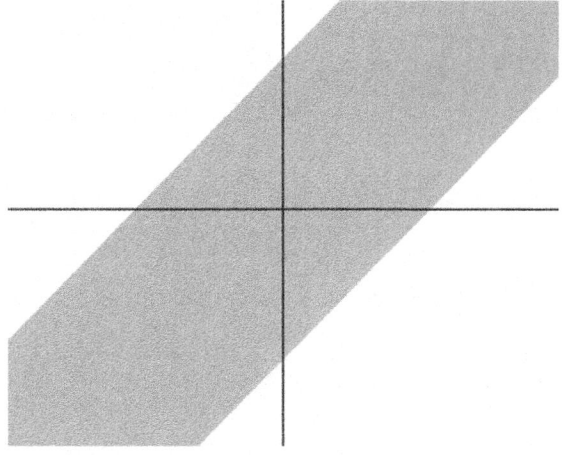

Fig. 3.19 Exercise 3.20: In grey the domain of F in the (s, t)-plane

Answer. (a) Because of the arcsine function

$$\mathrm{Dom}(F) = \left\{(s, t) \in \mathbb{R}^2 : -1 \leq s - t \leq 1\right\},$$

which is the stripe represented in Fig. 3.19.

(b) In order to achieve a description in cartesian coordinates of the image of F it is necessary to find $(x, y, z) \in \mathbb{R}^3$ for which there exist $(s, t) \in \mathrm{Dom}(F)$ that satisfy

$$\begin{cases} x = (s - t) \sin s \\ y = (s - t) \cos s \\ z = \arcsin(s - t). \end{cases}$$

In this case, $|z| \leq \pi/2$ and $x^2 + y^2 = (s - t)^2$. Furthermore, $(s - t) = \sin z$, so that the required equation is $x^2 + y^2 = \sin^2 z$. The geometry of this set is depicted in Fig. 3.20.

(c) The three components of F are of class C^1 in the interior of $\mathrm{Dom}(F)$, namely on $A = \left\{(s, t) \in \mathbb{R}^2 : -1 < s - t < 1\right\}$. Therefore $F \in C^1(A)$ is differentiable at $P_0 \in A$. What is required to show is (3.4), namely

$$\lim_{P \to P_0} \frac{f(P) - f(P_0) - JF(P_0) \cdot (P - P_0)}{\|P - P_0\|} = 0 \in \mathbb{R}^3.$$

Since

$$JF(s, t) = \begin{bmatrix} \sin s + (s - t) \cos s & -\sin s \\ \cos s - (s - t) \sin s & -\cos s \\ \frac{1}{\sqrt{1-(s-t)^2}} & -\frac{1}{\sqrt{1-(s-t)^2}} \end{bmatrix}$$

Fig. 3.20 The image of the function F of Exercise 3.20 exhibits an hourglass-type shape, with a sinusoidal profile

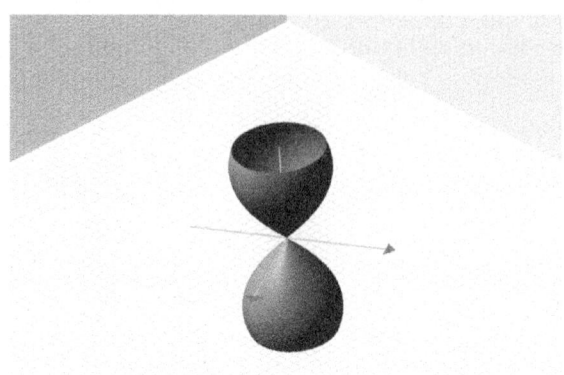

it follows that

$$JF(0,0) = \begin{bmatrix} 0 & 0 \\ 1 & -1 \\ 1 & -1 \end{bmatrix},$$

so that, in column notation

$$f(P) - f(P_0) - JF(P_0) \cdot (P - P_0) = \begin{bmatrix} (s-t)\sin s < \\ (s-t)\cos s \\ \arcsin(s-t) > \end{bmatrix} - \begin{bmatrix} 0 & 0 \\ 1 & -1 \\ 1 & -1 \end{bmatrix} \begin{bmatrix} s \\ t \end{bmatrix}$$

$$= \begin{bmatrix} (s-t)\sin s \\ (s-t)\cos s - s + t \\ \arcsin(s-t) - s + t \end{bmatrix}.$$

As for the limit (3.4), it is best to proceed componentwise. The first component is

$$\frac{(s-t)\sin s}{\sqrt{s^2+t^2}} = \frac{\sin s}{s} \frac{s(s-t)}{\sqrt{s^2+t^2}}.$$

The first factor is bounded in absolute value by 1 for any $s \neq 0$. The second can be estimated in polar coordinates

$$\left| \frac{\rho \cos \theta \rho (\cos \theta - \sin \theta)}{\rho} \right| \leq 2\rho.$$

It follows that the first limit is as required, namely

$$\lim_{(s,t) \to (0,0)} \frac{(s-t)\sin s}{\sqrt{s^2+t^2}} = 0.$$

By a Taylor expansion, the second component of the limit (3.4) can be written

$$\frac{(s-t)(\cos s - 1)}{\sqrt{s^2+t^2}} = \frac{\cos s - 1}{s^2} \frac{s^2(s-t)}{\sqrt{s^2+t^2}}.$$

Therefore, from $(1-\cos s)/s^2 \to 1/2$ and $|s|/\sqrt{s^2+t^2} \le 1$ it follows that

$$\left|\frac{(s-t)(\cos s - 1)}{\sqrt{s^2+t^2}}\right| \to 0$$

as $(s,t) \to (0,0)$, and the second request is met, too. Finally, using a Taylor expansion

$$\frac{\arcsin(s-t)-(s-t)}{\sqrt{s^2+t^2}} = \frac{\frac{1}{6}(s-t)^3 + (s-t)^3\omega(s-t)}{\sqrt{s^2+t^2}} = \frac{(s-t)^3\left(\frac{1}{6}+\omega(s-t)\right)}{\sqrt{s^2+t^2}}$$

where $\omega(w) \to 0$ as $w \to 0$. In polar coordinates $(s-t)^3/\sqrt{s^2+t^2}$ is estimated by

$$\frac{|\rho\cos\theta - \rho\sin\theta|^3}{\rho} \le \rho^2|\cos\theta - \sin\theta|^3 \le 8\rho^2,$$

so that in the end the third component of the limit (3.4) is also as required, i.e.

$$\lim_{(s,t)\to(0,0)} \frac{\arcsin(s-t)-(s-t)}{\sqrt{s^2+t^2}} = 0.$$

(d) First of all observe that $P_1 \in A$, so that F is certainly differentiable at P_1. The differential at P_1 is the linear map $L_{P_1}(P) = JF(P_1) \cdot P$, namely

$$L_{P_1}(s,t) = \begin{bmatrix} 1 & -1 \\ -\frac{\pi}{4} & 0 \\ \frac{4}{\sqrt{16-\pi^2}} & -\frac{4}{\sqrt{16-\pi^2}} \end{bmatrix} \begin{bmatrix} s \\ t \end{bmatrix} = \begin{bmatrix} s-t \\ -\frac{\pi}{4}s \\ \frac{4}{\sqrt{16-\pi^2}}(s-t) \end{bmatrix}.$$

This exercise is a combination of a little geometric ingenuity and analytic patience. Question (a) is more or less trivial and question (b) is not much more demanding: just express $(s-t)$ in terms of z. Question (c) boils down to understanding that a vector valued limit is equivalent to three limits, each of which is, in the case at hand, a standard limit of a scalar function of two variables. Question (d) is about the definition of differential.

3.21 Consider the function $f: [0,1] \times [0,1] \to \mathbb{R}$ defined by

$$f(x,t) = \begin{cases} 0 & x \le t \\ 1 & t < x \le 1 \end{cases}$$

and put $F(x) = \int_0^1 f(x,t)\,dt$.

(a) Is F continuous on $[0, 1]$?
(b) Is it possible to use differentiation under the integral sign for F in $[0, 1]$?
(c) Is it true that $F'(x) = \int_0^1 \frac{\partial}{\partial x} f(x, t) \, dt$ for those x at which F is differentiable?
(d) Decide if the equality $F'(x) = \int_0^1 g(x, t) \, dt$, where

$$g(x, t) = \begin{cases} \frac{\partial}{\partial x} f(x, t) & x \neq t \\ & x = t, \end{cases}$$

holds for some $x \in (0, 1)$ at which F is differentiable.

Answer. (a) Clearly,

$$F(x) = \int_0^1 f(x, t) \, dt = \int_0^x dt = x,$$

which is obviously continuous in $[0, 1]$.

(b) Evidently, F is differentiable everywhere and $F'(x) = 1$ for every $x \in [0, 1]$. However, the theorem cannot be used to infer this because the function f is not even. continuous, nor is it differentiable with respect to x.

(c) Clearly, the partial derivative $\frac{\partial}{\partial x} f(x, t)$ exists at every point of $(0, 1) \times (0, 1)$ where it vanishes, except along the diagonal $\{x = t\}$ where it does not exist. Strictly speaking, for every fixed $x \in (0, 1)$ the function $t \mapsto \frac{\partial}{\partial x} f(x, t)$ is not defined at $t = x$ and so Theorem 3.8 does not apply. Hence the answer is no.

(d) As already observed, for every fixed $x \in (0, 1)$ the function $t \mapsto \frac{\partial}{\partial x} f(x, t)$ is defined in $[0, x) \cup (x, 1]$ where it vanishes, because in each subinterval $f(x, \cdot)$ is constant, either 0 or 1. The extension of this function at x with value $f(x, x) = 0$ produces the continuous function $t \mapsto g(x, t) \equiv 0$, namely the zero function. Hence for every $x \in (0, 1)$

$$\int_0^1 g(x, t) \, dt = 0 \neq 1 = F'(x).$$

Hence the equality never holds true.

This is a basic example where differentiation under the integral sign does not hold true, because although perfectly integrable in each segment for any fixed x, the function $t \mapsto f(x, t)$ is not continuous at $t = x$. Being piecewise constant it is differentiable except at one point, with derivative vanishing wherever it is defined, namely in $[0, x) \cup (x, 1]$. There is a continuous extension of the derivative at the point x which is zero everywhere. The integral of this continuous extension vanishes and therefore differentiation under the integral sign fails.

3.22 Draw the graph of $F(x) = \int_0^1 \log(2 - x^2 t^2) \, dt$.

3.5 Guided Exercises

Answer. The function $f(x,t) = \log(2 - x^2 t^2)$ is defined and of class C^1 on its natural domain $A = \{(x, t) \in \mathbb{R}^2 : x^2 t^2 < 2\}$. Now, $(-\sqrt{2}, \sqrt{2}) \times [0, 1] \subset A$ and hence for any $\varepsilon > 0$ the rectangle $A_\varepsilon = [-\sqrt{2} + \varepsilon, \sqrt{2} - \varepsilon] \times [0, 1]$ is contained in A. Since $f \in C(A_\varepsilon)$, the funcion F is defined and continuous in $[-\sqrt{2} + \varepsilon, \sqrt{2} - \varepsilon]$ for every sufficiently small $\varepsilon > 0$, which amounts to saying that it is defined and continuous on $(-\sqrt{2}, \sqrt{2})$. Further, for $x = \pm\sqrt{2}$, the function $t \mapsto f(\pm\sqrt{2}, t) = \log(2 - 2t^2)$ admits improper integral at $t = 1$ and hence the full domain of F is actually $[-\sqrt{2}, \sqrt{2}]$. It is also useful to observe that F is an even function.

Now,
$$\frac{\partial f}{\partial x}(x, t) = -\frac{2xt^2}{2 - t^2 x^2}$$

is a continuous function on A_ε for every $\varepsilon > 0$ sufficiently small and it follows that F is differentiable in $[-\sqrt{2} + \varepsilon, \sqrt{2} - \varepsilon]$, hence in the end in $(-\sqrt{2}, \sqrt{2})$. Thus

$$F'(x) = -2x \int_0^1 \frac{t^2}{2 - t^2 x^2} \, dt$$

for $x \in (-\sqrt{2}, \sqrt{2})$. The integral appearing above is positive and so F' is positive in $(-\sqrt{2}, 0)$ and negative in $(0, \sqrt{2})$. Thus F is increasing in $(-\sqrt{2}, 0]$ and decreasing in $[0, \sqrt{2})$. Since $\log(2 - x^2 t^2) \geq \log(2 - 2t^2)$ for $x \in [\sqrt{2}, \sqrt{2}]$ and $t \in [0, 1]$, it holds

$$F(x) = \int_0^1 \log(2 - x^2 t^2) \, dt \geq \int_0^1 \log(2 - 2t^2) \, dt = F(-\sqrt{2}) = F(\sqrt{2})$$

for any $x \in [-\sqrt{2}, \sqrt{2}]$. Therefore F is actually increasing in $[-\sqrt{2}, 0]$ and decreasing in $[0, \sqrt{2}]$.

Notice that $F'(x)$ diverges to $\pm\infty$ for $x \to \mp\sqrt{2}$. Indeed, upon writing

$$F'(x) = -2x \int_0^1 \frac{1}{\left(\frac{\sqrt{2}}{t} - x\right)\left(\frac{\sqrt{2}}{t} + x\right)} \, dt$$

it is clear that the integral diverges if $x \to \pm\sqrt{2}$ because the denominator diverges of order exactly 1 for $t \to 1$. Finally,

$$F(0) = \int_0^1 \log 2 \, dt = \log 2,$$

and

$$F(\sqrt{2}) = F(-\sqrt{2}) = \int_0^1 \log(2 - 2t^2) \, dt$$

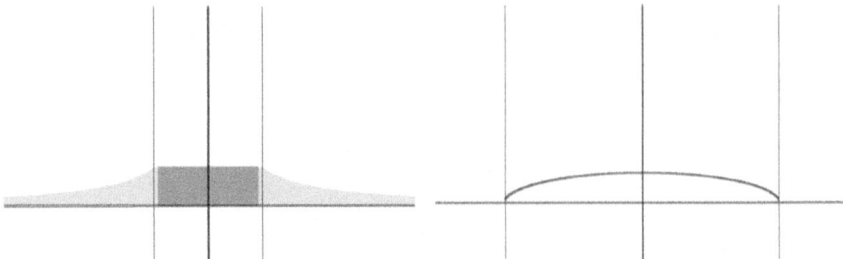

Fig. 3.21 On the left the domain of f in Exercise 3.48 in the (x, t) plane, where the darker region is a rectangle A_ε; on the right the graph of F

$$= \log 2 + \int_0^1 \log(1 - t^2)\, dt$$
$$= 3\log 2 - 2 > 0,$$

after integration by parts. The graph is depicted in Fig. 3.21.

This is a standard exercise, whereby a little care must be used to handle the domain. First, the integral is on the interval $t \in [0, 1]$, hence one must find for which x the segment $\{x\} \times [0, 1]$ is inside the domain of f. Secondly, one must also handle the case in which the integral is actually to be interpreted in the improper sense, which happens exactly for $x = \pm\sqrt{2}$. The rest is straightforward calculations allowed by derivation under the integral sign.

3.23 Let $F(x) = \displaystyle\int_1^2 \frac{e^{x+y} - 1}{\sqrt{x^2 + y^2}}\, dy$.

(a) Prove that $F \in C^1(\mathbb{R})$.
(b) Establish if F is invertible in some neighborhood of the origin.

Answer. (a) For any $a < b$, the function $f : [a, b] \times [1, 2] \to \mathbb{R}$ defined by

$$f(x, y) = \frac{e^{x+y} - 1}{\sqrt{x^2 + y^2}}$$

is continuous, and such is therefore F on $[a, b]$. Since $a < b$ are arbitrary, it follows that $F \in C^0(\mathbb{R})$. Next, the partial derivative

$$\frac{\partial f}{\partial x}(x, y) = \frac{e^{x+y}\sqrt{x^2 + y^2} - (e^{x+y} - 1)\frac{x}{\sqrt{x^2+y^2}}}{x^2 + y^2}$$

is continuous on $[a, b] \times [1, 2]$, which implies that F is differentiable in $[a, b]$ for any $a < b$ with continuous derivative given by

3.5 Guided Exercises

$$F'(x) = \int_1^2 \frac{\partial f}{\partial x}(x, y)\, dy.$$

Since $a < b$ are arbitrary, it follows that $F \in C^1(\mathbb{R})$.

(b) Observe that

$$F'(0) = \int_1^2 \frac{\partial f}{\partial x}(0, y)\, dy = \int_1^2 \frac{e^y}{y}\, dy > 0$$

and so, by continuity, $F'(x) > 0$ in a neighborhood of the origin. Therefore F is strictly increasing, hence invertible, in some neighborhood of the origin.

A standard exercise with no difficulty whatsoever. It is just a matter of checking that derivation under the integral sign is legitimate, and this is true because the function f_x is continuous. In part (b) it is enough to observe that $f_x(0, y)$ is positive in $[1, 2]$ and hence has a positive integral, so that $F'(0) > 0$.

3.24 Sia $F(t) = \int_0^\pi \frac{\sin x}{x} e^{xt}\, dx$.

(a) Determine $\mathrm{Dom}(F)$.
(b) Find where F is continuous and where it is differentiable, if anywhere.
(c) Compute, if existing, the limits $\lim_{t \to -\infty} F'(t)$, $\lim_{t \to +\infty} F'(t)$ and $\lim_{t \to +\infty} F(t)$.

Answer. (a) The function $(x, t) \mapsto \frac{\sin x}{x} e^{xt}$ is continuous on $(0, \pi] \times \mathbb{R}$ and extends naturally to a continuous function f on $[0, \pi] \times \mathbb{R}$ upon setting $f(0, t) = 1$. Clearly,

$$F(t) = \int_0^\pi f(x, t)\, dx$$

is well defined for every $t \in \mathbb{R}$ because $x \mapsto f(x, t)$ is continuous on $[0, \pi]$ for every $t \in \mathbb{R}$. Thus $\mathrm{Dom}(F) = \mathbb{R}$.

(b) The continuity of f on $[0, \pi] \times [c, d]$ for every $c < d$ implies that $F \in C^0(\mathbb{R})$. Furthermore,

$$\frac{\partial f}{\partial t}(x, t) = e^{xt} \sin x$$

for every $(x, t) \in \mathbb{R}^2$ and $\frac{\partial f}{\partial t} \in C^0([0, \pi] \times [c, d])$ for every $c < d$. Therefore F is differentiable in $\mathbb{R}$.

(c) An integration by parts gives

$$F'(t) = \int_0^\pi e^{xt} \sin x\, dx$$

$$= \left[-e^{xt} \cos x\right]_0^\pi + t \int_0^\pi e^{xt} \cos x\, dx$$

$$= \left[-e^{xt} \cos x\right]_0^\pi + t \left(\left[e^{xt} \sin x\right]_0^\pi - t \int_0^\pi e^{xt} \sin x\, dx\right),$$

that is
$$(1+t^2)F'(t) = e^{\pi t} + 1 \quad \Longrightarrow \quad F'(t) = \frac{e^{\pi t}+1}{1+t^2}.$$

Thus,
$$\lim_{t \to +\infty} F'(t) = \lim_{t \to +\infty} \frac{e^{\pi t}+1}{1+t^2} = +\infty, \quad \lim_{t \to -\infty} F'(t) = \lim_{t \to -\infty} \frac{e^{\pi t}+1}{1+t^2} = 0.$$

Observe that F is strictly increasing as $F' > 0$. A possible representation of F is
$$F(t) = F(0) + \int_0^t F'(s)\,ds = \int_0^\pi \frac{\sin x}{x}\,dx + \int_0^t \frac{e^{\pi s}}{1+s^2}\,ds + \arctan t$$

so that
$$\lim_{t \to +\infty} F(t) = +\infty.$$

Indeed, the first improper integral manifestly diverges while $\arctan t \to \pi/2$ and $\int_0^\pi \frac{\sin x}{x}\,dx$ has a finite value (obviously less than π because $|\sin x/x| \le 1$).

Part (a) is just establishing the integrability of a function depending on a parameter, and part (b) is a direct application of Theorem 3.8, item (i). Part (c) is the most interesting one. First, differentiation under the integral sign is applied to actually compute $F'(x)$ explicitly. Secondly, the Fundamental of Theorem of Calculus is applied in the form
$$F(t) = F(0) + \int_0^t F'(s)\,ds$$

in order to establish the behaviour of F at $+\infty$ using the explicit knowledge of F'.

3.25 Consider the function $F(x, y) = \int_0^1 \sqrt{1-(x^2+y^2)t^8}\,dt$.

(a) Determine $\mathrm{Dom}(F)$.
(b) Find, if existing, the minima and maxima of F in its domain.

Answer. (a) If $t \in [0, 1]$, then $f(x, y, t) = \sqrt{1-(x^2+y^2)t^8}$ is well defined and continuous provided that $1 - (x^2+y^2)t^8 \ge 0$, which is true for all (x, y) when $t = 0$ and equivalent to $x^2 + y^2 \le t^{-8}$ if $t \in (0, 1]$. In the latter case, $1 \le t^{-8}$, therefore $x^2 + y^2 \le t^{-8}$ for every $t \in (0, 1]$ if and only in $x^2 + y^2 \le 1$. It follows that $\mathrm{Dom}(F) = \{(x, y) \in \mathbb{R}^2 : x^2 + y^2 \le 1\}$.

(b) Since $\mathrm{Dom}(F)$ is compact, by Weierstrass' theorem F has both maxima and minima. Now, for $(x, y) \in D := \{(x, y) \in \mathbb{R}^2 : x^2 + y^2 < 1\} \subset \mathrm{Dom}(F)$, the function $f(x, y, t)$ has continuous partial derivatives f_x and f_y for all $t \in [0, 1]$, and differentiation under the integral sign gives

$$\frac{\partial F}{\partial x}(x, y) = \int_0^1 \frac{\partial}{\partial x}\sqrt{1-(x^2+y^2)t^8}\,dt = -x\int_0^1 \frac{t^8}{\sqrt{1-(x^2+y^2)t^8}}\,dt$$

3.5 Guided Exercises

$$\frac{\partial F}{\partial y}(x, y) = \int_0^1 \frac{\partial}{\partial y}\sqrt{1-(x^2+y^2)t^8}\,dt = -y\int_0^1 \frac{t^8}{\sqrt{1-(x^2+y^2)t^8}}\,dt.$$

Since the integral on the right hand side is strictly positive, the unique critical point of F in D is for $x = 0 = y$. In order to establish the nature of this critical point the second derivatives at $(0,0)$ are useful. Now, since $t^8/\sqrt{1-(x^2+y^2)t^8}$ has continuous partial derivatives, derivation under the integral sign may be performed again and yields, after a short computation

$$\frac{\partial^2 F}{\partial x^2}(x,y) = -\int_0^1 \frac{t^8}{\sqrt{1-(x^2+y^2)t^8}}\,dt - x^2\int_0^1 \frac{t^{16}}{\sqrt{[1-(x^2+y^2)t^8]^3}}\,dt$$

so that

$$\frac{\partial^2 F}{\partial x^2}(0,0) = -\int_0^1 t^8\,dt = -\frac{1}{9}.$$

Similarly, arguing by symmetry, $\frac{\partial^2 F}{\partial y^2}(0,0) = -1/9$. Finally,

$$\frac{\partial^2 F}{\partial x \partial y}(x,y) = -xy\int_0^1 \frac{t^8}{\sqrt{1-(x^2+y^2)t^8}}\,dt$$

which gives $\frac{\partial^2 F}{\partial x \partial y}(0,0) = 0$. It follows that the Hessian matrix at the origin is

$$HF(0,0) = \begin{bmatrix} -\frac{1}{9} & 0 \\ 0 & -\frac{1}{9} \end{bmatrix},$$

so that the origin is a (possibly local) maximum, with $F(0,0) = \int_0^1 dt = 1$. Finally, along the boundary of D, namely where $x^2 + y^2 = 1$, the function is constant and equal to

$$F\Big|_{\partial D} = \int_0^1 \sqrt{1-t^8}\,dt,$$

which is manifestly smaller than 1. It follows that all the points of the boundary are minima and the origin is the only maximum. This latter conclusion could also have been derived from the observation that $f(x, y, t) \leq 1$ in its domain, whence $F(x, y) \leq \int_0^1 dt = 1 = F(0,0)$.

This exercise is rather standard but interesting because derivation under the integral sign is used repeatedly. Here it has been solved using the usual machinery for functions of two variables, but it is worthwhile mentioning that it could also be approached by observing that $f(x, y, t)$, hence $F(x, y)$ itself, is radial in (x, y). So one could use polar coordinates and notice that $F(\rho\cos\theta, \rho\sin\theta) = G(\rho)$, with

$$G(\rho) = \int_0^1 \sqrt{1 - \rho^2 t^8}\, dt.$$

This simplifies matters because then, using derivation under the integral sign,

$$G'(\rho) = -\rho \int_0^1 \frac{t^8}{\sqrt{1 - \rho^2 t^8}}\, dt < 0,$$

so that, being decreasing, G has its maximum at $\rho = 0$, the origin, and its mininum at $\rho = 1$, the unit circle.

3.26 Consider the function $F(x, y) = \int_{\frac{1}{2x+3y}}^{x} \log(yt) e^{1-t^2}\, dt$.

(a) Determine $\text{Dom}(F)$.
(b) Find where F is continuous and where it is differentiable.
(c) Compute, if possible, F_x and F_y.

Answer. (a) It is necessary that $y \neq 0$ for the function $f(x, t) = \log(yt) e^{1-t^2}$ to be defined. Furthermore, it must be that $yt > 0$ for any fixed $y \neq 0$ along the integration interval. Therefore t must have the same sign as y along the integration interval. Finally, observe that the upper integration bound x might actually be taken to be 0 because the function $t \mapsto \log(yt) e^{1-t^2}$ has convergent improper integral. Indeed,

- if $y > 0$, then $\lim_{t \to 0^+} \log(yt) e^{1-t^2} = -\infty$,
- if $y < 0$, then $\lim_{t \to 0^-} \log(yt) e^{1-t^2} = +\infty$,

and in both cases the divergence order is smaller than any $\alpha < 1$, so that the integrals

$$\int_0^\varepsilon \log(yt) e^{1-t^2}\, dt, \quad \int_{-\varepsilon}^0 \log(yt) e^{1-t^2}\, dt$$

exist as improper integrals if $y > 0$ or $y < 0$, respectively. Summing up,

- if $y > 0$, and if $x \geq 0$, then the integral defining $F(x, y)$ is meaningful because then $2x + 3y > 0$, the integration interval I is inside $[0, +\infty)$ and the function $t \mapsto \log(yt) e^{1-t^2}$ is integrable in I, possibly in the improper sense at 0 if $x = 0$.

- if $y < 0$, and if $x \leq 0$, then the integral defining $F(x, y)$ is meaningful because then $2x + 3y < 0$, the integration interval I is inside $(-\infty, 0]$ and the function $t \mapsto \log(yt) e^{1-t^2}$ is integrable in I, possibly in the improper sense at 0 if $x = 0$.

Therefore $\text{Dom}(F) = \{(x, y) \in \mathbb{R}^2 : xy \geq 0,\ y \neq 0\}$ (Fig. 3.22).

(b) Since f is continuous with continuous partial derivatives at all points in the interior $D = \{(x, y) \in \mathbb{R}^2 : xy > 0,\}$ of $\text{Dom}(F)$, the same holds for F in D.

Fig. 3.22 In grey the interior of the domain of F in Exercise 3.26; the (black) y axis corresponds to the upper integration bound $x = 0$, and also belongs to the domain

(c) Using formulae (3.12),

$$\frac{\partial F}{\partial x}(x, y) = e^{1-x^2} \log(xy) + e^{1-\frac{1}{(2x+3y)^2}} \frac{2}{(2x+3y)^2} \log\left(\frac{y}{2x+3y}\right)$$

$$\frac{\partial F}{\partial y}(x, y) = e^{1-\frac{1}{(2x+3y)^2}} \frac{3}{(2x+3y)^2} \log\left(\frac{y}{2x+3y}\right) + \int_{\frac{1}{2x+3y}}^{x} \frac{e^{1-t^2}}{y} \, dt.$$

Here the most serious issue is the determination of the domain. The difficulty consists in the fact that since yt must be positive, then for any fixed y the variable t must have the same sign as y and, consequently, the integration bounds also must have the same sign, at least "weekly". This means that since the function $t \mapsto f(x, y, t)$ to be integrated is integrable in the improper sense at $t = 0$, the (upper) integration bound is allowed to vanish.

3.27 Consider the function $F(x) = \int_0^x \frac{\log(1+tx)}{1+t^2} \, dt$ to prove that

$$\int_0^1 \frac{\log(1+t)}{1+t^2} \, dt = \frac{\pi}{8} \log 2.$$

Answer. The function

$$f(x, t) = \frac{\log(1+tx)}{1+t^2}$$

is defined if $1 + tx > 0$, which is always true for $x \geq 0$ and $t \in [0, 1]$. Since f is continuous and differentiable in $[0, 1] \times [0, 1]$, and since

$$\frac{\partial f}{\partial x}(x, t) = \frac{t}{(1+t^2)(1+tx)}$$

differentiation under the integral sign gives

$$F'(x) = \int_0^x \frac{\partial f}{\partial x}(x,t)\,dt + \frac{\log(1+x^2)}{1+x^2} = \int_0^x \frac{t}{(1+t^2)(1+tx)}\,dt + \frac{\log(1+x^2)}{1+x^2}.$$

Solving the decomposition into partial fractions problem

$$\frac{t}{(1+t^2)(1+tx)} = \frac{A}{1+tx} + \frac{Bt+C}{1+t^2}$$

for the rational function of t on the left, with coefficients depending on x, gives

$$A = -\frac{x}{1+x^2}, \quad B = \frac{1}{1+x^2}, \quad C = \frac{x}{1+x^2}$$

and yields

$$\int_0^x \frac{t}{(1+t^2)(1+tx)}\,dt = -\frac{x}{1+x^2}\int_0^x \frac{1}{1+tx}\,dt + \frac{1}{1+x^2}\int_0^x \frac{t+x}{1+t^2}\,dt$$

$$= \frac{x \arctan x}{1+x^2} - \frac{\log(1+x^2)}{2(1+x^2)}$$

after direct calculation of the indicated integrals of rational functions. Thus,

$$F'(x) = \int_0^x \frac{t}{(1+t^2)(1+tx)}\,dt + \frac{\log(1+x^2)}{1+x^2} = \frac{x \arctan x}{1+x^2} + \frac{\log(1+x^2)}{2(1+x^2)}.$$

A final integration by parts gives

$$F(x) = \int_0^x F'(u)\,du$$

$$= \int_0^x \left[\frac{u \arctan u}{1+u^2} + \frac{\log(1+u^2)}{2(1+u^2)}\right]du$$

$$= \left(\frac{1}{2}\log(1+u^2)\arctan u\Big|_0^x - \frac{1}{2}\int_0^x \frac{\log(1+u^2)}{1+u^2}\,du\right) + \int_0^x \frac{\log(1+u^2)}{2(1+u^2)}\,du$$

$$= \frac{1}{2}\log(1+x^2)\arctan x.$$

Therefore

$$F(1) = \frac{1}{2}\log(2)\arctan 1 = \frac{\pi}{8}\log 2.$$

This exercise is by no means standard. It is one of the many existing examples of the so-called Feynman technique for solving various types of integrals. The main idea is to introduce a second variable, or parameter, in an integral that contains a single variable and to then derive differential equations by taking one or several derivatives with respect to the parameter. In this case the parameter is x and gives rise to the integral function $F(x)$, the value of interest being $x = 1$.

3.6 Problems

3.28 Determine the regions in which the following functions are continuous.

(a) $\begin{cases} \dfrac{x^2 y^2}{x^4 + y^4} & (x, y) \neq (0, 0) \\ 0 & (x, y) = (0, 0) \end{cases}$

(b) $\begin{cases} \dfrac{x^2 y}{x^4 + y^2} & (x, y) \neq (0, 0) \\ 0 & (x, y) = (0, 0) \end{cases}$

(c) $\begin{cases} \dfrac{e^{xy} - e^{-xy}}{2(x^2 + y^2)} & (x, y) \neq (0, 0) \\ 0 & (x, y) = (0, 0) \end{cases}$

(d) $\begin{cases} \dfrac{x^2 \sin xy}{x^2 + y^2} & (x, y) \neq (0, 0) \\ 0 & (x, y) = (0, 0) \end{cases}$

(e) $\begin{cases} \arctan \dfrac{x}{x^2 + y^2} & (x, y) \neq (0, 0) \\ 0 & (x, y) = (0, 0). \end{cases}$

3.29 Compute, where existing, the partial derivatives of the following functions.

(a) $\log \tan(x/y)$

(b) x^y

(c) $x^3 y^2 z + 2x - 3y + z + 5$

(d) $e^{\sin(y/x)}$

(e) $\arctan(y/x)$

(f) $\log(x + \sqrt{x^2 + y^2})$

(g) $\log \sin(x/\sqrt{y})$

(h) $x^{(y^z)}$

(i) $y\sqrt{|x||y|}$

(l) $\log(x^2 + y^2)$

(m) $\sum_{i=1}^{n} a_i x_i$, $(a_1, \ldots, a_n) \in \mathbb{R}^n$.

3.30 For each of the following functions establish if they are continuous, if they admit partial derivatives at $(0, 0)$ and if they are differentiable $(0, 0)$.

(a) $\begin{cases} \dfrac{xy}{x^2+y^2} & (x,y) \neq (0,0) \\ 0 & (x,y) = (0,0) \end{cases}$

(b) $\begin{cases} \dfrac{x+y^3}{x^2+y^2} & (x,y) \neq (0,0) \\ 0 & (x,y) = (0,0) \end{cases}$

(c) $\begin{cases} \dfrac{x(1-\cos y)}{y} & |y| > x^4 \\ 0 & (x,y) = (0,0) \\ \dfrac{\sin y^2}{x^4} & |y| \leq x^4,\ (x,y) \neq (0,0). \end{cases}$

3.31 Compute the limit
$$\lim_{t \to 0} \frac{f(P_0 + tQ) - f(P_0)}{t}$$
in each of the following cases (in neither of which it is a directional derivative):

(a) $f(x,y) = \sqrt{2 - 3x^2 - y^2 + x}$, $P_0 = (1,1)$, $Q = (3,4)$.
(b) $f(x,y) = x^{(y^2)}$, $P_0 = (1,1)$, $Q = (3,4)$.
(c) $f(x,y) = \arctan\left(\dfrac{x+y}{1-xy}\right)$, $P_0 = (0,0)$, $Q = (-1,1)$.
(d) $f(x,y,z) = x^2 + 2y^2 + 3z^2$, $P_0 = (1,1,0)$, $Q = (1,-1,2)$.
(e) $f(x,y,z) = (x/y)^z$, $P_0 = (1,1,1)$, $Q = (2,1,-1)$.
(f) $f(x,y) = y\sqrt{|x||y|}$, $P_0 = (1,1)$, $Q = (-1,2)$.

3.32 Consider the function $f(x,y) = \sqrt{x^2 + y^2}$.
(a) Is f continuous at the origin?
(b) Does f admit partial derivatives at the origin?
(c) Is f differentiable at the origin?
(d) For which unit vectors Q does there exist the directional derivative $\partial f/\partial Q$ at the origin?

3.33 Consider the function $f(x,y) = \sqrt{|xy|}$.
(a) Is f continuous at the origin?
(b) Does f admit partial derivatives at the origin?
(c) Is f differentiable at the origin?
(d) For which unit vectors Q does there exist the directional derivative $\partial f/\partial Q$ at the origin?

3.34 Determine for which values of the real parameters $\alpha > 0$ and $\beta, \gamma \in \mathbb{R}$, if any, the following functions

$$f(x,y) = \begin{cases} \dfrac{\log(1 + (x^2)^\alpha) - |y|^\alpha}{x^2 + y^2} & (x,y) \neq (0,0) \\ 0 & (x,y) = (0,0) \end{cases},$$

3.6 Problems

$$g(x, y) = \begin{cases} \dfrac{xy}{|\sin x - y|^\beta} & (x, y) \neq (0, 0) \\ \gamma & (x, y) = (0, 0) \end{cases}$$

are continuous or differentiable at the origin.

3.35 Let $\alpha > 0$ be a parameter and consider the function

$$f(x, y) = \begin{cases} \dfrac{(1 - \cos(xy)) \log(1 + x^2 + y^2)}{(x^2 + y^2)^\alpha} & (x, y) \neq (0, 0) \\ 0 & (x, y) = (0, 0). \end{cases}$$

(a) Establish for which values of α, if any, f is continuous at $(0, 0)$.
(b) Establish for which values of α, if any, f is differentiable at $(0, 0)$.

3.36 Let $\alpha > 0$ be a parameter and consider the function

$$f(x, y) = \begin{cases} \dfrac{|y - 1|^\alpha \cos x}{\sqrt{x^2 + (y - 1)^2}} & (x, y) \neq (0, 1) \\ 0 & (x, y) = (0, 1) \end{cases}$$

(a) Establish for which values of α, if any, f is continuous at $(0, 1)$.
(b) Establish for which values of α, if any, f is differentiable at $(0, 1)$.

3.37 Let $\alpha > 0$ be a parameter and consider the function

$$f(x, y) = \begin{cases} (x^2 + 2y^2)^\alpha \sin\left(\dfrac{1}{2x^2 + y^2}\right) & (x, y) \neq (0, 0) \\ 0 & (x, y) = (0, 0). \end{cases}$$

(a) Establish for which values of α, if any, f is continuous at $(0, 0)$.
(b) Establish for which values of α, if any, f admits partial derivatives at $(0, 0)$.
(c) Establish for which values of α, if any, f is differentiable at $(0, 0)$.

3.38 Let $\alpha \in \mathbb{R}$ be a parameter and consider the function

$$f(x, y) = \begin{cases} (x^2|y - 1|)^\alpha & (x, y) \neq (0, 1) \\ 0 & (x, y) = (0, 1). \end{cases}$$

(a) Establish for which values of α, if any, f is differentiable at $(0, 1)$.
(b) Fix $\alpha = 1/3$. Establish for which unit vectors Q, if any, the directional derivative $(\partial f / \partial Q)(0, 1)$ exists and, for those values, compute it.

3.39 Show that $f(x, y) = x^{2/3} y$ is differentiable at the origin but f_x is not continuous at the origin.

3.40 Show that the mixed partials at the origin of

$$f(x, y) = \begin{cases} y^2 \sin(1/y) & y \neq 0 \\ 0 & y = 0 \end{cases}$$

are equal although f has a non continuous partial derivative at $(0, 0)$.

3.41 Consider the function

$$f(x, y) = \begin{cases} x^2 y^2 \cos\left(\dfrac{1}{x^2 y^2}\right) & xy \neq 0 \\ 0 & xy = 0. \end{cases}$$

(a) Is f continuous in $\mathbb{R}^2$?
(b) Does f admit partial derivatives in $\mathbb{R}^2$?
(c) Are the partial derivatives of f continuous in $\mathbb{R}^2$?
(d) Is f differentiable in $\mathbb{R}^2$?

3.42 Consider the function

$$f(x, y) = \begin{cases} \dfrac{xy(x^2 - y^2)}{x^2 + y^2} & (x, y) \neq (0, 0) \\ 0 & (x, y) = (0, 0). \end{cases}$$

(a) Is f continuous at $(0, 0)$?
(b) Does f admit partial derivatives at $(0, 0)$?
(c) Are the partial derivatives of f continuous at $(0, 0)$?
(d) Is it true that the mixed partials of f at $(0, 0)$ are equal?

3.43 Consider the function

$$f(x, y) = \begin{cases} \dfrac{x^2(y^2 - 1)}{x^2 + y^2 - 1} & x^2 + y^2 \neq 1 \\ 0 & x^2 + y^2 = 1. \end{cases}$$

(a) Compute the maximum directional derivative of f at $(0, 1)$.
(b) Establish if f is continuous at $(1, 0)$.

3.44 Consider the function $F(x, y) = \left((x + y)\sqrt{x^2 + y^2}, (x - y)\sqrt{x^2 + y^2}\right)$.

(a) Is F differentiable at $(0, 0)$?
(b) Find, if existing, the differential of F at $(1, 1)$.

(c) Determine, where existing, the Jacobian determinant of F and specify where it is differentiable.

3.45 Consider the vector valued functions $F\colon \mathbb{R}^2 \to \mathbb{R}^2$ and $G\colon \mathbb{R}^2 \to \mathbb{R}^2$ defined by $F(x, y) = (xe^y, ye^x)$ and $G(x, y) = (\sin x, \cos y)$, respectively.

(a) Write $G \circ F$.
(b) Write $F \circ G$.
(c) Compute $J(G \circ F)$.
(d) Compute $J(F \circ G)$.

3.46 What is the order with which $F(x) = \int_0^{\sqrt[4]{x}} e^{xt^4}\, dt$ vanishes for $x \to 0^+$?

3.47 Find the McLaurin polynomial of order 5 of $F(x) = \int_0^x e^{xt^3}\, dt$.

3.48 Consider $F(x) = \int_0^1 \dfrac{\log(1 + x + y)}{\sqrt[4]{x^2 + y^2}}\, dy$.

(a) Determine the domain of F.
(b) Is F invertible in a neighborhood of $x_0 = 0$?

3.49 Consider $F(x) = \int_0^1 \dfrac{e^y - x}{y + 1 - x}\, dy$.

(a) Determine the domain of F.
(b) Compute $F'(0)$.

3.50 Compute the integral $\displaystyle\int_{-\infty}^{+\infty} \dfrac{\sin x}{x}\, dx$ à la Feynman by considering the function
$$H(t) = \int_0^{+\infty} \dfrac{\sin x}{x} e^{-tx}\, dx. \text{ (Hint: Show that } H(t) \to 0 \text{ as } t \to +\infty\text{).}$$

Reference

1. Baronti, M., De Mari, F., van der Putten, R., Venturi, I.: Calculus problems, Unitext, 101, La Matematica per il 3+2. Springer, Berlin (2016)

Chapter 4
Minima and Maxima, Implicit Functions

The analysis of local or global extreme points of a scalar valued function $f: A \to \mathbb{R}$ is rather different according as one looks at the interior points of A or if the subset in which one searches the extreme points of f is a "constraint", that is a set of smaller dimension, in a sense to be made precise. In the first case one speaks of *unconstrained* minima or maxima because it is possible to compare the behaviour of f in the neighboring points of a given point without constraints or restrictions, in the sense that all directions are available for comparison if a full ball around the inspected point falls within A, albeit small. In the second case one speaks of *constrained* minima or maxima because the analysis is constrained by a proper, and typically lower dimensional, subset, on which f is studied. A typical example is when $A = \mathring{A} \cup \partial A$, the union of the interior points and the boundary, for example when $A = \overline{B(O, r)}$, the closure of a ball.

Functions of one or several variables arise in various ways, and the expectation that their graphs somehow exhaust the list of relevant geometric objects is manifestly wrong. Think for example at the very elementary example of a circle in the plane. According to any reasonable notion of dimension, a circle in $\mathbb{R}^2$ is a one dimensional object and one might therefore expect that it may be represented as the graph of a function of a single variable. This is obviously false, but one may emend this statement in two ways. The first is to consider the image of the curve $t \mapsto (\cos t, \sin t)$, whose graph though lies in $\mathbb{R}^3$. The second is to weaken the statement from a global to a local one, that is, by asking that any "sufficiently small" portion of the circle (to wit, at most half of it) may be viewed as the graph of a *bona fide* function of a single variable. For example the graph of $x \mapsto \sqrt{1-x^2}$ represents the upper half of the circle $x^2 + y^2 = 1$, and that of $y \mapsto \sqrt{1-y^2}$ the rightmost half of it. This fact is one occurrence of a general phenomenon, whereby under sufficiently regularity assumptions the set of zeroes of a function $(x, y) \mapsto F(x, y)$ may be locally represented as a graph. This actually holds true in any dimension, and is the content of the famous Dini's theorem, also referred to as the Implicit Function Theorem.

4.1 Unconstrained Minima and Maxima

Definition 4.1 (*Global and local extreme points*) A point P_0 in the domain A of the function $f: A \subseteq \mathbb{R}^n \to \mathbb{R}$ is called:

(i) a *global maximum point* for f if $f(P_0) \geq f(P)$ for every $P \in A$;
(ii) a *global minimum point* for f if $f(P_0) \leq f(P)$ for every $P \in A$;
(iii) *local maximum point* for f if there exists $\delta > 0$ such that $f(P_0) \geq f(P)$ for every $P \in B(P_0, \delta) \cap A$;
(iv) *local minimum point* for f if there exists $\delta > 0$ such that $f(P_0) \leq f(P)$ for every $P \in B(P_0, \delta) \cap A$.

If P_0 satisfies either (i) or (ii), then it is called a *global extreme point* for f, whereas if it satisfies either (iii) or (iv), then it is called a *local extreme point* for f. It is worthwhile observing that the word "point" is often omitted and one speaks of global or local maxima and minima. Furthermore, if the inequalities hold in the strong sense for $P \neq P_0$, then, say, a local minimum is called a *strong local minimum*, and similarly in the other cases.

As in the case of functions of a single variable, any global extreme point is also a local extreme point, but the converse is false, as the reader can verify with simple examples. The next result is of fundamental importance.

Theorem 4.1 (Weierstrass) *If $A \subseteq \mathbb{R}^n$ is compact, that is closed and bounded, and if $f: A \to \mathbb{R}$ is continuous, then f admits global maxima and minima, that is, there exist $P_0, Q_0 \in A$ such that $f(P_0) \leq f(P) \leq f(Q_0)$ for every $P \in A$.*

It is worthwhile recalling that neither minima nor maxima need be unique, neither in the above result nor in the following natural extension.

Theorem 4.2 (Weierstrass, extended version) *If $A \subseteq \mathbb{R}^n$ is closed and unbounded, if $f: A \to \mathbb{R}$ is continuous and for $\|P\| \to +\infty$ it satisfies $f(P) \to +\infty$ (or $f(P) \to -\infty$, respectively), then f admits global minima (global maxima, respectively).*

The next result is analogous to the one-variable version.

Theorem 4.3 (Fermat) *Suppose that $A \subseteq \mathbb{R}^n$ is open and that $P_0 \in A$ is a local extreme point for $f: A \to \mathbb{R}$. If f admits partial derivatives at P_0, then $\nabla f(P_0) = 0$.*

Definition 4.2 (*Critical points*) Suppose that $A \subseteq \mathbb{R}^n$ is open, that $f: A \to \mathbb{R}$ adimts partial derivatives at P_0 and that $\nabla f(P_0) = 0$. Then P_0 is called a *critical point* for f.

Fermat's theorem asserts that extreme points that are interior points of the domain of a function that has partial derivatives are necessarily critical points. The converse, however is false. For example, the origin is a critical point for $f(x, y) = xy$, which is differentiable everywhere, but it is neither a local minimum nor a local maximum.

4.1 Unconstrained Minima and Maxima

Definition 4.3 (*Saddle point*) An interior point P_0 of the domain $A \subseteq \mathbb{R}^n$ of the function $f : A \to \mathbb{R}$ is called a *saddle* point for f if there exist two different lines $\ell_1 = \{P_0 + tP_1 : t \in \mathbb{R}\}$ and $\ell_2 = \{P_0 + tP_2 : t \in \mathbb{R}\}$, where $P_1, P_2 \in \mathbb{R}^n \setminus \{0\}$, such that the restriction $f|_{\ell_1 \cap A}$ has a local maximum at P_0 and the restriction $f|_{\ell_2 \cap A}$ has a local minimum at P_0.

Coming back to the example $f(x, y) = xy$, clearly, the origin is a saddle point because $f(x, -x) = -x^2$ has a (global) maximum at the origin and $f(x, x) = x^2$ has a (global) minimum at the origin. It is important to observe that not all critical points are either local extrema or saddle points, like for example the origin for the function

$$f(x, y) = \begin{cases} (x^2 + y^2)^2 \sin\left(\frac{1}{\sqrt{x^2+y^2}}\right) & (x, y) \neq (0, 0) \\ 0 & (x, y) = (0, 0), \end{cases}$$

as the reader is urged to check in detail. Notice that along all lines through the origin f has the same behaviour because it is radial, namely it behaves like the function $\rho \mapsto \rho^4 \sin(1/\rho)$ which oscillates infinitely many times near the origin.

The analysis of critical points for scalar functions involves higher order derivatives, and more precisely the quadratic form determined by the Hessian matrix of the function at the critical point. This basic notion is recalled below.

Definition 4.4 (*Quadratic forms*) A *quadratic form* is a homogeneous polynomial of degree two in n variables. Any quadratic form is associated to a symmetric $n \times n$ matrix $A = (a_{ij})$, namely, it is of the form

$$Q_A(x_1, \ldots, x_n) = \sum_{i,j=1}^n a_{ij} x_i x_j,$$

where $a_{ij} = a_{ji}$ if $i \neq j$. The quadratic form Q_A is said to be :

(i) *positive definite* if $Q_A(P) > 0$ for every $P \in \mathbb{R}^n \setminus \{O\}$;
(ii) *negative definite* if $Q_A(P) < 0$ for every $P \in \mathbb{R}^n \setminus \{O\}$;
(iii) *positive semidefinite* if $Q_A(P) \geq 0$ for every $P \in \mathbb{R}^n$ and $Q_A(P_0) = 0$ for some $P_0 \in \mathbb{R}^n \setminus \{O\}$;
(iv) *negative semidefinite* if $Q_A(P) \leq 0$ for every $P \in \mathbb{R}^n$ and $Q_A(P_0) = 0$ for some $P_0 \in \mathbb{R}^n \setminus \{O\}$;
(v) *indefinite* if there exist $P_1, P_2 \in \mathbb{R}^n$ such that $Q_A(P_1) < 0 < Q_A(P_2)$.

Notice that the quadratic form may also be expressed in the form

$$Q_A(P) = {}^t P A P,$$

where P is the column vector whoise transpose is the row vector ${}^t P = (x_1, \ldots, x_n)$. Sufficient conditions for the quadratic form Q_A to be either positive or negative definite may be expressed by means of the so-called principal minors. If $A = (a_{ij})$

is an $n \times n$ matrix, its principal submatrix A_j is the $j \times j$ matrix given by its first j rows and j columns, that is

$$A_j = \begin{bmatrix} a_{11} & \ldots & a_{1j} \\ \vdots & & \vdots \\ a_{j1} & \ldots & a_{jj} \end{bmatrix}. \tag{4.1}$$

Theorem 4.4 (Principal minors criterion) *The quadratic form Q_A associated to the symmetric $n \times n$ matrix A is:*

(i) *positive definite if and only if* $\det A_j > 0$ *for every* $j = 1, \ldots, n$;
(ii) *negative definite if and only if* $(-1)^j \det A_j > 0$ *for every* $j = 1, \ldots, n$.

Finer information can be derived from knowledge of the spectrum of A, or, more precisely, from the sign of its eigenvalues. Remember that any symmetric real matrix has n real eigenvalues, counted with multiplicities.

Theorem 4.5 (Eigenvalue criterion) *The quadratic form Q_A associated to the symmetric $n \times n$ matrix A is:*

(i) *positive definite if and only if all the eigenvalues of A are positive;*
(ii) *positive semidefinite if and only if all the eigenvalues of A are non negative;*
(iii) *negative definite if and only if all the eigenvalues of A are negative;*
(iv) *negative semidefinite if and only if all the eigenvalues of A are non positive;*
(v) *indefinite if there exist a negative and a positive eigenvalue of A.*

Of particular relevance in the analysis of the nature of the critical point P_0 for the function $f : A \to \mathbb{R}$ is the quadratic form associated to the Hessian matrix at P_0.

Definition 4.5 (*Hessian form*) Let $A \subset \mathbb{R}^n$ be open, suppose that $P_0 \in A$ and that $f \in C^2(A)$. The *Hessian form* of f at P_0 is the quadratic form associated to the Hessian matrix

$$Hf(P_0) = \left[\frac{\partial^2 f}{\partial x_i \partial x_j}(P_0) \right]_{i,j},$$

namely

$$Q_{Hf(P_0)}(x_1, \ldots, x_n) = \sum_{i,j=1}^n \frac{\partial^2 f}{\partial x_i \partial x_j}(P_0) x_i x_j.$$

Observe that the assumption $f \in C^2(A)$ entails that $Hf(P_0)$ is symmetric.

Sufficient conditions for deciding the nature of a critical point, that is, assessing if a critical point is a local extreme for f or not and, if so, whether it is a local minimum or a local maximum, may be derived from the Hessian form. This fact hinges on the second order Taylor expansion (3.11).

Theorem 4.6 (*Nature of a critical point*) *Let $A \subset \mathbb{R}^n$ be open, take $f \in C^2(A)$ and suppose that $P_0 \in A$ is a critical point for f. Then:*

(i) *if $Hf(P_0)$ is positive definite, then P_0 is a local mimimum;*
(ii) *if $Hf(P_0)$ is negative definite, then P_0 is a local maximum;*
(ii) *if $Hf(P_0)$ is indefinite, then P_0 is a saddle point.*

The above general results take a more practical form in the most common cases, namely when $n = 2$ or $n = 3$. In the next theorem the notation introduced in (4.1) will be used.

Theorem 4.7 *Let $A \subset \mathbb{R}^2$ be open, suppose that $f \in C^2(A)$ and that $P_0 \in A$ is a critical point for f. Then:*

(i) *if $f_{xx}(P_0) > 0$ and $\det Hf(P_0) > 0$ then P_0 is a local minimum point;*
(ii) *if $f_{xx}(P_0) < 0$ and $\det Hf(P_0) > 0$ then P_0 is a local maximum point;*
(iii) *$\det Hf(P_0) < 0$ then P_0 is a saddle point.*

Let $A \subset \mathbb{R}^3$ be open, suppose that $f \in C^2(A)$ and that $P_0 \in A$ is a critical point for f. Then:

(i) *if $f_{xx}(P_0) > 0$, $\det[Hf(P_0)]_2 > 0$ and $\det Hf(P_0) > 0$ then P_0 is a local minimum point;*
(ii) *if $f_{xx}(P_0) < 0$, $\det[Hf(P_0)]_2 > 0$ and $\det Hf(P_0) < 0$ then P_0 is a local maximum point.*

Observe that in three variables $\det Hf(P_0) < 0$ does not imply that the Hessian is indefinite an hence in this case P_0 is not necessarily a saddle point, as it could well be a maximum, such as $P \mapsto -\|P\|^2$.

4.2 Constrained Minima and Maxima

It is customary to study constrained minima and maxima along "nice" closed sets, which typically arise as zeroes of continuous functions. Although more general scenarios are certainly possible, this is a standard assumption. Actually, often the functions that describe the constraints are even more regular than just continuous.

Definition 4.6 (*Constraint*) Let $A \subset \mathbb{R}^n$ be open and suppose that $g_1, \ldots g_k : A \to \mathbb{R}$ are continuous functions, with $1 \leq k < n$. The closed set

$$C = C_{g_1,\ldots,g_k} = \{P \in A : g_1(P) = \cdots = g_k(P) = 0\}$$

is called the *constraint* determined by the functions $g_1, \ldots g_k$.

Evidently, $C_{g_1,\ldots,g_k} = \cap_{j=1}^k g_j^{-1}(\{0\})$ is closed. The dependency on $g_1, \ldots, g_k$ will often be omitted.

Definition 4.7 (*Constrained extreme point*) Let $A \subset \mathbb{R}^n$ be open. Suppose that $g_1, \ldots g_k \colon A \to \mathbb{R}$ are continuous functions and let C denote the constraint determined by them. The point $P_0 \in C$ is called:

(i) a local maximum for $f \colon A \to \mathbb{R}$ *constrained* on C if there exists $\delta > 0$ such that $f(P_0) \geq f(P)$ for every $P \in B(P_0, \delta) \cap C$;
(ii) a local minimum for $f \colon A \to \mathbb{R}$ *constrained* on C if there exists $\delta > 0$ such that $f(P_0) \leq f(P)$ for every $P \in B(P_0, \delta) \cap C$.

In order to gain geometric intuition, suppose now that $n = 2$ and consider a constraint C given as the zero set $\{P : g(P) = 0\}$ of a nice smooth function. For visualization, think of C as a nice curve. Consider a point $P_0 \in C$. It is well known that the gradient $\nabla g(P_0)$, whenever non vanishing, is orthogonal to the curve C at P_0. Consider now a scalar field $f \colon A \to \mathbb{R}$ defined in some neighborhood of P_0 and suppose that P_0 is, say, a constrained minimum for f along C. Keep in mind that the gradient $\nabla f(P_0)$ is orthogonal to the level set of f passing through P_0. If the gradients $\nabla f(P_0)$ and $\nabla g(P_0)$ were linearly independent, then the level set $\{P \in A : f(P) = f(P_0)\}$ would cross the constraint C transversally at P_0 and there would be points along the line C crossing the region where the values of f are bigger than $f(P_0)$ but also the region where the values of f are smaller than $f(P_0)$, in contradiction with the assumption that P_0 is a contrained local minimum. The conclusion is that the two gradients must be aligned (see Fig. 4.1). This is the geometric content of the next result.

Theorem 4.8 (*Lagrange multipliers*) *Let $A \subset \mathbb{R}^n$ be open. Let C be the constraint determined by the scalar functions $g_1, \ldots, g_k \in C^1(A)$, with $k < n$, and suppose that the Jacobian matrix of the vector valued map $G \colon A \to \mathbb{R}^k$ given by*

$$G(P) = {}^t(g_1(P), \ldots, g_k(P))$$

has maximum rank k at every point $P \in C$. If $f \in C^1(A)$ and if $P_0 \in A$ is a constrained critical point for f on C, then there exist real numbers $\lambda_1, \ldots, \lambda_k$ such that

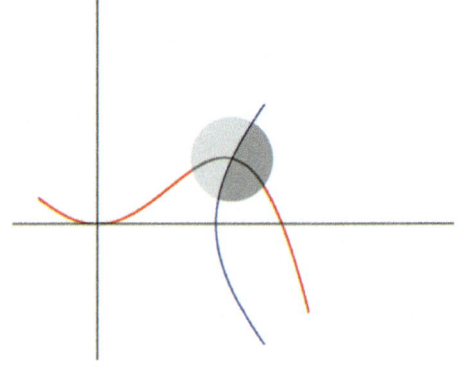

Fig. 4.1 In red the constraint, in blue a level curve of f transversal to the constraint; in different tones of grey parts of the regions where f is smaller or bigger than along the level curve

4.2 Constrained Minima and Maxima

P_0 *is a critical point for the function*

$$L(P) = f(P) + \sum_{j=1}^{k} \lambda_j g_j(P),$$

that is

$$\nabla f(P_0) + \sum_{j=1}^{k} \lambda_j \nabla g_j(P_0) = 0.$$

The real numbers $\lambda_1, \ldots, \lambda_k$ are called the Lagrange multipliers *associated with the constrained problem, and the scalar field $L: A \to \mathbb{R}$ is called the associated* Lagrangian function.

It is worthwhile observing that in the case of a single function $g: A \to \mathbb{R}$ there is a single Lagrange multiplier λ, and any constrained critical point P_0 satisfies the system of $n+1$ equations given by

$$\begin{cases} \nabla f(P_0) + \lambda \nabla g(P_0) = 0 \\ g(P_0) = 0. \end{cases}$$

In the simplest situation when $n = 2$, this is the system of three equations:

$$\begin{cases} \dfrac{\partial f}{\partial x}(x_0, y_0) + \lambda \dfrac{\partial g}{\partial x}(x_0, y_0) = 0 \\[2mm] \dfrac{\partial f}{\partial y}(x_0, y_0) + \lambda \dfrac{\partial g}{\partial y}(x_0, y_0) = 0 \\[2mm] g(x_0, y_0) = 0. \end{cases}$$

Recall that, in general, the Jacobian of the vector valued map $F: A \subseteq \mathbb{R}^n \to \mathbb{R}^m$ is an $m \times n$ matrix valued function on A (gradients are always row vectors)

$$JF(P_0) = \begin{bmatrix} \nabla f_1(P_0) \\ \vdots \\ \vdots \\ \nabla f_m(P_0) \end{bmatrix} = \begin{bmatrix} \dfrac{\partial f_1}{\partial x_1}(P_0) & \cdots & \dfrac{\partial f_1}{\partial x_n}(P_0) \\ \vdots & & \vdots \\ \dfrac{\partial f_m}{\partial x_1}(P_0) & \cdots & \dfrac{\partial f_m}{\partial x_n}(P_0) \end{bmatrix}. \quad (4.2)$$

4.3 The Implicit Function Theorem

As outlined in the beginning of this chapter, the notion of implicit function is related to the possibility of expressing the solutions of the equation $F(x_1, \ldots, x_n) = 0$ as the graph of a function, at least locally. The general picture is that of a set of m equations in the n variables $x_1, \ldots, x_n$. In order for the system

$$\begin{cases} f_1(x_1, \ldots, x_n) = 0 \\ \ldots \\ f_m(x_1, \ldots, x_n) = 0 \end{cases}$$

to admit, in general, a meaningful solution set it is sensible to assume $m < n$, that is, fewer equations than variables. A quick way to write the above system is to consider the possibly vector valued map $F = (f_1, \ldots, f_m)$ defined in some open subset A of $\mathbb{R}^n$ and to consider its zero set

$$Z(F) = \{P \in \mathbb{R}^n : F(P) = 0\}.$$

The most natural examples to bear in mind are as follows. When $n = 2$ and $m = 1$ then the set of zeroes of a scalar function f, namely the solutions of $f(x, y) = 0$, are expected to give rise to what might be informally referred to as a "curve" in the plane. When $n = 3$ and $m = 1$ then the set of zeroes of a scalar function f, namely the solutions of $f(x, y, z) = 0$, should single out a "surface". Finally, when $n = 3$ and $m = 2$ then the solutions of two equations, namely $f_1(x, y, z) = 0 = f_2(x, y, z)$, being the intersection of two "surfaces", will in general represent a "curve" in three dimensional space. It is also important to observe that in general the set $Z(F)$ needs not be connected.

The next definition formalizes the notion of implicit function.

Definition 4.8 (*Implicit function*) Let $n \geq 2$ and consider an open set $A \subset \mathbb{R}^n$. Fix $1 \leq m < n$, write the points of A as $P = (X, Y) \in \mathbb{R}^{n-m} \times \mathbb{R}^m$ and take a map $F : A \to \mathbb{R}^m$. The equation $F(P) = 0$ is said to define an *implicit function* of X locally around $P_0 = (X_0, Y_0) \in A$ if

(i) $P_0 \in Z(F)$;
(ii) there exist a neighborhood $\mathscr{U} \subseteq \mathbb{R}^{n-m}$ of X_0, an open neighborhood $\mathscr{V} \subseteq \mathbb{R}^n$ of P_0 and a function $\Phi : \mathscr{U} \to \mathbb{R}^m$ such that $Z(F) \cap \mathscr{V} = \{(X, \Phi(X)) : X \in \mathscr{U}\}$.

In this case, Φ is the implicit function of X defined by $F(X, Y) = 0$, in the sense that the points of the zero set $Z(F)$ close enough to P_0, namely those in $\mathscr{V}$, are of the form $(X, \Phi(X))$, hence if $F(X, Y) = 0$ then $Y = \Phi(X)$, and conversely if $Y = \Phi(X)$, then $F(X, Y) = 0$.

In order to properly formulate the most important result on the subject, it is convenient to write the Jacobian matrix in accordance with the splitting (X, Y) of coordinates. With reference to (4.2), it is natural to write $JF(P_0)$ in the form

4.3 The Implicit Function Theorem

$$JF(P_0) = \begin{bmatrix} J_X F(P_0) & J_Y F(P_0) \end{bmatrix}$$

whereby $J_X F(P_0)$ is the $m \times (n-m)$ submatrix containing all the derivatives with respect to the variables $X = (x_1, \ldots, x_{n-m})$, and $J_Y F(P_0)$ is the square $m \times m$ submatrix containing all the derivatives with respect to the variables $Y = (y_1, \ldots, y_m)$. Explicitly:

$$J_Y F(P_0) = \begin{bmatrix} \frac{\partial f_1}{\partial y_1}(P_0) & \cdots & \frac{\partial f_1}{\partial y_m}(P_0) \\ \vdots & & \vdots \\ \frac{\partial f_m}{\partial y_1}(P_0) & \cdots & \frac{\partial f_m}{\partial y_m}(P_0) \end{bmatrix}.$$

When $n = 2$ and $m = 1$ and the map whose zeroes are under investigation is simply a scalar function $f : A \to \mathbb{R}$, then

$$Jf(P_0) = \begin{bmatrix} f_x(P_0) & f_y(P_0) \end{bmatrix}$$

and this clarifies the notation introduced above.

Theorem 4.9 (Implicit Function Theorem, or Dini's theorem) *Let $n \geq 2$ and let A be an open subset of $\mathbb{R}^n$. Suppose that $F : A \to \mathbb{R}^m$, with $1 \leq m < n$, is of class $C^1(A)$. Suppose further that $P_0 = (X_0, Y_0) \in A \cap Z(F)$ and that $\det J_Y F(P_0) \neq 0$. Then the equation $F(P) = 0$ defines an implicit function of X locally around P_0. Precisely:*

(i) there exist a neighborhood $\mathcal{U} \subseteq \mathbb{R}^{n-m}$ of X_0, an open neighborhood $\mathcal{V} \subseteq \mathbb{R}^n$ of P_0 and a function $\Phi : \mathcal{U} \to \mathbb{R}^m$ such that $Z(F) \cap \mathcal{V} = \{(X, \Phi(X)) : X \in \mathcal{U}\}$;
(ii) $\Phi \in C^1(\mathcal{U})$ and $J\Phi(X) = -(J_Y F(X, \Phi(X)))^{-1} J_X F(X, \Phi(X))$.

It is worthwhile spelling out Dini's theorem in the case when $n = 2$ and $m = 1$ because it takes a simpler form and it is perhaps the most common case. See Fig. 4.2 for an illustration.

Theorem 4.10 (2D Dini's theorem) *Let A be an open subset of $\mathbb{R}^2$, suppose that $f \in C^1(A)$ and that $f(x_0, y_0) = 0$. Suppose further that*

$$f_y(x_0, y_0) \neq 0.$$

Then the equation $f(x, y) = 0$ defines an implicit function of x locally around (x_0, y_0). Precisely:

(i) there exist a neighborhood $\mathcal{U} \subseteq \mathbb{R}$ of x_0, an open neighborhood $\mathcal{V} \subseteq \mathbb{R}^2$ of (x_0, y_0) and a function $\varphi : \mathcal{U} \to \mathbb{R}$ such that $Z(F) \cap \mathcal{V} = \{(x, \varphi(x)) : x \in \mathcal{U}\}$;
(ii) $\varphi \in C^1(\mathcal{U})$ and

$$\varphi'(x) = -\frac{f_x(x, \varphi(x))}{f_y(x, \varphi(x))}.$$

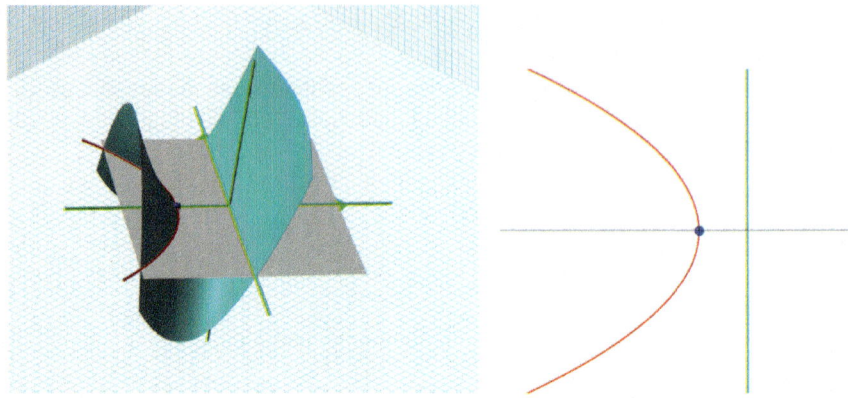

Fig. 4.2 An illustration of Dini's theorem: the zero level set of a scalar function of two variables consists of two curves (left). Both of them may be described by expressing one variable as a function of the other (right)

Clearly, a statement analogous to that of Theorem 4.10 holds if $f_x(x_0, y_0) \neq 0$, in which case one may express x as a function ψ of y and obtain the formula

$$\psi'(y) = -\frac{f_y(\psi(y), y)}{f_x(\psi(y), y)}.$$

4.4 Guided Exercises

4.1 Find, if existing, the critical points of the function $f(x, y) = x^2 + 3xy + y^2 + x - y$ and discuss their nature.

Answer. Clearly, $f \in C^\infty(\mathbb{R}^2)$ because it is a polynomial. The critical points are the solutions of equation $\nabla f(x, y) = 0$, namely of the system

$$\begin{cases} f_x(x, y) = 2x + 3y + 1 = 0 \\ f_y(x, y) = 3x + 2y - 1 = 0 \end{cases} \iff \begin{cases} 5x + 5y = 0 \\ 3x + 2y - 1 = 0 \end{cases}$$

which has the only solution $P_0 = (1, -1)$. Now, $f_{xx}(x, y) = f_{yy}(x, y) = 2$ and $f_{xy}(x, y) = f_{yx}(x, y) = 3$, so that the Hessian matrix is constant and is equal to

$$Hf(x, y) = \begin{bmatrix} 2 & 3 \\ 3 & 2 \end{bmatrix} = Hf(P_0).$$

Clearly, $\det Hf(P_0) < 0$ and hence P_0 is a saddle point.

4.4 Guided Exercises

Warm-up exercise on critical points: a polynomial with a single saddle point. It could haves being guessed upon writing $f(x, y) = (x - 1)(y + 1) + (x + y)^2$, which is the sum of the prototypical polynomial with a single saddle point at $P_0 = (1, -1)$ and the squared distance function, which is not going to mess things up.

4.2 Find, if existing, all local extrema of the function

$$f(x, y) = x^2 + 4xy + 5y^2 + x - y.$$

Answer. Clearly, $f \in C^\infty(\mathbb{R}^2)$ because it is a polynomial. Hence the local maxima or minima are necessarily critical poins and the latter are the solutions of equation $\nabla f(x, y) = 0$, namely of the system

$$\begin{cases} f_x(x, y) = 2x + 4y + 1 = 0 \\ f_y(x, y) = 4x + 10y - 1 = 0 \end{cases} \iff \begin{cases} 6x + 14y = 0 \\ 2x + 4y + 1 = 0 \end{cases}$$

which has the only solution $P_0 = (-7/2, 3/2)$. Now, $f_{xx}(x, y) = 2$, $f_{yy}(x, y) = 10$ and $f_{xy}(x, y) = f_{yx}(x, y) = 4$, so that the Hessian matrix is constant and is equal to

$$Hf(x, y) = \begin{bmatrix} 2 & 4 \\ 4 & 10 \end{bmatrix} = Hf(P_0).$$

Clearly, $Hf(P_0)$ is positive definite and hence P_0 is a local minimum.

A second warm-up exercise on critical points: a polynomial with a single critical point which is actually a local minimum. It is easy to show that $f(P) \to +\infty$ when $\|P\| \to +\infty$, so that by Weierstrass generalized theorem f has a global minimum. As such, it is a local minimum hence a critical point, hence P_0. Notice the similarity of the polynomials of this exercise and that in Exercise 4.1, which nonetheless exhibit very different behaviours.

4.3 Find, if existing, the global extrema of the function $f(x, y) = \sqrt{1 - x^2 - y^2} + y$.

Answer. Evidently, the function f is defined on $\{(x, y) \in \mathbb{R}^2 : x^2 + y^2 \leq 1\}$, which is a compact subset of the plane. Since f is the composition of continuous functions, it is continuous on its domain. By Weierstrass' theorem, f has both maxima and minima on its domain. Now, $f \in C^2(A)$, where $A = \{(x, y) \in \mathbb{R}^2 : x^2 + y^2 < 1\}$ is the interior of $\mathrm{Dom}(f)$. If a global maximum or minimum of f lies in A, it is certainly a critical point of f. Now, at all points of A the partial derivatives are

$$f_x(x, y) = -\frac{x}{\sqrt{1 - x^2 - y^2}}, \quad f_y(x, y) = \frac{-y + \sqrt{1 - x^2 - y^2}}{\sqrt{1 - x^2 - y^2}},$$

so that

$$\nabla f(x,y) = (0,0) \iff \begin{cases} x = 0 \\ -y + \sqrt{1-x^2-y^2} = 0 \end{cases} \iff \begin{cases} x = 0 \\ y \geq 0 \\ 1 - y^2 = y^2. \end{cases}$$

Therefore $P_0 = (0, \sqrt{2}/2)$ is the only critical point of f in A. The second-order partial derivatives in A are given by

$$f_{xx}(x,y) = -\frac{1-y^2}{(1-x^2-y^2)^{3/2}}, \quad f_{yy}(x,y) = -\frac{1-x^2}{(1-x^2-y^2)^{3/2}}$$

$$f_{xy}(x,y) = f_{yx}(x,y) = -\frac{xy}{(1-x^2-y^2)^{3/2}}.$$

Hence the Hessian matrix at P_0 is

$$Hf(P_0) = \begin{bmatrix} -\sqrt{2} & 0 \\ 0 & -2\sqrt{2} \end{bmatrix},$$

and is clearly negative definite. Therefore, P_0 is a local maximum, and $f(P_0) = \sqrt{2}$.

As for the boundary $B = \partial(\text{Dom}(f)) = \{(x,y) \in \mathbb{R}^2 : x^2 + y^2 = 1\}$, it is clear that the restriction $f\big|_B(x,y) = y$ is an increasing function of y. Hence $P_1 = (-1, 0)$ is the global minimum of $f\big|_B$, and $f(P_1) = -1$, while $P_2 = (0, 1)$ is the global maximum of $f\big|_B$, and $f(P_2) = 1$. In conclusion, the unique global maximum is P_0 and P_1 is the unique global minimum.

It is worth observing that $P_2 = (0, 1)$ is not a local maximum. Indeed, the restriction of f on the y-axis is

$$\varphi(y) := f(0, y) = \sqrt{1-y^2} + y, \quad y \in [-1, 1]$$

and a quick inspection reveals that φ is strictly decreasing in $[\sqrt{2}/2, 1]$. Therefore, there are points P along the y-axis which are arbitrarily close to P_2 and for which $f(P) > f(P_2)$. Since P_2 is a global maximum of $f\big|_B$, however, there are points Q on the unit circle B which are arbitrarily close to P_2 and for which $f(Q) < f(P_2)$. Therefore, in a small neighborhood $\mathscr{U}$ of P_2 it holds that either $f(P_2) \geq f(P)$ for all $P \in \mathscr{U} \cap \text{Dom}(f)$ or $f(P_2) \leq f(P)$ for all $P \in \mathscr{U} \cap \text{Dom}(f)$. Hence P_2 is not a local maximum.

This exercise is somewhat standard, but interesting. A continuous function on a compact set certainly has maxima and minima, so one is certain to be looking for points that do indeed exist. In order to find them, it is advisable to break the problem into a search in the interior A of the domain first, and then along the boundary B. On the open set A one finds the critical points and then uses the Hessian matrix to study their nature. On the boundary B it is often the case that the restriction of f may be studied as a function of a single variable, say $t \mapsto \psi(t)$, so that the analysis simplifies (in the case at hand t may actually be taken as y). There is, however, a

4.4 Guided Exercises

subtle issue. When studying $t \mapsto \psi(t)$, one is not considering points of the domain of f that are very close to the critical points of ψ, because they belong to A. It is like walking on a mountain track: the surrounding terrain might have all kind of irregularities while the track itself steadily raises upwards. Further, one may reach the highest point of the track which is not the summit of the mountain.

4.4 Find, if existing, all local extrema of the function $f(x, y) = x \sin y - y$.

Answer. Clearly $f \in C^\infty(\mathbb{R}^2)$. The critical points are the solutions of the equation $\nabla f(x, y) = (0, 0)$, namely

$$\begin{cases} f_x(x, y) = \sin y = 0 \\ f_y(x, y) = x \cos y - 1 = 0 \end{cases} \iff \begin{cases} y = k\pi \\ x(-1)^k - 1 = 0 \end{cases} \iff \begin{cases} y = k\pi \\ x = (-1)^k \end{cases}$$

with $k \in \mathbb{Z}$. There are infinitely many critical points $P_k = ((-1)^k, k\pi)$, with $k \in \mathbb{Z}$. Since

$$f_{xx}(x, y) = 0 \quad f_{yy}(x, y) = -x \sin y, \quad f_{xy}(x, y) = f_{yx}(x, y) = \cos y,$$

the Hessian matrix at the critical point P_k is

$$Hf(P_k) = \begin{bmatrix} 0 & (-1)^k \\ (-1)^k & 0 \end{bmatrix}.$$

Since $\det Hf(P_k)) = -1 < 0$ for every $k \in \mathbb{Z}$, every P_k is a saddle point.

An interesting function which oscillates with respect to y for any fixed $x \neq 0$ and is a straight line with respect to x for any fixed y. The critical points occur when the the straight line is horizontal and crosses an oscillation at a peak or at a valley (see Fig. 4.3).

4.5 Find, if existing, all local extrema of the function $f(x, y) = \dfrac{x^2 y^2}{2} - xy$ on the set

$$X = \{(x, y) \in \mathbb{R}^2 : -2 \leq x \leq 0, -2 \leq y \leq 0, xy \leq 1\}.$$

Answer. The set X is compact (see Fig. 4.4). Indeed, it is closed because it is the intersection of the inverse images of several continuous functions and it is bounded because it is contained in the ball $B(O, 4)$. The function f is a polynomial and hence global extrema exist by Weierstrass' theorem.

A interior point $(x, y) \in \mathring{X}$ is critical for f if and only if

$$\begin{cases} f_x(x, y) = xy^2 - y = 0 \\ f_y(x, y) = x^2 y - x = 0 \end{cases} \iff \begin{cases} y(xy - 1) = 0 \\ x(xy - 1) = 0, \end{cases}$$

which may happen if and only if (x, y) solves at least one of the following

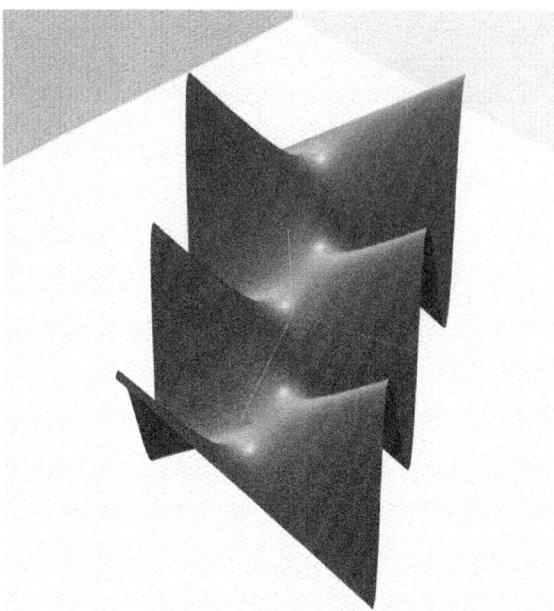

Fig. 4.3 Sequence of saddle points for the function $f(x, y) = x \sin y - y$ of Exercise 4.4

Fig. 4.4 The domain of Exercise 4.5, left, and the extension of f to its natural domain, the full plane, right

$$\begin{cases} x = 0 \\ y = 0 \end{cases} , \quad \begin{cases} y = 0 \\ xy = 1 \end{cases} , \quad \begin{cases} x = 0 \\ xy = 1 \end{cases} , \quad xy = 1.$$

The second and third systems have no solutions, the first and and the fourth equation have solutions only on the boundary ∂X, hence are not interior. Therefore there are no critical points.

4.4 Guided Exercises

Now, on the portion of X lying on the axes the function vanishes, and on the hyperbolic arc $\{(x, y) \in \mathbb{R}^2 : xy = 1\} \cap X$ the function is constant and equal to $-1/2$. Finally, on the vertical segment joining $(-2, -1/2)$ to $(-2, 0)$ it is

$$f(-2, y) = 2y^2 + 2y,$$

with $y \in [-1/2, 0]$, a function which increases from $-1/2$ to 0. Similarly, on the horizontal segment joining $(-1/2, -2)$ to $(0, -2)$ it is $f(x, -2) = 2x^2 + 2x$, with $x \in [-1/2, 0]$, which increases from $-1/2$ to 0. This shows that all the points on the axes are global maxima and all the points on the hyperbolic arc are global minima.

This function attains its extrema on the boundary of the set X, and no point in the interior of X is a critical point. It must be stressed that this happens because the domain X is not the natural domain of f, which is the whole plane. Indeed, if one removes the restriction to look at f on X, and considers f as a globally defined function g, then g has indeed many critical points, namely all the points along the hyperbola $xy = 1$ and the origin. A minute thought reveals that in fact the hyperbola $xy = 1$ consists of global minima and the origin is a saddle point (see Fig. 4.4). Indeed, the function is the composition of $t \mapsto \frac{t^2}{2} - t$ with $(x, y) \mapsto xy$. The former attains the global minimum $-1/2$ at $t = 1$ and the latter has a saddle point at the origin.

4.6 Find, if existing, all local or global extrema of the function

$$f(x, y, z) = (x^2 + y^2)(1 - z^2) + z^2.$$

Answer. Evidently f is unbounded both from below and from above. Indeed, on the coordinate plane $z = 0$ it attains the values $f(x, y, 0) = x^2 + y^2$ which fill the set $[0, +\infty)$ whereas on the cylinder $\{(x, y, z) \in \mathbb{R}^3 : x^2 + y^2 = 2\}$ it takes the values $2 - z^2$ which fill the set $(-\infty, 2]$.

Since $f \in C^\infty(\mathbb{R}^3)$ the critical points of f will contain all possible local extrema. The gradient of f vanishes at (x, y, z) if and only if

$$\begin{cases} f_x(x, y, z) = 2x(1 - z^2) = 0 \\ f_y(x, y, z) = 2y(1 - z^2) = 0 \\ f_z(x, y, z) = 2z[1 - (x^2 + y^2)] = 0 \end{cases}$$

so that the critical poins are the elements of the set

$$C = O \cup \{(x, y, z) \in \mathbb{R}^3 : x^2 + y^2 = 1, \text{ and } z = \pm 1\},$$

namely a point (the origin) and two circles which will be denoted by C_+, at height $z = 1$, and C_- at height $z = -1$, respectively. The second derivatives at $P = (x, y, z)$ are

$$f_{xx}(P) = f_{yy}(P) = 2(1 - z^2), \qquad f_{zz}(P) = 2[1 - (x^2 + y^2)]$$

$$f_{xy}(P) = f_{yx}(P) = 0, \quad f_{xz}(P) = f_{zx}(P) = -4xz, \quad f_{yz}(P) = f_{zy}(P) = -4yz$$

so that the Hessian matrix is

$$Hf(P) = \begin{bmatrix} 2(1-z^2) & 0 & -4xz \\ 0 & 2(1-z^2) & -4yz \\ -4xz & -4yz & 2(1-(x^2+y^2)) \end{bmatrix}.$$

Therefore, $Hf(O) = \text{diag}(2, 2, 2)$ which is clearly positive definite, so that O is a local minimum. Next, let $P_\pm(\theta) = (\cos\theta, \sin\theta, \pm 1)$ parametrize the points on the circles $C_\pm$ in cylindrical coordinates centered at the origin, with $\theta \in [0, 2\pi)$. Thus,

$$Hf(P_\pm(\theta)) = \begin{bmatrix} 0 & 0 & \mp 4\cos\theta \\ 0 & 0 & \mp 4\sin\theta \\ \mp 4\cos\theta & \mp 4\sin\theta & 0 \end{bmatrix}.$$

The eigenvalues of $Hf(P_\pm(\theta))$ are the roots of the polynomial

$$\det(\lambda I - Hf(P_\pm(\theta))) = \det \begin{bmatrix} \lambda & 0 & \pm 4\cos\theta \\ 0 & \lambda & \pm 4\sin\theta \\ \pm 4\cos\theta & \pm 4\sin\theta & \lambda \end{bmatrix} = \lambda^3 - 16\lambda.$$

They are $0, 4, -4$, which shows that each point on each of the circles $C_\pm$ is a saddle point.

This is a standard exercise. Non constant polynomials are never bounded, but may well be either bounded from below or from above. This is neither, and exhibits a non trivial set $\mathscr{C}$ of critical points. The analysis of the Hessian matrix reveals at once the presence of a local minimum and infinitely many saddle points. Being a function of three variables, the intuition gets a little lost, and it is of course impossible to appeal to some kind of drawing for its graph. That's when analytical methods provide a powerful tool.

4.7 Find, if existing, all local or global extrema of the function

$$f(x, y) = \frac{1}{2}y^2 + x^2 - 3x^4 - 2x^2y^2 - y^4.$$

Answer. Observe that $f \in C^\infty(\mathbb{R}^2)$. Further, $f(0, y) = \frac{y^2}{2} - y^4 \to -\infty$ when $y \to \pm\infty$, which indicates that perhaps $f(P) \to -\infty$ as $P \to \infty$. Using polar coordinates,

$$f(\rho\cos\theta, \rho\sin\theta) = \rho^2(\frac{1}{2}\sin^2\theta + \cos^2\theta) - \rho^4(3\cos^4\theta + 2\cos^2\theta\sin^2\theta + \sin^4\theta)$$
$$\leq \rho^2 - \rho^4\varphi(\theta),$$

4.4 Guided Exercises

where evidently $\varphi(\theta) := 3\cos^4\theta + 2\cos^2\theta \sin^2\theta + \sin^4\theta$. Now, since φ is continuous on the compact interval $[0, 2\pi]$ and strictly positive, it attains a positive minimum at, say, $\theta^* \in [0, 2\pi]$. It follows that

$$f(\rho\cos\theta, \rho\sin\theta) \leq \rho^2 - \rho^4\varphi(\theta^*) \to -\infty$$

as $\rho \to +\infty$. This uniform limit shows that indeed $f(P) \to -\infty$ as $P \to \infty$. Hence, by Weierstrass' theorem in generalized form f has a global maximum, certainly among the critical points. The gradient of f vanishes at (x, y, z) if and only if

$$\begin{cases} f_x(x, y) = 2x - 12x^3 - 4xy^2 = 2x(1 - 6x^2 - 2y^2) = 0 \\ f_y(x, y) = y - 4x^2y - 4y^3 = y(1 - 4x^2 - 4y^2) = 0. \end{cases}$$

The zero-set of f_x is the union of the ellipse of equation $1 - 6x^2 - 2y^2 = 0$ and the y-axis and, whereas the zero-set of f_y is the union of the x-axis and the circle of equation $1 - 4x^2 - 4y^2 = 0$ (see Fig. 4.5). These two sets intersect at the following nine points:

$$A = (0, \tfrac{1}{2}), \quad B = (0, -\tfrac{1}{2}), \quad C = (0, 0), \quad D = (\tfrac{\sqrt{6}}{6}, 0), \quad E = (-\tfrac{\sqrt{6}}{6}, 0),$$

$$F = (\tfrac{\sqrt{2}}{4}, \tfrac{\sqrt{2}}{4}), \quad G = (-\tfrac{\sqrt{2}}{4}, \tfrac{\sqrt{2}}{4}), \quad H = (\tfrac{\sqrt{2}}{4}, -\tfrac{\sqrt{2}}{4}), \quad L = (-\tfrac{\sqrt{2}}{4}, -\tfrac{\sqrt{2}}{4}).$$

The second derivatives at $P = (x, y)$ are

$$f_{xx}(P) = 2 - 36x^2 - 4y^2, \quad f_{xy}(P) = f_{yx}(P) = -8xy, \quad f_{yy}(P) = 1 - 4x^2 - 12y^2$$

so that the Hessian matrix is

$$Hf(P) = \begin{bmatrix} 2 - 36x^2 - 4y^2 & -8xy \\ -8xy & 1 - 4x^2 - 12y^2 \end{bmatrix}.$$

A direct computation gives:

$$Hf(A) = Hf(B) = \begin{bmatrix} 1 & 0 \\ 0 & -2 \end{bmatrix} \implies A \text{ and } B \text{ are saddle points};$$

$$Hf(C) = \begin{bmatrix} 2 & 0 \\ 0 & 1 \end{bmatrix} \implies C \text{ is a local minimum};$$

$$Hf(D) = Hf(E) = \begin{bmatrix} -4 & 0 \\ 0 & \tfrac{1}{3} \end{bmatrix} \implies D \text{ and } E \text{ are saddle points};$$

Fig. 4.5 To the left the critical points of the function in Exercise 4.7: in blue the zero-set of f_x and in red the zero-set of f_y. On the right the graph of f, with a plane at an intermediate level between the minimum and the lower saddles

$$Hf(F) = Hf(L) = \begin{bmatrix} -3 & -1 \\ -1 & -1 \end{bmatrix} \implies F \text{ and } L \text{ are local maxima;}$$

$$Hf(G) = Hf(H) = \begin{bmatrix} -3 & 1 \\ 1 & -1 \end{bmatrix} \implies G \text{ and } H \text{ are local maxima.}$$

Finally, an immediate calculation gives $f(F) = f(G) = f(H) = f(L) = 3/32$, so that the local maxima are all global maxima.

The graph of f in a ball around the origin resembles a volcanic crater, featuring four equally high peaks around its rim, and four cols, two higher and two lower, see Fig. 4.5. Together with the local minimum, the pit of the crater at the origin, they account for the nine critical points of f. The depicted plane simulates a lake filling the crater, at a lower elevation than the lower cols.

4.8 Find, if existing, all local or global extrema of the function

$$f(x, y) = \frac{a}{2}(x - y)^2 + \frac{(a+3)}{2}(x+y)^2$$

as a ranges in $\mathbb{R}$.

Answer. Clearly, $f \in C^\infty(\mathbb{R}^2)$. Therefore, the possible local or global extrema are critical points. The gradient of f vanishes at (x, y) if and only if

$$\begin{cases} f_x(x, y) = a(x - y) + (a+3)(x+y) = (2a+3)x + 3y = 0 \\ f_y(x, y) = -a(x - y) + (a+3)(x+y) = (2a+3)y + 3x = 0 \end{cases}$$

which is equivalent to

$$\begin{cases} y = -\left(\frac{2a+3}{3}\right)x \\ a(a+3)x = 0. \end{cases}$$

4.4 Guided Exercises

This system yields the following mutually exclusive possibilities:

(i) if $a(a+3) \neq 0$, then $O = (0, 0)$ is the only critical point;
(ii) if $a = -3$, then the critical points are precisely those of the line $y = x$;
(iii) if $a = 0$, then the critical points are precisely those of the line $y = -x$.

The nature of the critical points is therefore the following:

(i) In this case, since the second derivatives are $f_{xx}(x, y) = f_{yy}(x, y) = 2a + 3$, $f_{xy}(x, y) = f_{yx}(x, y) = 3$, the Hessian matrix is constant and equals

$$Hf(O) = \begin{bmatrix} 2a+3 & 3 \\ 3 & 2a+3 \end{bmatrix}.$$

The eigenvalues of $Hf(O)$ are the roots of the characteristic polynomial, namely $\det(\lambda I - Hf(O)) = \lambda^2 - 2(2a+3)\lambda + 4a(a+3)$, and therefore are:

- both positive if $a > 0$, in which case O is a local minimum;
- both negative if $a < -3$, in which case O is a local maximum;
- with opposite signs if $-3 < a < 0$, in which case O is a saddle point.

(ii) If $a = -3$, then $f(x, y) = -\frac{3}{2}(x-y)^2$ has global maxima (equal to 0) along the line $y = x$ and no other local extrema.
(iii) If $a = 0$, then $f(x, y) = \frac{3}{2}(x+y)^2$ has global minima (equal to 0) along the line $y = -x$ and no other local extrema.

As the parameter a slides, the set of critical points changes, and their nature also changes. Technically speaking, the exercise presents no difficulty, but some attention in the analysis of the Hessian matrix, and its eigenvalues, must be paid. One may think of a "movie" representing the evolution of a surface as time passes, where the parameter a represents time. The qualitative changes occur in the time span $(-3 - \varepsilon_1, \varepsilon_2)$, with small and positive ε_1 and ε_2. In Fig. 4.6, a snapshot is taken of each of the qualitatively different situations occurring.

4.9 Find, if existing, all local or global extrema of the function

$$f(x, y) = (x+a)^2 - 2(x-1)y^2$$

as a ranges in $\mathbb{R}$.

Answer. Clearly, $f \in C^\infty(\mathbb{R}^2)$. Therefore, the possible local or global extrema are critical points. The gradient of f vanishes at (x, y) if and only if

$$\begin{cases} f_x(x, y) = 2(x+a) - 2y^2 = 0 \\ f_y(x, y) = -4(x-1)y = 0. \end{cases}$$

Looking first at f_y, it must either be $x = 1$ or $y = 0$. If $y = 0$ then $f_x(x, 0) = 2(x+a) = 0$ happens if and only if $x = -a$. Therefore all points of the form $(-a, 0)$

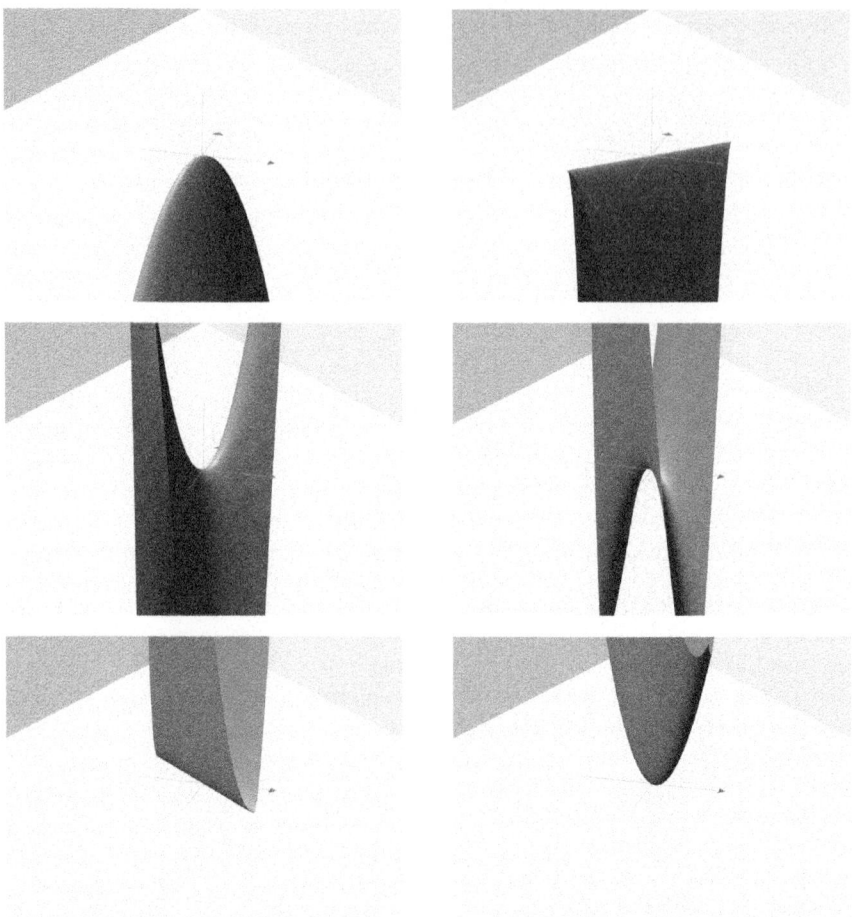

Fig. 4.6 The various graphs of the polynomial functions that correspond to different values of the parameter a in Exercise 4.8. From top to bottom and from left to right, the snapshots are relative to $a < -3$, $a = -3$, $-3 < a < -3/2$, $-3/2 < a < 0$, $a = 0$ and $a > 0$

are critical points. If $x = 1$ then $f_x(1, y) = 2(1 + a) - 2y^2 = 0$ happens if and only if $1 + a \geq 0$ and $y = \pm\sqrt{1 + a}$. Thus, for $a \geq -1$ the points $(1, \pm\sqrt{1 + a})$ are critical points. In order to establish the nature of the various critical points, it is sensible to look at the second derivatives. Since

$$f_{xx}(x, y) = 2, \qquad f_{xy}(x, y) = f_{yx}(x, y) = -4y, \qquad f_{yy}(x, y) = -4(x - 1),$$

the Hessian matrix is

$$Hf(x, y) = \begin{bmatrix} 2 & -4y \\ -4y & -4(x - 1) \end{bmatrix}.$$

(i) Now, for the critical points $(-a, 0)$ it is

$$Hf(-a, 0) = \begin{bmatrix} 2 & 0 \\ 0 & 4(a+1) \end{bmatrix}.$$

Therefore:
- if $a > -1$ then both eigenvalues are positive and $(-a, 0)$ is a local minimum, which cannot be a global minimum because for any fixed a

$$\lim_{y \to +\infty} f(2, y) = \lim_{y \to +\infty} (a+2)^2 - 2y^2 = -\infty;$$

- if $a < -1$ then $Hf(-a, 0)$ is indefinite and $(-a, 0)$ is a saddle point;
- if $a = -1$ then $(1, 0)$ is a saddle point as well. Indeed, if $a = -1$ then the restriction of f to the x-axis is $f(x, 0) = (x-1)^2$, which has a minimum at $x = 1$ where it vanishes. However, on the parabola $x = 1 + y^2$ it is $f(1 + y^2, y) = -y^4$. Therefore, there are points arbitrarily close to $(1, 0)$ where the function takes positive values and points arbitrarily close to $(1, 0)$ where the function takes negative values.

(ii) As for the remaining critical points, which only occur when $a > -1$, namely the points $(1, \pm\sqrt{1+a})$, the Hessian is

$$Hf(1, \pm\sqrt{1+a}) = \begin{bmatrix} 2 & \mp 4\sqrt{1+a} \\ \mp 4\sqrt{1+a} & 0 \end{bmatrix}$$

and has determinant $-16(1+a) < 0$. Thus, these are all saddle points.

In conclusion:
- if $a > -1$ there is a local minimum at $(-a, 0)$ and two saddle points at $(1, \pm\sqrt{1+a})$;
- if $a = -1$ the above critical points merge to a single saddle point at $(1, 0)$;
- if $a < -1$ there is a single saddle point at $(-a, 0)$.

This exercise only requires a parameter discussion. Many of the comments relative to Exercise 4.8 apply in this case as well. Here the transition occurs at $a = -1$.

4.10 Find, if existing, all local or global extrema of the function

$$f(x, y) = (k-1)x^2 + \frac{1}{3}y^3 - y^2$$

as k ranges in $\mathbb{R}$.

Answer. Evidently, $f \in C^\infty(\mathbb{R}^2)$. Observe that $\varphi(y) := f(0, y) = \frac{1}{3}y^3 - y^2$ diverges to $\pm\infty$ as $y \to \pm\infty$ and hence f has neither lower nor upper bounds and so it has neither global maxima nor global minima. The possible local extrema are the critical points, namely the solutions of the system

$$\begin{cases} f_x(x, y) = 2(k - 1)x = 0 \\ f_y(x, y) = y^2 - 2y = 0. \end{cases}$$

Now, if $k \neq 1$ then the critical points are $O = (0, 0)$ and $P_1 = (0, 2)$, whereas if $k = 1$ then all the points of the lines $\ell_0 = \{(x, 0) : x \in \mathbb{R}\}$ and $\ell_2 = \{(x, 2) : x \in \mathbb{R}\}$ are critical. Since

$$f_{xx}(x, y) = 2(k - 1), \qquad f_{xy}(x, y) = f_{yx}(x, y) = 0, \qquad f_{yy}(x, y) = 2y - 2,$$

the Hessian matrix is

$$Hf(x, y) = \begin{bmatrix} 2(k - 1) & 0 \\ 0 & 2y - 2 \end{bmatrix}.$$

- If $k \neq 1$, then $\det Hf(O) = -4(k - 1)$ and hence if $k < 1$ then O is a local maximum since $f_{xx}(O) < 0$. If $k > 1$, then O is a saddle point. As for the point P_1, if $k \neq 1$, then $\det Hf(P_1) = 4(k - 1)$ and hence if $k < 1$ then P_1 is a saddle point and if $k > 1$, then P_1 is a local minimum because $f_{xx}(P_1) > 0$.

- If $k = 1$, then $\det(Hf)$ vanishes both on ℓ_0 and on ℓ_2. The Hessian alone does not allow to conclude but the function f for $k = 1$ only depends on y and takes the form $\varphi(y) = \frac{1}{3}y^3 - y^2$. This is easily understood as a function of a single variable: it has a local maximum at $y = 0$ and a local minimum at $y = 2$, and diverges to $-\infty$ for $y \to -\infty$ and to $+\infty$ for $y \to +\infty$ (see Fig. 4.7). It follows that the points of ℓ_0 are all local maxima and the points of ℓ_2 are all local minima.

This is a standard parametric exercise, whereby the critical points give rise to geometrically different scenarios. Here for all values of the parameter except $k = 1$ the critical points are exactly two and their nature is understood via the Hessian matrix. For $k = 1$ the critical poins form two parallel lines, and this is a reflection of the fact that for that value of k the function really depends on y alone and is therefore constant along all lines parallel to the x axis. This feature allows to easily derive the nature of critical points although the Hessian matrix gives no useful information.

4.11 Find, if existing, all local or global extrema of the function

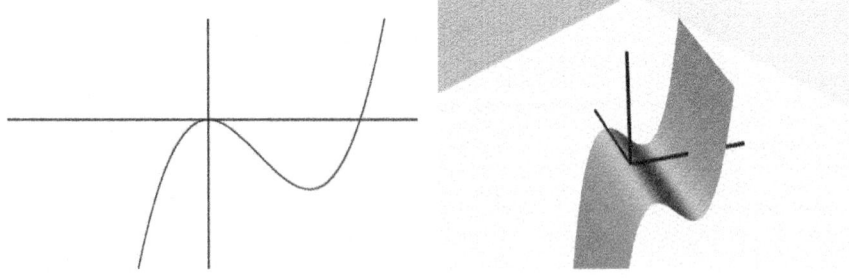

Fig. 4.7 The functions $y \mapsto \frac{1}{3}y^3 - y^2$ and $(x, y) \mapsto \frac{1}{3}y^3 - y^2$ appearing in Exercise 4.10 when $k = 1$

4.4 Guided Exercises

$$f(x, y) = \frac{x^2 - y}{x^2 + y^2 + 1}.$$

Answer. Clearly, $f \in C^\infty(\mathbb{R}^2)$. Therefore, the possible local or global extrema are critical points. The gradient of f vanishes at (x, y) if and only if

$$\begin{cases} f_x(x, y) = \dfrac{2xy^2 + 2x + 2xy}{(x^2 + y^2 + 1)^2} = 0 \\ f_y(x, y) = \dfrac{-x^2 + y^2 - 1 - 2x^2 y}{(x^2 + y^2 + 1)^2} = 0, \end{cases}$$

which is equivalent to

$$\begin{cases} x(y^2 + y + 1) = 0 \\ -x^2 + y^2 - 1 - 2x^2 y = 0 \end{cases} \iff \begin{cases} x = 0 \\ y^2 - 1 = 0. \end{cases}$$

Therefore there are exactly two critical points $P_0 = (0, -1)$ and $P_1 = (0, 1)$. Observe that $f(P_0) = 1/2$ and $f(P_1) = -1/2$. Now,

$$\begin{aligned} f(x, y) - f(P_0) &= \frac{x^2 - y}{x^2 + y^2 + 1} - \frac{1}{2} \\ &= \frac{2x^2 - 2y - x^2 - y^2 - 1}{2(x^2 + y^2 + 1)} \\ &= \frac{x^2 - (y + 1)^2}{2(x^2 + y^2 + 1)}. \end{aligned}$$

It follows that $f(x, y) - f(P_0) \geq 0$ if and only if $x^2 - (y + 1)^2 \geq 0$. Clearly this is true on the line $\{(x, -1) : x \in \mathbb{R}\}$ and false on the line $\{(0, y) : y \in \mathbb{R}\}$. Therefore there are points P arbitrarily close to P_0 on which $f(P) > f(P_0)$ and points Q arbitrarily close to P_0 on which $f(Q) < f(P_0)$. This shows that P_0 is a saddle point. Analogously,

$$f(x, y) - f(P_1) = \frac{3x^2 + (y - 1)^2}{2(x^2 + y^2 + 1)}$$

which is non negative for every $(x, y) \in \mathbb{R}^2$. Therefore P_1 is a global minimum.

In this exercise it is shown that at times it is more convenient to study the local behaviour of the function near a critical point rather than using second derivatives. Indeed, these are somewhat complicated in this case and easily lead to mistakes in the calculations, whereas the explicit form of $f(x, y) - f(P_0)$ and of $f(x, y) - f(P_1)$ is easily established and, in both cases, straightforward to analyze.

4.12 Find, if existing, all global extrema of the function

Fig. 4.8 The interesting points arising in Exercise 4.12; Q is outside the domain of f (the grey region), P_3 is an interior point (the maximum point) and P_1, P_2, P_4, P_5 are on the boundary, one of which, P_5, is the minimum point

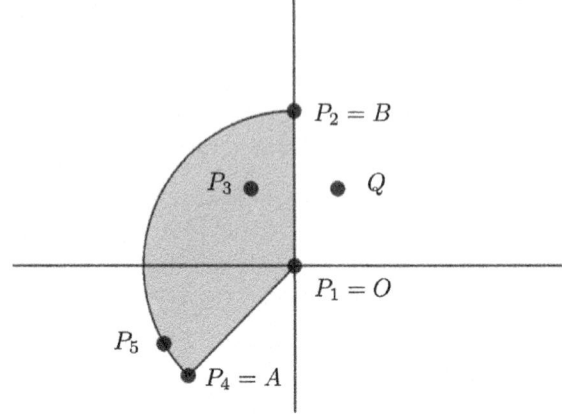

$$f(x, y) = x^3 + xy^2 - 2xy$$

on the set $X = \{(x, y) \in \mathbb{R}^2 : x^2 + y^2 \leq 4,\ x \leq 0,\ y \geq x\}$.

Answer. Evidently, f is of class C^∞ on $\mathbb{R}^2$, hence in particular continuous on the compact set X. By Weierstrass' theorem, it therefore admits both maximum and minimum points. The critical points are the solutions that lie in the interior of X of the system of equations

$$\begin{cases} f_x(x, y) = 3x^2 + y^2 - 2y = 0 \\ f_y(x, y) = 2xy - 2x = 0 \end{cases} \iff \begin{cases} 3x^2 + y^2 - 2y = 0 \\ x(y - 1) = 0. \end{cases}$$

Looking at the second equation, either $x = 0$ or $y = 1$. In the former case the first equation yields $y(y - 2) = 0$ and the points $P_1 = (0, 0)$ and $P_2 = (0, 2)$ satisfy the equations. In the latter case, the first equation yields $3x^2 - 1 = 0$ and the points $Q = (\sqrt{3}/3, 1)$ and $P_3 = (-\sqrt{3}/3, 1)$ satisfy the equations. However, P_1 and P_2 are not interior points (in fact they lie on the boundary of X) and $Q \notin X$. Thus, only P_3 is a critical point. Observe that

$$f(P_1) = f(P_2) = 0, \qquad f(P_3) = \frac{2}{3\sqrt{3}}.$$

As for the other points on the boundary, it is natural to consider the following three subsets of ∂X: the segment $\overline{AO}$ joining $A = (-\sqrt{2}, -\sqrt{2})$ and the origin O, the segment $\overline{OB}$ joining O with $B = (0, 2)$ and the circular arc $\overset{\frown}{AB}$ joining A to B along the circle centered at the origin of radius 2 (see Fig. 4.8).

- The segment $\overline{AO}$. Upon setting $x = y = t$ with $t \in [-\sqrt{2}, 0]$, the restriction of f to this segment is given by $f(t, t) = 2t^3 - 2t^2 =: g(t)$. Now, the derivative is

4.4 Guided Exercises

$g'(t) = 6t^2 - 4t = 2t(3t - 2)$ and hence $g'(t) > 0$ for $t < 0$ or $t > 2/3$. It follows that g is increasing in $[-\sqrt{2}, 0]$ and hence if $P_4 = (-\sqrt{2}, -\sqrt{2})$ then

$$f(P_4) = -4(1 + \sqrt{2})$$

is the minimum of f along the segment and $f(0, 0) = f(P_1) = 0$ is the maximum.
- The segment $\overline{OB}$. The restriction of f to this segment is given by $f(0, y) = 0$. Thus here the function vanishes identically.
- The circular arc $\widehat{AB}$. Upon setting $(x, y) = (2 \cos t, 2 \sin t)$ with $t \in [\pi/2, 5\pi/4]$ the restriction of f to this arc is

$$\begin{aligned} h(t) :&= f(2\cos t, 2 \sin t) \\ &= 8 \cos^3 t + 8 \cos t \sin^2 t - 8 \cos t \sin t \\ &= 8 \cos t (1 - \sin t). \end{aligned}$$

Now, differentiating

$$h'(t) = -8 \sin t + 8 \sin^2 t - 8 \cos^2 t = 0 \iff 2 \sin^2 t - \sin t - 1 = 0$$

where $\sin t \in [-1/\sqrt{2}, 1]$. The latter equation has the two solutions $\sin t = 1$ or $\sin t = -1/2$. For $t \in [\pi/2, 5\pi/4]$ these correspond to $t = \pi/2$ and $t = 7\pi/6$, namely to the points $(0, 2) = P_2$ and $P_5 = (-\sqrt{3}, -1)$. It is immediately checked that $t = \pi/2$ is a maximum for h (where it vanishes) and $t = 7\pi/6$ is the minimum. Furthermore

$$f(P_5) = -6\sqrt{3} < -4(1 + \sqrt{2}) = f(P_4) < f(P_2) = 0.$$

Hence P_2 is the maximum point for the restriction of f to $\widehat{AB}$ and P_5 is the minimum point. Therefore, $P_5 = (-\sqrt{3}, -1)$ is the (unique) minimum point and lies on the boundary whereas $P_3 = (-\sqrt{3}/3, 1)$ is the (unique) maximum point and is an interior point.

This is somewhat a prototypical example of a situation where it is necessary to separate the analysis of the interior points from the boundary points of the assigned domain X. Since the function is naturally defined on a much larger set (in this case f is actually a polynomial and hence it is defined everywhere in $\mathbb{R}^2$) there are solutions of the equation $\nabla f(x, y) = 0$ that do not lie in X, much less within its interior. Next, ∂X consists of three very nice sets, and on each of them the restriction of f is quite easy to analyze. In the end one collects all the candidates and compares the values that f attains on each of them, finding the maxima and the minima. In this case there is one maximum point and one minimum point.

4.13 Find, if existing, all global extrema of the function

$$f(x, y) = \log(5 - x^2 - y^2)$$

Fig. 4.9 The interesting points arising in Exercise 4.13; there are no critical interior points and several points on the boundary where the restriction of f has local minima or maxima

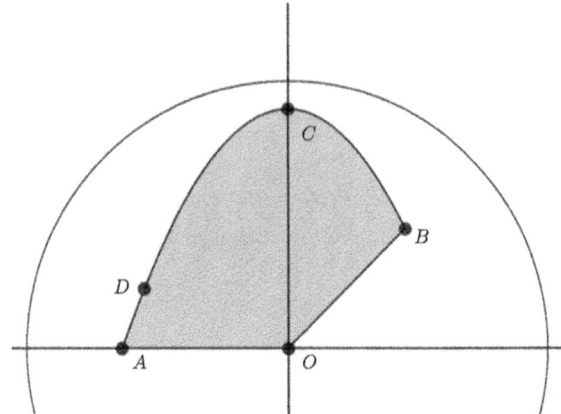

on the set $X = \{(x, y) \in \mathbb{R}^2 : y \leq 2 - x^2,\ y \geq 0,\ y \geq x\}$.

Answer. The function f is of class C^∞ on its domain $\{(x, y) \in \mathbb{R}^2 : x^2 + y^2 < 5\}$, namely the circle of radius $\sqrt{5}$ centered at the origin, which contains the set X (see Fig. 4.9). Therefore, f is continuous on the compact set X and by Weierstrass' theorem it admits both maximum and minimum points. The critical points are the solutions that lie in the interior of X of the system of equations

$$\begin{cases} f_x(x, y) = \dfrac{-2x}{5 - x^2 - y^2} = 0 \\[1ex] f_y(x, y) = \dfrac{-2y}{5 - x^2 - y^2} = 0. \end{cases}$$

The system has the only solution $(x, y) = (0, 0) = O$ which is not an interior point of X. Actually, $O \in \partial X$.

As for the other points on the boundary, it is natural to consider the following three subsets of ∂X: the segment $\overline{AO}$ joining $A = (-\sqrt{2}, 0)$ to the origin O, the segment $\overline{OB}$ joining O to $B = (1, 1)$ and the arc $\widehat{AB}$ joining A to B along the parabola $y = 2 - x^2$ (see Fig. 4.9).

- The segment $\overline{AO}$. The restriction of f to this segment is given by the function of one variable $g(x) := f(x, 0) = \log(5 - x^2)$ with $x \in [-\sqrt{2}, 0]$. Since $g'(x) = -2x/(5 - x^2)$ and $x < 0$, the function g is increasing and thus has a minimum at $x = -\sqrt{2}$, which corresponds to A, where $f(A) = \log 3$, and a maximum at $x = 0$, which corresponds to O, where $f(O) = \log 5$.
- The segment $\overline{OB}$. The restriction of f to this segment is given by the function $h(t) := f(t, t) = \log(5 - 2t^2)$ with $t \in [0, 1]$. Since $h'(t) = -4t/(5 - 2t^2)$ and $t \geq 0$, the function h has a maximum at $t = 0$, which corresponds to O, and a minimum at $t = 1$, which corresponds to B, where $f(B) = \log 3$.

4.4 Guided Exercises

- The parabolic arc $\widehat{AB}$. The restriction of f to this set is given by the function $p(x) := f(x, 2 - x^2) = \log(1 + 3x^2 - x^4)$ with $x \in [-\sqrt{2}, 1]$. Now,

$$p'(x) = \frac{6x - 4x^3}{1 + 3x^2 - x^4}$$

and since the denominator is positive on $[-\sqrt{2}, 1]$, it follows that $p'(x) > 0$ when $x(3 - 2x^2) > 0$. This happens only in $[-\sqrt{2}, -\sqrt{(3/2)}] \cup [0, 1]$. Hence the restriction of f to $\widehat{AB}$ has a local minimum at A, a local maximum at $D = (-\sqrt{(3/2)}, 1/2)$ where $f(D) = \log(13/4)$, another local minimum at $C = (0, 2)$ where $f(C) = 0$ and another local maximum at B. The conclusion is that the minimum point is C and the maximum point is O.

This exercise is very similar in spirit to Exercise 4.12. In this case there are no interior critical points and the analysis reduces to the boundary, which consists of two segments and a piece of a parabola. All these are easily described and the restriction of f along each of these sets is an elementary function of one variable.

4.14 Find, if existing, all global extrema of the function

$$f(x, y) = x^3 + x^2 + y^3 - y^2$$

on the set $X = \{(x, y) \in \mathbb{R}^2 : 2x^2 + 3y^2 \leq \frac{40}{21}\}$.

Answer. Clearly, X is compact. Indeed, it is closed because it is the inverse image of a continuous function and it is bounded because it is contained in the ball $B(O, \sqrt{\frac{20}{21}})$:

$$2x^2 + 2y^2 \leq 2x^2 + 3y^2 \leq \frac{40}{21}.$$

By Weierstrass' theorem, the continuous function f on $\mathbb{R}^2$ restricted to the compact set X has global extrema. The interior points of X are those for which the inequality defining X is strict, which form an open set. The gradient of f vanishes at an interior point (x, y) if and only if

$$\begin{cases} f_x(x, y) = 3x^2 + 2x = x(3x + 2) = 0 \\ f_y(x, y) = 3y^2 - 2y = y(3y - 2) = 0. \end{cases}$$

There are thus four critical points of the polynomial $x^3 + y^2 + y^3 - y^2$, namely $O = (0, 0)$, $A = (0, 2/3)$, $B = (-2/3, 0)$ and $C = (-2/3, 2/3)$. The point C, however, is not in X.

The second derivatives at $P = (x, y)$ are

$$f_{xx}(P) = 6x + 2, \quad f_{xy}(P) = f_{yx}(P) = 0, \quad f_{yy}(P) = 6y - 2.$$

It follows that
$$Hf(O) = \begin{bmatrix} 2 & 0 \\ 0 & -2 \end{bmatrix},$$
so that O is a saddle point. Furthermore,
$$Hf(A) = \begin{bmatrix} 2 & 0 \\ 0 & 2 \end{bmatrix}, \quad Hf(B) = \begin{bmatrix} -2 & 0 \\ 0 & -2 \end{bmatrix},$$
so that A is a local minimum and B is a local maximum. As for the boundary behaviour, the problem is to find the extrema of f under the constraint
$$\partial X = \{(x, y) \in \mathbb{R}^2 : g(x, y) = 0\}$$
where $g(x, y) = 2x^2 + 3y^2 - \frac{40}{21}$. Observe that $f, g \in C^1(\mathbb{R}^2)$. Furthermore, since $\nabla g(x, y) = (4x, 6y)$, the gradient of g never vanishes on its zero-set. The extrema for $f|_{\partial X}$ may then be analyzed by means of Langrange multipliers. This means that if (x, y) is an extremum of $f|_{\partial X}$ then there exists a Lagrange multiplier $\lambda \in \mathbb{R}$ such that $\nabla f(x, y) + \lambda \nabla g(x, y) = (0, 0)$, namely, there must be $\lambda \in \mathbb{R}$ such that
$$\begin{cases} 3x^2 + 2x + 4\lambda x = 0 \\ 3y^2 - 2y + 6\lambda y = 0 \\ 2x^2 + 3y^2 = \frac{40}{21} \end{cases} \iff \begin{cases} x(3x + 2 + 4\lambda) = 0 \\ y(3y - 2 + 6\lambda) = 0 \\ 2x^2 + 3y^2 = \frac{40}{21}. \end{cases}$$

Now, if $x = 0$, then $(0, y) \in \partial X$ only if $y = \pm 2\sqrt{10}/3\sqrt{7}$ and the two points
$$E = (0, \frac{2\sqrt{10}}{3\sqrt{7}}), \quad F = (0, -\frac{2\sqrt{10}}{3\sqrt{7}})$$
are constrained critical points. Next, if $y = 0$, then $(x, 0) \in \partial X$ only if $x = \pm 2\sqrt{5}/\sqrt{21}$ and the two points
$$G = (\frac{2\sqrt{5}}{\sqrt{21}}, 0), \quad H = (-\frac{2\sqrt{5}}{\sqrt{21}}, 0)$$
are constrained critical points. Finally, if $x \neq 0$ and $y \neq 0$, the system of equations becomes
$$\begin{cases} 3x + 2 + 4\lambda = 0 \\ 3y - 2 + 6\lambda = 0 \\ 2x^2 + 3y^2 = \frac{40}{21}. \end{cases}$$

From the first equation $x = -(2 + 4\lambda)/3$ and $y = (2 - 6\lambda)/3$, so that the third equation becomes $49\lambda^2 - 14\lambda + 1 = 0$ and has the only solution $\lambda = 1/7$. The last constrained critical point is then

$$L = \left(-\frac{6}{7}, \frac{8}{21}\right).$$

Finally, a tedious computation gives:

$$f(E) = \frac{40}{63}\left(\frac{2\sqrt{10}}{3\sqrt{7}} - 1\right) = -0,1290\ldots, \quad f(F) = \frac{40}{63}\left(-\frac{2\sqrt{10}}{3\sqrt{7}} - 1\right) = -1.1408\ldots$$

$$f(G) = \frac{20}{21}\left(\frac{2\sqrt{5}}{\sqrt{21}} + 1\right) = 1,881\ldots \quad f(H) = \frac{20}{21}\left(-\frac{2\sqrt{5}}{\sqrt{21}} + 1\right) = 0,0229\ldots$$

$$f(L) = \frac{20}{1323} = 0.0151\cdots$$

Hence F is the minimum of f on ∂X and G is the maximum. Finally,

$$f(A) = -\frac{4}{27} = -0.14814\cdots > f(F), \quad f(B) = 0.14814\cdots = \frac{4}{27} < f(G).$$

In conclusion, F is the minimum of f on X and G is the maximum.

This exercise is a straight illustration on how to look for local extrema of a function on a domain which has both a nonempty interior and a nonempty boundary. The latter situation may require using the methods of Lagrange multipliers. The final part shows that once all the constrained critical points are found, further analysis is required.

4.15 Find, if existing, all extrema of the function

$$f(x, y) = 2x^2 - 4y^2$$

on the set $X = \{(x, y) \in \mathbb{R}^2 : x^2 + y^4 - 4 = 0\}$.

Answer. Put $g(x, y) = x^2 + y^4 - 4$. Obviously, $f, g \in C^\infty(\mathbb{R}^2)$ and X is therefore closed. Furthermore, on X it is $0 \leq y^4 = 4 - x^2$, so that $-2 \leq x \leq 2$ and consequently $-\sqrt{2} \leq y \leq \sqrt{2}$. Therefore X is compact and f admits global extrema on X by Weierstrass' theorem. Also, the gradient $\nabla g(x, y) = (2x, 4y^3)$ never vanishes on X. The extrema for $f|_{\partial X}$ may then be analyzed by means of Langrange multipliers. This means that if (x, y) is an extremum of $f|_{\partial X}$ then there exists a Lagrange multiplier $\lambda \in \mathbb{R}$ such that $\nabla f(x, y) + \lambda \nabla g(x, y) = (0, 0)$, namely, there must be $\lambda \in \mathbb{R}$ such that

$$\begin{cases} f_x + \lambda g_x = 4x + 2\lambda x = 2x(2 + \lambda) = 0 \\ f_y + \lambda g_y = -8y + 4\lambda y^3 = 4y(\lambda y^2 - 2) = 0 \\ x^2 + y^4 = 4. \end{cases}$$

The origin $O = (0, 0)$ satisfies the first two equations but is not in X. If $x = 0$, then the constraint equation yields $y = \pm\sqrt{2}$ which, from the second equation, gives $\lambda = 1$. Hence $(0, \pm\sqrt{2})$ are constrained critical points. Finally, if $x \neq 0$, then $\lambda = -2$. Hence from the second equation $y = 0$ and the constraint forces $x = \pm 2$. In conclusion, there are exactly four constrained critical points, $(0, \pm\sqrt{2})$ and $(\pm 2, 0)$. Finally $f(0, \pm\sqrt{2}) = -8 < 0 < 8 = f(\pm 2, 0)$. Thus $(0, \pm\sqrt{2})$ are the global minima and $(\pm 2, 0)$ are the global maxima.

This exercise is a straightforward, clean application of the method of Lagrange multipliers. All is required is a thorough analysis of the appropriate system of equations.

4.16 Find, if existing, all global extrema of the function

$$f(x, y) = x^2 + 2y^2 - 2xy + 4x$$

on the set $X = \{(x, y) \in \mathbb{R}^2 : x^2 + 2y^2 - 1 = 0\}$.

Answer. Clearly, X is an ellipse hence a compact set. Since $f \in C^\infty(\mathbb{R}^2)$, its restriction $f|_X$ is continuous and it therefore admits global extrema by Weierstrass' theorem. Write $g(x, y) = x^2 + 2y^2 - 1$, a C^1-function with gradient $\nabla g(x, y) = (2x, 4y)$ which never vanishes on X. The extrema for $f|_{\partial X}$ may then be analyzed by means of Langrange multipliers. Observe that on X it is $f|_{\partial X}(x, y) = 1 - 2xy + 4x$, hence there is no harm in considering $h(x, y) = 1 - 2xy + 4x$ in place of f. This means that if (x, y) is an extremum of $f|_{\partial X}$ then there exists a Lagrange multiplier $\lambda \in \mathbb{R}$ such that $\nabla h(x, y) + \lambda \nabla g(x, y) = (0, 0)$, namely, there must be $\lambda \in \mathbb{R}$ such that

$$\begin{cases} h_x + \lambda g_x = -2y + 4 + 2\lambda x = 0 \\ h_y + \lambda g_y = -2x + 4\lambda y = 0 \\ x^2 + 2y^2 = 1. \end{cases}$$

Now, if $y = 0$, then the second equation gives $x = 0$, but $(0, 0) \notin X$. Thus $y \neq 0$, and the second equation implies $\lambda = x/2y$, whence

$$\begin{cases} 0 = -2y + 4 + 2\dfrac{x}{2y}x = \dfrac{1}{y}(-2y^2 + 4y + x^2) \\ x^2 + 2y^2 = 1. \end{cases} \iff \begin{cases} x^2 = 2y^2 - 4y \\ 1 = (2y^2 - 4y) + 2y^2. \end{cases}$$

The last equation has the solutions $y = (1 \pm \sqrt{2})/2$, which plugged into the first one gives $x^2 = (-1 \mp 2\sqrt{2})/2$. Obviously, $x^2 = (-1 - 2\sqrt{2})/2$ has no solution, whereas $x^2 = (-1 + 2\sqrt{2})/2$ yields two solutions, thereby giving rise to the two constrained critical points:

$$P_1 = \left(\sqrt{\dfrac{2\sqrt{2} - 1}{2}}, \dfrac{1 - \sqrt{2}}{2}\right), \quad P_2 = \left(-\sqrt{\dfrac{2\sqrt{2} - 1}{2}}, \dfrac{1 - \sqrt{2}}{2}\right)$$

Finally, observe that $f(P_1) = f(x_1, y_1) = h(x_1, y_1) > 1$ because $h(x, y) = 1 - 2xy + 4x$ and $x_1 > 0 > y_1$. Similarly, $f(P_2) = f(x_2, y_2) = h(x_2, y_2) < 1$ because $x_2 < 0$ and $y_2 < 0$. In conclusion, P_1 is the unique global maximum and P_2 is the unique global minimum.

The main observation here is that when looking for constrained extrema of f, it is of course legitimate to substitute f with a function h which coincides with f on the constraint. This may simplify the calculations somewhat.

4.17 Draw, in a neighborhood of the point $P_0 = (0, 1)$, the graph of the function implicitly defined by the equation $y^3 + \log(x + y) - xy - 1 = 0$.

Answer. Put $f(x, y) = y^3 + \log(x + y) - xy - 1 = 0$. Evidently, $f \in C^\infty(A)$, where the set $A = \{(x, y) \in \mathbb{R}^2 : x + y > 0\}$ contains the point $P_0 = (0, 1)$, and $f(P_0) = 0$. Furthermore, since

$$\begin{cases} f_x(x, y) = \dfrac{1}{x+y} - y \\ f_y(x, y) = 3y^2 + \dfrac{1}{x+y} - x \end{cases}$$

it follows that $\nabla f(P_0) = (0, 4) \neq (0, 0)$. By Dini's Theorem there exist a neighborhood $\mathscr{U}$ of $x_0 = 0$ and a unique function $y \colon \mathscr{U} \to \mathbb{R}$ such that $y(0) = 1$ and $f(x, y(x)) = 0$ for every $x \in \mathscr{U}$. Furthermore,

$$\begin{aligned} y'(x) &= -\frac{f_x(x, y(x))}{f_y(x, y(x))} \\ &= -\frac{1 - xy(x) - y^2(x)}{3xy^2(x) + 3y^3(x) - x^2 - xy(x) + 1}. \end{aligned}$$

Put for simplicity

$$N(x) := 1 - xy(x) - y^2(x), \qquad D(x) := 3xy^2(x) + 3y^3(x) - x^2 - xy(x) + 1.$$

As $f_x(P_0) = 0$ had already being established, certainly $y'(0) = 0$, that is $N(0) = 0$. Therefore:

$$y''(0) = -\frac{N'(0)D(0) - N(0)D'(0)}{D^2(0)} = -\frac{N'(0)}{D(0)}.$$

Now, $N'(x) = -y(x) - xy'(x) - 2y(x)y'(x)$ and hence $N'(0) = -1$. Further, $D(0) = 4$. It follows that $y''(0) = 1/4$. Appealing to Taylor's expansion, in a neighborhood of $x_0 = 0$ one may write $y(x) = 1 + \frac{1}{8}x^2 + o(x^2)$.

This is a very standard use of Dini's theorem. Local information on the implicit function may be derived from knowledge of the derivatives, which may be computed successively. From a (rough) qualitative point of view, one or two derivatives often suffice. In Fig. 4.10, a comparison between the actual implicit function and its second order McLaurin polynomial.

4.18 Draw, in a neighborhood of the point $P_0 = (0, 0)$, the graph of the function implicitely defined by the equation $x + e^y - e^x - 2y = 0$.

Answer. Put $f(x, y) = x + e^y - e^x - 2y$. Clearly, $f \in C^\infty(\mathbb{R}^2)$ and an immediate computation gives $\nabla f(0, 0) = (0, -1) \neq (0, 0)$. By Dini's Theorem there exist a neighborhood $\mathscr{U}$ of $x_0 = 0$ and a unique function $y \colon \mathscr{U} \to \mathbb{R}$ such that $y(0) = 0$ and $f(x, y(x)) = 0$ for every $x \in \mathscr{U}$. First of all,

$$y'(0) = -\frac{f_x(0, y(0))}{f_y(0, y(0))} = 0.$$

A way to compute $y''(0)$ is to start from the equation

$$x + e^{y(x)} - e^x - 2y(x) = 0,$$

which holds true in $\mathscr{U}$, and to take two derivatives with respect to x, obtaining first

$$1 + e^{y(x)} y'(x) - e^x - 2y'(x) = 0,$$

and then

$$e^{y(x)} (y'(x))^2 + e^{y(x)} y''(x) - e^x - 2y''(x) = 0.$$

This yields

$$e^{y(0)} (y'(0))^2 + e^{y(0)} y''(0) - 1 - 2y''(0) = 0.$$

and since $'(0)$,

$$y''(0) - 1 - 2y''(0) = 0 \quad \Longrightarrow \quad y''(0) = -1.$$

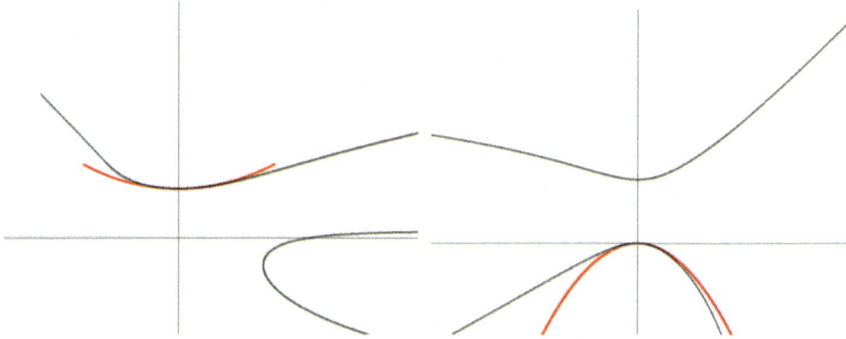

Fig. 4.10 On the left, in black, the level set $f(x, y) = 0$ of Exercise 4.17, determining the implicit function in a neighborhood of P_0; in red its second order McLaurin polynomial; on the right the analogous pictures for Exercise 4.18

4.4 Guided Exercises

Therefore the second order McLaurin expansion is $y(x) = -\frac{1}{2}x^2 + o(x^2)$.

In this exercise, very similar to the previous one, an alternative method for the computation of derivatives is used. In Fig. 4.10, a comparison between the actual implicit function and its second order McLaurin polynomial.

4.19 Verify that the equation $y^7 + 3y - 2xe^{3x} = 0$ defines a function $y = \varphi(x)$ for all $x \in \mathbb{R}$, and determine its behaviour for $x \to \pm\infty$.

Answer. Put $f(x, y) = y^7 + 3y - 2xe^{3x}$. Since

$$f_y(x, y) = 7y^6 + 3,$$

it follows $f_y(P) > 0$ for every P, and so in particular this happens at every P for which $f(P) = 0$. This means that Dini's Theorem applies at every P for which $f(P) = 0$, in the sense that for each such $P = (x_P, y_P)$ there exists a neighborhood $\mathcal{U}_P$ of x_P and a function $\varphi_P : \mathcal{U}_P \to \mathbb{R}$ such that $f_P(x, \varphi_P(x)) = 0$ for every $x \in \mathcal{U}_P$. The idea is now to "glue" all these open sets and define a unique function of x by patching together all the local functions given by Dini's Theorem. Precisely, denote by $Z(f) = \{P \in \mathbb{R}^2 : f(P) = 0\}$ the zeroes of f and define the function

$$\varphi : \bigcup_{P \in Z(f)} \mathcal{U}_P \to \mathbb{R}$$

by $\varphi(Q) = \varphi_P(Q)$ if $Q \in \mathcal{U}_P$. This makes sense because if Q is in the intersection of two open sets $\mathcal{U}_{P_1} \cap \mathcal{U}_{P_2}$, then the functions φ_{P_1} and φ_{P_2} agree on $\mathcal{U}_{P_1} \cap \mathcal{U}_{P_2}$, by Dini's Theorem and hence $\varphi_{P_1}(Q) = \varphi_{P_2}(Q)$. It must be shown that

$$\bigcup_{P \in Z(f)} \mathcal{U}_P = \mathbb{R}.$$

This follows from the observation that the function $y \mapsto y^7 + 3y$ is of course surjective, that is, it attains every real value. Hence, for any given $x \in \mathbb{R}$ there exists y such that

$$2xe^{3x} = y^7 + 3y,$$

that is $(x, y) \in Z(f)$. Next, observe that by Dini's Theorem

$$\varphi'(x) = -\frac{f_x(x, \varphi(x))}{f_y(x, \varphi(x))} = \frac{2e^{3x}(1 + 3x)}{7\varphi(x)^6 + 3}.$$

Therefore $\varphi'(x) > 0$ for $x > -1/3$, $\varphi'(x) < 0$ for $x < -1/3$ and $\varphi'(x) = 0$ for $x = -1/3$. Hence φ decreases in $(-\infty, -1/3]$ and increases in $[-1/3, +\infty)$. Therefore φ does admit limits for $x \to \pm\infty$. From the equation

$$2xe^{3x} = \varphi(x)^7 + 3\varphi(x),$$

it follows that

$$+\infty = \lim_{x \to +\infty} 2xe^{3x} = \lim_{x \to +\infty} (\varphi(x)^6 + 3)\varphi(x)$$
$$0 = \lim_{x \to -\infty} 2xe^{3x} = \lim_{x \to -\infty} (\varphi(x)^6 + 3)\varphi(x).$$

Hence, if $\lim_{x \to -\infty} \varphi(x) = \ell$ then $(\ell^6 + 3)\ell = 0$ implies $\ell = 0$. Similarly, if $\lim_{x \to -\infty} \varphi(x) = m$ then necessarily $m = +\infty$.

4.5 Problems

4.20 Find the local extrema of $f(x, y) = x + y^2 + x|y|$, if existing.

4.21 Find the global extrema on $\{(x, y) \in \mathbb{R}^2 : 0 \leq y \leq |x|, -1 \leq x \leq 1\}$, if existing, of $f(x, y) = 8x + 4y - 1$.

4.22 Find the global extrema of $f(x, y) = x^2 + 2xy$ on $\{(x, y) \in \mathbb{R}^2 : 2x^2 + y^2 = 3\}$, if existing.

4.23 Find the global extrema of $f(x, y) = \log(3 - xy)$ inside and on the triangle with vertices at $(0, 0)$, $(1, 0)$, $(0, 1)$, if existing.

4.24 Find the global extrema of $f(x, y) = x^3 - 3xy^2$ inside and on the square with vertices at $(0, 0)$, $(1, 0)$, $(0, 1)$ and $(1, 1)$, if existing.

4.25 Find the global extrema on $\{(x, y) \in \mathbb{R}^2 : 0 \geq y \geq x^2 - 1\}$, if existing, of $f(x, y) = \sqrt{4 - x^2 - y^2}$.

4.26 Find the global extrema on $\{(x, y) \in \mathbb{R}^2 : 2x^2 + y^2 \leq 2\}$ of $f(x, y) = 3x^2y - y^3$, if existing.

4.27 Find the global extrema, if existing, of $f(x, y) = x^2 + y^2 - 4\log(x^2 + y + 1)$ on $\{(x, y) \in \mathbb{R}^2 : y \geq -x^2\}$.

4.28 Find the global extrema on $\{(x, y, z) \in \mathbb{R}^3 : x^2 + y^2 + z^2 = 9, x^2 + (y - 1)^2 = 4\}$ of $f(x, y, z) = z$, if existing.

4.29 Find the local and global extrema, if existing, of $f(x, y) = x^2 + kxy + y^2$ as k varies in $\mathbb{R}$.

4.30 Find the global and local extrema on $\mathbb{R}^2$ of $f(x, y) = x^3y^2(1 - x - y)$, if existing. Then find its global extrema, if existing, on $Q = \{(x, y) \in \mathbb{R}^2 : |x| + |y| \leq 1\}$.

4.31 Find the local and global extrema of $f(x, y) = \log(1 + x^2 + ky^2)$ as k varies in $\mathbb{R}$, if existing.

4.5 Problems

4.32 Find the local and global extrema of $f(x, y) = x^4 - (1+k)xy$ as k varies in $\mathbb{R}$, if existing.

4.33 Verify that the equation $\cos(xy) - e^{2x-y} - 3y^2 + x = 0$ defines an implicit function $x = x(y)$ in a neighborhood of the point $O = (0, 0)$ and sketch a graph of it.

4.34 Verify that the equation $\sin(x - y^2) + e^x - y^2 - y - 1 = 0$ defines an implicit function $x = x(y)$ in a neighborhood of the point $O = (0, 0)$ and sketch a graph of it.

4.35 Find the global extrema on $\{(x, y) \in \mathbb{R}^2 : x \geq 0, y \geq 0, y \leq -3x + 3\}$ of $f(x, y) = 2x^2 + xy^2 - 2xy$, if existing.

4.36 Find the global extrema on $\{(x, y) \in \mathbb{R}^2 : 1 \leq x \leq 2, \frac{\pi}{4} \leq y \leq \frac{3}{4}\pi\}$, if existing, of $f(x, y) = \frac{|\cos y|}{\sin y}\sqrt{x}$.

4.37 Determine for which values of the parameter $k \in \mathbb{R} \setminus \{1/2\}$ the level set of the function $f(x, y) = \frac{x + ky}{x^2 - ky + 1}$ through the point $(1, 4)$ defines locally a function of one variable.

4.38 Show that the equation $x \sin y + y \cos x - e^x + e^y = 0$ defines locally around $(0, 0)$ an implicit function $y = y(x)$ and write its second order Taylor expansion.

4.39 Find the local and global extrema of

$$f(x, y) = \begin{cases} \dfrac{x^2 y^3}{x^4 + y^6} & (x, y) \neq (0, 0) \\ 0 & (x, y) = (0, 0) \end{cases}$$

and then find the global ones on $\{(x, y) \in \mathbb{R}^2 : x^4 + y^6 = 1\}$.

4.40 Find the global extrema on $\{(x, y, z) \in \mathbb{R}^3 : x^2 + y^2 + e^{-1} \leq z \leq y + 1\}$, if existing, of $f(x, y, z) = \log(z - x^2 - y^2) + y - z$.

4.41 Show that the level set through $(1, 0)$ of $f(x, y) = x^2 e^y + y^2 e^x - y$ defines implicitely a function φ of a single variable. Prove that φ has an extremum and find the zeroes of φ. Find an interval I contained in $\text{Dom}(f)$.

4.42 Find the global extrema on $\{(x, y) \in \mathbb{R}^2 : x^2 + y^2 \leq 1\}$, if existing, of

$$f(x, y) = \begin{cases} \dfrac{x^2 + \arctan(y^2)}{\sqrt{x^2 + y^2}} & (x, y) \neq (0, 0) \\ 0 & (x, y) = (0, 0). \end{cases}$$

4.43 Show that the zeroes of $f(x, y, z) = \arctan(z + xy) + e^z + \log(1 + x^2 + y^2) - 1$ in a neighborhood of $(0, 0, 0)$ define a function $z = z(x, y)$. Find at least one critical point of this function, prove that it is bounded above and find an upper bound.

Chapter 5
Multiple Integrals

Recall that, as explained in [1], the Riemann integral $\int_a^b f(x)\,dx$ of a non negative bounded function $f : [a, b] \to \mathbb{R}$ is meant to capture the area under its graph. This number in not well defined for all positive functions, but for a rich supply of them, including the piecewise continuous ones. An analogous concept may be defined for functions of several variables, and it is of particular interest in the case of two or three variables, in which cases the corresponding notions are called *double integrals* and *triple integrals*. In dimension two or three intuitive geometric interpretations are available, and are certainly of great help, but the meaning of the various definitions and results remains in all dimensions and is in fact of fundamental importance both from the theoretical point of view and for applications. It should be mentioned, however, that although very powerful and satisfactory for most practical computations, the Riemann integral is not the best possible notion of (definite) integral, and is in fact improved in many important respects by the Lebesgue integral, most notably because the class of integrable functions according to the latter notion is larger and because it allows for better properties of natural function spaces. This fact must be kept in mind but is of no harm in the process of learning basic integration techniques, which is the main purpose of the present chapter.

5.1 Measure and Measurable Sets

Definition 5.1 (*n-rectangles*) If $I_1,\ldots, I_n$ are bounded non empty intervals, the Cartesan product $R = I_1 \times \cdots \times I_n \subset \mathbb{R}^n$ is called an *n-rectangle*. If $a_j = \inf I_j$ and $b_j = \sup I_j$, the *measure* of R is the positive number.

$$|R| = (b_1 - a_1)(b_2 - a_2)\ldots(b_n - a_n).$$

A 2-rectangle will be called a *rectangle* (a square if of the form $R = I \times I$) and a 3-rectangle will be called a parallelepiped (a cube if of the form $R = I \times I \times I$).

Suppose that $P \subset \mathbb{R}^n$ is a finite union

$$P = R_1 \cup R_2 \cup \cdots \cup R_m$$

of n-rectangles meeting at most along their boundaries, that is, assume that for all $1 \leq j, k \leq m$ with $j \neq k$ it holds

$$(R_j \cap R_k) \subset (\partial R_j \cup \partial R_k).$$

Then the measure of P is defined as the positive number

$$|P| = \sum_{j=1}^{m} |R_j|.$$

Any such finite union P is called an *almost disjoint* union of n-rectangles, and the corresponding set $\{R_1, \ldots, R_m\}$ an almost disjoint family of n-rectangles.

Notice that the bounded intervals that appear in the definition of n-rectangle can only be of the following kinds: $[a, b], [a, b), (a, b], (a, b)$. What are therefore the possible rectangles? The reader is urged to draw them all, and should notice that the measure is, by definition, independent of the portion of boundary included in $I_1 \times I_2$.

Definition 5.2 (*Peano-Jordan measurable set*) If Ω is a bounded non empty subset of $\mathbb{R}^n$, then.

$$|\Omega|_{-} := \sup\{|P| : P \subset \Omega\}$$

$$|\Omega|_{+} := \inf\{|P| : P \supset \Omega\},$$

where P varies among the almost disjoint unions of n-rectangles, are called the *inner Peano-Jordan measure* and *outer Peano-Jordan measure* of Ω, respectively. Since Ω is bounded, both $|\Omega|_{-}$ and $|\Omega|_{+}$ are finite. The set Ω is called *Peano-Jordan measurable* if the inner and outer Peano-Jordan measures coincide, and in this case their common value

$$|\Omega| = |\Omega|_{-} = |\Omega|_{+}$$

is called the *Peano-Jordan measure* of Ω. It is customary to assign zero measure to the empty set.

It is worth observing that not all bounded sets are Peano-Jordan measurable. For example, the Dirichlet set

$$D = \{(x, y) \in [0, 1] \times [0, 1] : x, y \in \mathbb{Q}\} \subset \mathbb{R}^2$$

has outer measure 1 and inner measure 0, as the reader may easily check. Hence it is not Peano-Jordan measurable. Notice that $\partial D = \overline{D} = [0, 1] \times [0, 1]$, so that $|\partial D| = 1$. This fact is just an example of the very general statement in the next result.

Theorem 5.1 *A bounded non empty subset of $\mathbb{R}^n$ is Peano-Jordan measurable if and only if its boundary is Peano-Jordan measurable and has zero Peano-Jordan measure.*

The Peano-Jordan measure enjoys the basic properties described below.

Proposition 5.1 *Suppose that $A, B \subset \mathbb{R}^n$ are Peano-Jordan measurable. Then*

(i) *$A \cup B$, $A \cap B$ and $A \setminus B$ are Peano-Jordan measurable and $|A \cup B| = |A| + |B| - |A \cap B|$;*
(ii) *if $A \subseteq B$, then $|A| \leq |B|$.*

The notion of *characteristic function* (also known as *indicator function*) is very useful and practical.

Definition 5.3 Given any subset A of $\mathbb{R}^n$ its characteristic function χ_A is defined by

$$\chi_A(x) = \begin{cases} 1 & x \in A \\ 0 & x \notin A. \end{cases}$$

It plays an important role in integration theory because, as explained below, the Peano-Jordan measure of A may be expressed as the Rieman integral of χ_A.

5.2 Double Integrals

Throughout this section $n = 2$. Most of what holds in the case n = 2 also holds, *mutatis mutandis*, in higher dimension, but it is easiest to formulate things in the lowest possible dimension. For simplicity, the wording "Peano-Jordan measurable" will be simplified to "measurable".

Definition 5.4 (*Integral of a step function*) A *step function* on $Q = [a, b] \times [c, d]$ is a function $s : Q \to \mathbb{R}$ of the form

$$s(x, y) = \sum_{j=1}^{m} c_j \chi_j(x, y)$$

where $c_j \in \mathbb{R}$, $\{R_1, \ldots, R_m\}$ is an almost disjoint family of rectangles contained in Q and where $\chi_j = \chi_{R_j}$, for short. The set of all step funcions on Q will be denoted by $S(Q)$. The *double integral* of $s \in S(Q)$ is the real number

$$\iint_Q s = \sum_{j=1}^{m} c_j |R_j|. \tag{5.1}$$

Step functions can be used to approximate bounded functions.

Proposition 5.2 *Fix a rectangle $Q = [a, b] \times [c, d]$.*

(i) *If $s, t \in S(Q)$ are such that $s(x, y) \leq t(x, y)$ for all $(x, y) \in Q$, then $\iint_Q s \leq \iint_Q t$.*
(ii) *If $f: Q \to \mathbb{R}$ is bounded, then there exist $s, t \in S(Q)$ such that for all $(x, y) \in Q$*
    $s(x, y) \leq f(x, y) \leq t(x, y).$

Definition 5.5 (*Riemann integral over rectangles*) Let $f: Q \to \mathbb{R}$ be a bounded function on the rectangle Q. The real numbers:

$$\overline{\iint_Q} f = \inf \left\{ \iint_Q t : t \in S(Q), t \geq f \text{ on } Q \right\}$$

$$\underline{\iint_Q} f = \sup \left\{ \iint_Q s : s \in S(Q), s \leq f \text{ on } Q \right\}$$

are called the upper and lower Riemann integrals of f, respectively. The function f is said to be *Riemann integrable* on Q if its lower and upper integrals coincide, and in this case their common value

$$\iint_Q f := \overline{\iint_Q} f = \underline{\iint_Q} f$$

is called the *Riemann double integral* of f on Q. The set of all Riemann integrable functions on Q is denoted by $R(Q)$.

Not all bounded functions are integrable. For example, χ_D, the characteristic function of t the Dirichlet set $D = \{(x, y) \in [0, 1] \times [0, 1] : x, y \in \mathbb{Q}\}$, has upper integral equal to 1 and lower integral equal to 0 and is therefore not integrable.

Proposition 5.3 *Let Q be a rectangle. Then $f \in R(Q)$ if and only if for every $\varepsilon > 0$ there exist $s_\varepsilon, t_\varepsilon \in S(Q)$ such that $s_\varepsilon \leq f \leq t_\varepsilon$ on Q and $\iint_Q t_\varepsilon - \iint_Q s_\varepsilon < \varepsilon$.*

It is worth observing that of course step functions are integrable in the sense of Definition 5.5 and that their integral is precisely given by (5.1). Furthermore, it should be obvious that for any rectangle $R \subset Q$ it is

$$|R| = \iint_Q \chi_R$$

This fact inspires the natural extension of Riemann integral to sets A that are more general than rectangles. The right choice is that of measurable sets, because one of

5.2 Double Integrals

the reasonable requirements is that the constant function $x \to 1$ be integrable on A, and its integral should then give the measure of A.

Definition 5.6 (*Riemann integral over measurable sets*) Let Ω be a bounded measurable set and let $f: \Omega \to \mathbb{R}$ be a bounded function on Ω. If Q is any rectangle containing Ω,[1] put.

$$\tilde{f}(x) = \begin{cases} f(x) & x \in \Omega \\ 0 & x \in Q \setminus \Omega, \end{cases}$$

the so-called extension by zero of f to Q. The funcion f is said to be Riemann integrable on Ω if $\tilde{f}$ is Riemann integrable on Q. In such case, the real number

$$\iint_\Omega f := \iint_Q \tilde{f}$$

is called the Riemann integral of f on Ω, and is independent of the choice of Q. The set of all integrable functions on the set Ω is denoted by $R(\Omega)$.

Proposition 5.4 *Let $\Omega \subset \mathbb{R}^2$ be a bounded measurable set and let $f, g \in R(\Omega)$, $\alpha, \beta \in \mathbb{R}$. Then*

(i) $\alpha f + \beta g \in R(\Omega)$ and $\iint_\Omega (\alpha f + \beta g) = \alpha \iint_\Omega f + \beta \iint_\Omega g$;
(ii) if $f \leq g$ then $\iint_\Omega f \leq \iint_\Omega g$;
(iii) $|f| \in R(\Omega)$ and $\left|\iint_\Omega f\right| \leq \iint_\Omega |f|$;
(iv) if $|\Omega| = 0$, then $\iint_\Omega f = 0$;
(v) if $\Omega_1, \Omega_2 \subseteq \Omega$ are measurable and $|\Omega_1 \cap \Omega_2| = 0$, then $f \in R(\Omega_1) \cap R(\Omega_2)$ and

$$\iint_{\Omega_1 \cup \Omega_2} f = \iint_{\Omega_1} f + \iint_{\Omega_2} f.$$

It is important to observe that from (iv) and (v) above it follows that if the region of interest is of the form $\Omega_1 \cup \Omega_2$ with both Ω_1 and Ω_2 measurable, then

$$|\Omega_2| = 0 \quad \Rightarrow \quad \iint_{\Omega_1 \cup \Omega_2} f = \iint_{\Omega_1} f. \tag{5.2}$$

This can be spelled by saying that measurable sets of measure zero are negligible in the calculation of integrals.

Theorem 5.2 (*Integrability of continuous functions*) *Let $A \subset \mathbb{R}^2$ be a bounded measurable set and let $f: A \to \mathbb{R}$ be a bounded function on A. If f is continuous on $A \setminus \Omega$ where Ω is a measurable subset of A with $|\Omega| = 0$, then f is integrable on A.*

---

[1] Such rectangles do exist because Ω is bounded.

It is of course of interest to obtain explicit formulae for the calculation of double integrals of, say, continuous functions defined over reasonbable measurable sets. A very useful class of domains for which practical procedures can indeed be implemented is the class of sets that are obtained as unions of segments parallel to eachother and orthogonal to one of the axes, provided that their "profile" is described by graphs of (piecewise) continuous functions. This is the content of the next definition.

Definition 5.7 (*Normal domains*) If g_1, $g_2 \colon [a, b] \to \mathbb{R}$ are continuous functions, then the set.

$$\Omega = \{(x, y) \in \mathbb{R}^2 : a \leq x \leq b, g_1(x) \leq y \leq g_2(x)\} \tag{5.3}$$

is called a *domain normal to the x-axis*. Similarly, if h_1, $h_2 : [c, d] \to \mathbb{R}$ are continuous functions, then the set

$$B = \{(x, y) \in \mathbb{R}^2 : c \leq y \leq d, h_1(y) \leq x \leq h_2(y)\} \tag{5.4}$$

is called a *domain normal to the y-axis*.

Below in Fig. 5.1 two normal domains are depicted. The primary property of a normal domain Ω in integration theory is that the integral of any $f \in R(\Omega)$ can be reduced to a computation that involves two integrals of a single variable, as described next. The outcoming formulae are called *reduction formulae*.

Theorem 5.3 (*Reduction formulae*) *Any normal domain is measurable. Furthermore,*

(1) *if Ω is a domain normal to the x-axis as in (5.3) and $f \in R(\Omega)$, then*

$$\iint_\Omega f = \int_a^b \left(\int_{g_1(x)}^{g_2(x)} f(x, y) \mathrm{d}y \right) \mathrm{d}x$$

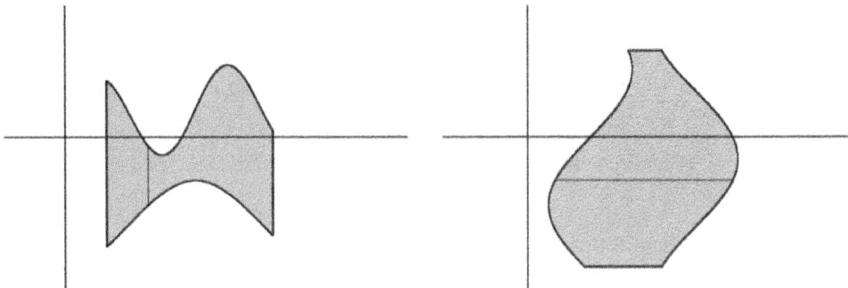

Fig. 5.1 Two normal domains; the domain on the left is normal to the *x*-axis, the domain on the right is normal to the *y*-axis. In each picture it is depicted one of the parallel segments of which the domain is the union, all of which are normal to the appropriate axis

5.2 Double Integrals

(2) *if B is a domain normal to the y-axis as in (5.4) and $f \in R(B)$, then*

$$\iint_B f = \int_c^d \left(\int_{h_1(y)}^{h_2(y)} f(x,y) dx \right) dy$$

In integration theory, and in particular when computing a double integral $\iint_\Omega f$, the geometry or the specific symmetries of Ω play an important role and can be at times best exploited by using appropriate coordinates. For example, the circular ring

$$\Omega = \{(x,y) \in \mathbb{R}^2 : r^2 \leq x^2 + y^2 \leq R^2\} \tag{5.5}$$

can be described in the polar coordinates $F : [0, 2\pi) \times (0, +\infty) \to \mathbb{R}^2 \setminus \{(0,0)\}$ given, as in (1.18), by

$$\begin{cases} x = \rho \cos \theta \\ y = \rho \sin \theta \end{cases}$$

in the simpler form

$$F^{-1}(\Omega) = \{(\theta, \rho) \in [0, 2\pi) \times (0, +\infty) : r \leq \rho \leq R\},$$

which is a rectangle in the strip $[0, 2\pi) \times (0, +\infty)$, see Fig. 5.2. Sometimes using the appropriate coordinates allows for much faster computations. For this reason, it is customary to ephasize the role of coordinates by writing

$$\iint_\Omega f(x,y) dx dy$$

instead of $\iint_\Omega f$, which may also be interpreted naturally when Ω is a normal domain, see Theorem 5.3. In this sense, the question arises how to write, for example, the integral of a function defined on a circular ring in the plane in terms of the variables ρ and θ. This is a particular instance of a general situation, as formalized next in the general context of $\mathbb{R}^n$.

Definition 5.8 Suppose that $\Omega, \Omega' \subset \mathbb{R}^n$ are open sets. A map $\Phi : \Omega' \to \Omega$ is called *a regular coordinate transformation* if it satisfies the following properties:

(i) Φ is bijective;
(ii) $\Phi \in C^i(\Omega')$;
(i) Φ is regular in the sense that the Jacobian matrix $J\Phi(P)$ is non singular at every point $P \in \Omega'$.

Recall that when $d = 2$ the Jacobian matrix $J\Phi$ of the map $\Phi : \Omega' \to \Omega$ with components (f, g), that is, of the mapping $\Phi(x,y) = (f(x,y), g(x,y))$ is given by

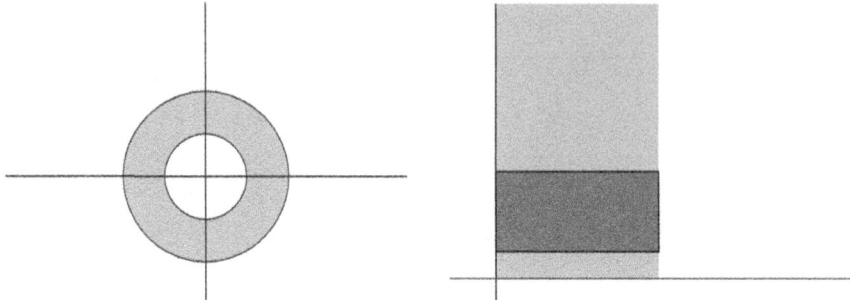

Fig. 5.2 A ring in the plane of Cartesian coordinates becomes a rectangle in polar coordinates

$$J\Phi(x, y) = \begin{bmatrix} \frac{\partial f}{\partial x}(x, y) & \frac{\partial f}{\partial y}(x, y) \\ \frac{\partial g}{\partial x}(x, y) & \frac{\partial g}{\partial y}(x, y) \end{bmatrix}.$$

In the special case of polar coordinates, the Jacobian is computed as follows. Recall from (1.18) that the map

$$\Phi : [0, 2\pi) \times (0, +\infty) \to \mathbb{R}^2, \quad (\theta, \rho) \mapsto (\rho \cos \theta, \rho \sin \theta),$$

is a bijection of the strip $[0, 2\pi) \times (0, +\infty)$ onto the punctured plane $\mathbb{R}^2 \setminus \{(0, 0)\}$, which restricts to a bijection of the open strip $\Omega' = (0, 2\pi) \times (0, +\infty)$ onto the open slit plane $\Omega = \mathbb{R}^2 \setminus \{(x, 0) : x \geq 0\}$. The Jacobian matrix is

$$J\Phi(\theta, \rho) = \begin{bmatrix} \frac{\partial(\rho \cos \theta)}{\partial \theta} & \frac{\partial(\rho \cos \theta)}{\partial \rho} \\ \frac{\partial(\rho \sin \theta)}{\partial \theta} & \frac{\partial(\rho \sin \theta)}{\partial \rho} \end{bmatrix} (\theta, \rho) = \begin{bmatrix} -\rho \sin \theta & \cos \theta \\ \rho \cos \theta & \sin \theta \end{bmatrix}.$$

Observe that for every $(\theta, \rho) \in \Omega^t$

$$\det(J\Phi(\theta, \rho)) = -\rho \neq 0$$

so that the polar coordinates map defines a regular coordinate transformation on the aforementioned open sets.

By means of regular coordinate transformations it is possible to compute double integrals using coordinates that are not the usual Cartesian coordinates, namely it is possible to "change the variables" as in the case of functions of one variable. This is often suggested by the symmetries of the problem at hand.

Theorem 5.4 (*Change of variables for double integrals*) *Suppose that Ω, Ω' are open subsets of $\mathbb{R}^2$ and that $\Phi : \Omega' \to \Omega$ is a regular coordinate transformation. If $A \subset \Omega$ is a measurable subset, then $\Phi^{-1}(A)$ is also measurable and for every continuous function $f : A \to \mathbb{R}$ it holds that.*

$$\iint_A f(x, y) dx dy = \iint_{\Phi^{-1}(A)} f(\Phi(u, v)) |\det J\Phi(u, v)| du dv \qquad (5.6)$$

Formula (5.6) in the case of polar coordinates works as follows. If $f : A \subset \mathbb{R}^2 \to \mathbb{R}$ is a continuous function and if $F(\theta, \rho) = f(\rho \cos\theta, \rho \sin\theta)$, then

$$\iint_A f(x, y) dx dy = \iint_{\Phi^{-1}(A)} F(\theta, \rho) \rho \, d\rho \, d\theta$$

Keep in mind (5.2), that is, that sets of measure zero are negligible in the computation of integrals. Hence, the change of variables formula written above holds for all measurable subsets A of $\mathbb{R}^2$ and it is not necessary to assume that A is contained in the slit plane $\Omega = \mathbb{R}^2 \setminus \{(x, 0) : x \geq 0\}$. In the very specific case of the circular ring described above in (5.5), and using the reduction formulae, this becomes

$$\iint_A f(x, y) dx dy = \int_0^{2\pi} \left(\int_r^R F(\theta, \rho) \rho \, d\rho \right) d\theta = \int_r^R \left(\int_0^{2\pi} F(\theta, \rho) d\theta \right) \rho \, d\rho$$

because the ring is a rectangle in the (θ, ρ)-plane, hence a domain that is normal to both axes, so that both reduction formulae are valid.

5.3 Triple Integrals

Much of what has been established for double integrals holds with little or no change for integrals of (scalar) functions over subsets of $\mathbb{R}^3$, that are named *triple integrals* for manifest reasons. In particular, the notions of step function on sets of the form $Q = [a, b] \times [c, d] \times [e, f]$ and of Riemann integral of a function defined over such a region are completely analogous to the two-dimensional case, as well as the notion of Riemann integral over a bounded (Peano-Jordan) measurable set. The explicit wording is therefore left as a useful exercise. As far as notation is concerned, the triple integral over the measurable set $\Omega \subset \mathbb{R}^3$ of $f \in \mathcal{R}(\Omega)$ will be written with either of the following symbols

$$\iiint_\Omega f, \quad \iiint_\Omega f(x, y, z) dx dy dz.$$

The general properties of triple integrals, i.e. the content of Proposition 5.4, holds *verbatim* for $\Omega \subset \mathbb{R}^3$ without any change at all, and of course Theorem 5.2 continues to hold with the same statement provided that $\Omega \subset \mathbb{R}^3$. It also goes without saying that for $d > 3$ there are no additional difficulties.

Things look technically different when it comes to reduction formulae. This is primarily due to the fact that for three-dimensional sets there are two ways to approach

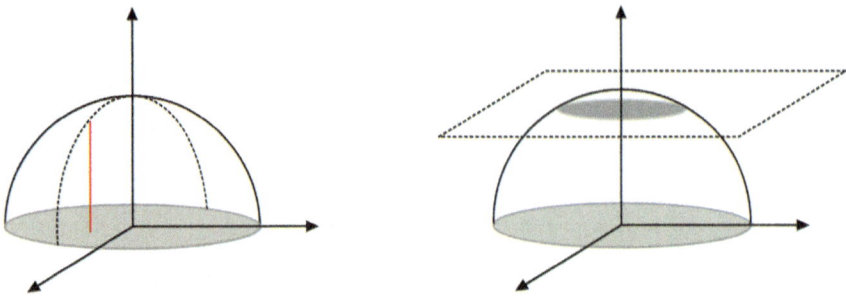

Fig. 5.3 Segments (left) and sections (right) filling the same solid region

the iterative argument at the heart of reduction formulae: either one computes first a 1D integral and then a 2D integral, or viceversa. The resulting 2D integrals can then be treated using the techniques introduced in the previous section, but the first step requires a comment.

To make the point clearer, suppose that the region of interest Ω is the closed solid 3D region above the plane $z = 0$ and below the unit sphere $x^2 + y^2 + z^2 = 1$. One may think of Ω in two ways, either as union of parallel vertical segments or as union of horizontal sections. More precisely (Fig. 5.3):

- either as the union of vertical segments parametrized by the unit disk $D \subset \mathbb{R}^2$

$$\Omega = \bigcup_{(x,y) \in D} PQ(x, y),$$

where for fixed $(x_0, y_0) \in D$ the first endpoint of the vertical segment $PQ(x_0, y_0)$ is $P = (x_0, y_0, 0)$ and the second, $Q = (x_0, y_0, z)$, is on the surface of the unit sphere in $\mathbb{R}^3$, that is, $PQ(x_0, y_0) = \left\{ (x_0, y_0, z) \in \mathbb{R}^3 : 0 \le z \le \sqrt{1 - x_0^2 - y_0^2} \right\}$,

- or as the union of horizontal sections parametrized by the segment $[0, 1] \subset \mathbb{R}$

$$\Omega = \bigcup_{0 \le z \le 1} \Omega_z$$

where for fixed $z_0 \in [0, 1]$ the section $\Omega_{z_0} = \left\{ (x, y, z_0) : x^2 + y^2 + z_0^2 \le 1 \right\}$ is the intersection of Ω with the plane Π_{z_0} parallel to the xy-plane at height z_0.

The reduction formula in the first case consists in integrating along each segment and then integrating the resulting function of two variables on D, whereas in the second case one integrates first on each section and then the resulting function of one variable on $[0, 1]$.

It is important to observe that the region Ω treated in the above example admits both descriptions, but it is very easy to find domains in $\mathbb{R}^3$ that can be desecribed in one way but not the other, or viceversa.

5.3 Triple Integrals

Below is a general version of the reduction formulae just discussed.

Theorem 5.5 (Integration by segments or by sections) *Suppose that $\Omega \subset \mathbb{R}^3$ is measurable and that $f: \Omega \to \mathbb{R}$ is continuous.*

(1) **(Integration by segments).** *If there exist a bounded measurable set $D \subset \mathbb{R}^2$ and two continuous funcions $\varphi, \psi : D \to \mathbb{R}$ such that*

$$\Omega = \{(x, y, z) \in \mathbb{R}^3 : (x, y) \in D, \phi(x, y) \leq z \leq \psi(x, y)\}$$

Then

$$\iiint_\Omega f = \iint_D \left(\int_{\phi(x,y)}^{\psi(x,y)} f(x, y, z) dz \right) dxdy$$

(2) **(Integration by sections).** *If there exist an interval $[a, b] \subset \mathbb{R}$ and a family of bounded measurable sets $\{S_z \subset \mathbb{R}^2 : z \in [a, b]\}$ such that*

$$\Omega = \{(x, y, z) \in \mathbb{R}^3 : z \in [a, b], (x, y) \in S_z\},$$

then

$$\iiint_\Omega f = \int_a^b \left(\iint_{S_z} f(x, y, z) dxdy \right) dz.$$

Changes of variables for triple integrals work just like those for double integrals, in the sense that Theorem 5.4 holds with the analogous statement, whereby evidently Ω, Ω' are open subsets of $\mathbb{R}^3$, $\Phi : \Omega' \to \Omega$ is a regular coordinate transformation and $f: \Omega \to \mathbb{R}$ is continuous function where $\tilde{\Omega} \subset \Omega$ is a measurable set. The change of variables formula (5.6) becomes then

$$\iiint_{\tilde{\Omega}} f(x, y, z) dxdydz = \iiint_{\Phi^{-1}(\tilde{\Omega})} f(\Phi(u, v, w)) |\det J\Phi(u, v, w)| dudvdw \quad (5.7)$$

An important special case is that of spherical coordinates. Recall from (1.20) that the spherical coordinate map is $\Phi : [0, \pi] \times [0, 2\pi) \times (0, +\infty) \to \mathbb{R}^3$, sending the triple (θ, ϕ, ρ) to

$$\begin{cases} x = \rho \sin \theta \cos \varphi \\ y = \rho \sin \theta \sin \varphi \\ z = \rho \cos \theta, \end{cases}$$

a bijection of the cylinder $(0, \pi) \times [0, 2\pi) \times (0, +\infty)$ onto the slit space $\mathbb{R}^3 \setminus \{z\text{-axis}\}$. The Jacobian matrix turns out to be non singular for $\theta \notin \{0, \pi\}$, hence for all values (θ, ϕ, ρ) in the cylinder $(0, \pi) \times [0, 2\pi) \times (0, +\infty)$. Indeed, it is easy to see that

$$|\det J\Phi(\theta,\varphi,\rho)| = \rho^2 \sin\theta$$

Therefore, if $F(\theta,\varphi,\rho) = f(\rho\sin\theta\cos\varphi, \rho\sin\theta\sin\varphi, \rho\cos\theta)$, then

$$\iiint_\Omega f(x,y,z)dxdydz = \iiint_{\Phi^{-1}(\Omega)} F(\theta,\varphi,\rho)\rho^2 \sin\theta d\theta d\varphi d\rho$$

An other useful example is that of the cylindrical coordinates

$$\begin{cases} x = x_0 + \rho\cos\theta \\ y = y_0 + \rho\sin\theta \\ z = z_0 + z \end{cases}$$

whereby the relevant map is $\Phi : [0, 2\pi) \times (0, +\infty) \times \mathbb{R} \to \mathbb{R}^3$ that gives a bijection of $[0, 2\pi) \times (0, +\infty) \times \mathbb{R}$ onto $\mathbb{R}^3 \setminus \{z\text{ - axis }\}$. It is left as an exercise to check that in this case the absolute value of the Jacobian determinant is

$$|\det J\Phi(\theta,\rho,z)| = \rho.$$

5.4 Barycenters and Moments of Inertia

The notions of *barycenter*, or *center of mass*, and of *moment of inertia* are central in Mechanics and in Engineering. For this reason they are referred to with their physical terminology although they are mathematical objects that find their natural formulations in the context of integration theory and actually occur as particular instances of multiple integrals, with no need of additional mathematical concepts. It is natural to define them simultaneously in the cases n = 2 and $n = 3$, but it would be just as easy to treat the general case $\mathbb{R}^n$ for any $n \geq 2$.

Thus, hereafter n = 2 or n = 3, $\Omega \subset \mathbb{R}^n$ is a (Peano-Jordan) measurable set and $\mu \in R(\Omega)$ is a *density function*, that is, a non negative integrable function which is assumed to be strictly positive at least on an open subset of Ω. The *mass* of Ω relative to the density μ is the positive number

$$m(\Omega) = \begin{cases} \iint_\Omega \mu(x,y)dxdy & n = 2 \\ \iiint_\Omega \mu(x,y,z)dxdydz & n = 3. \end{cases}$$

The barycenter, or center of mass, of Ω relative to the density μ is defined:

- if $n = 2$ as $B = (x_B, y_B) \in \mathbb{R}^2$, where

5.4 Barycenters and Moments of Inertia

$$x_B = \frac{1}{m(\Omega)} \iint_\Omega x\mu(x,y)dxdy,$$

$$y_B = \frac{1}{m(\Omega)} \iint_\Omega y\mu(x,y)dxdy;$$

- if $n=3$ as $B = (x_B, y_B, z_B) \in \mathbb{R}^3$, where

$$x_B = \frac{1}{m(\Omega)} \iiint_\Omega x\mu(x,y,z)dxdydz,$$

$$y_B = \frac{1}{m(\Omega)} \iiint_\Omega y\mu(x,y,z)dxdydz,$$

$$z_B = \frac{1}{m(\Omega)} \iiint_\Omega z\mu(x,y,z)dxdydz.$$

In Fig. 5.4 the barycenter of a planar triangular region with constant density is depicted.

A moment of inertia involves the relative position of the region Ω and a reference geometric object, usually a line or point. Assume first that l is a line in $\mathbb{R}^n$, $n = 2$ or $n = 3$, and denote by $d_l(P)$ the distance of $P \in \Omega$ from l, that is

$$d_\ell(P) = \min\{d(P,Q) : Q \in \ell\}.$$

It is useful to recall two standard cases. The first is when $n = 2$ and the line l is given in Cartesian form, namely

$$\ell = \{(x,y) \in \mathbb{R}^2 : ax + by + c = 0\}.$$

In this case the distance of $P = (x,y)$ from l is given by

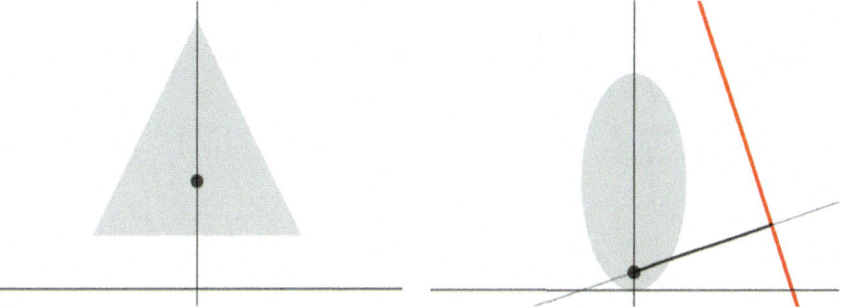

Fig. 5.4 On the left: the black dot is the barycenter of the grey triangular region; on the right the length of the thicker black segment is the distance of the black dot (one of the points of the elliptical grey region) from the red line

$$d_\ell(x, y) = \frac{|ax + by + c|}{\sqrt{a^2 + b^2}}.$$

In Fig. 5.4 on the right are depicted a point P and a line l: their distance $d_f(P)$ is the length of the segment lying on the perpendicular line to l with endpoints P and the intersection of the two lines.

The second, which is mostly useful when $n = 3$, is when the line is given in parametric form as

$$\ell = \{P_0 + tQ_0 : t \in \mathbb{R}\}$$

where $P_0, Q_0 \in \mathbb{R}^3$ are two given vectors. In this case the distance of P from l is given by

$$d_\ell(P) = \frac{\|(P - P_0) \wedge Q_0\|}{\|Q_0\|}$$

where $(P - P_0) \wedge Q_0$ is the vector product between $P - P_0$ and Q_0, which is a vector orthogonal to both. In the case where $P_0 = O$ is the origin, its expression in coordinates for the points $P = (x, y, z)$, and $Q_0 = (x_0, y_0, z_0)$, is

$$P \wedge Q_0 = (yz_0 - zy_0, zx_0 - xz_0, xy_0 - yx_0).$$

The *moment of inertia* of Ω with respect to f is defined:

- if $n = 2$ as the non negative number

$$I_\ell = \iint_\Omega d_\ell^2(x, y) \mu(x, y) dx dy \tag{5.8}$$

- if $n = 3$ as the non negative number

$$I_\ell = \iiint_\Omega d_\ell^2(x, y, z) \mu(x, y, z) dx dy dz \tag{5.9}$$

Of particular relevance are the moments of inertia with respect to the coordinate axes. More precisely,

- if $n = 2$, the quantities

$$I_x = \iint_\Omega y^2 \mu(x, y) dx dy, \quad I_y = \iint_\Omega x^2 \mu(x, y) dx dy$$

are the moments of inertia of Ω with respect to the x-axis and the y-axis, respectively.

- If $n = 3$, the quantities

5.4 Barycenters and Moments of Inertia

$$I_x = \iiint_\Omega (y^2 + z^2)\mu(x, y, z)dxdydz$$

$$I_x = \iiint_\Omega (y^2 + z^2)\mu(x, y, z)dxdydz$$

$$I_y = \iiint_\Omega (x^2 + z^2)\mu(x, y, z)dxdydz$$

are the moments of inertia of Ω with respect to the x-axis, the y-axis and the z-axis, respectively.

Finally, the moment of inertia of Ω with respect to the point Q_0 is given by the formulae (5.8) and (5.9) where the $d_l(P)$ is substituted by the distance $d(P, Q_0)$ of the point $P \in \Omega$ from Q_0, see formula (1.14). Observe that in the particular case of $Q_0 = O$, the moment of inertia with respect to the origin I_O can be computed from the moments of inertia with respect to the coordinate axes by

$$I_O = \begin{cases} I_x + I_y & n = 2 \\ \frac{1}{2}(I_x + I_y + I_z) & n = 3. \end{cases}$$

5.4.1 Solids of Revolution

Many interesting regions in $\mathbb{R}^3$ exhibit particular symmetries. One of the most relevant ones is when $\Omega \subset \mathbb{R}^3$ is a *solid of revolution*, that is, when it is obtained by rotating a planar region S, called the *meridian section* of Ω, around an axis contained in the plane where S lies. A prototypical situation is when S lies in the yz-plane and the axis of rotation is the z-axis. In the cylindrical coordinates $\Phi : (0, +\infty) \times [0, 2\pi) \times \mathbb{R} \to \mathbb{R}^3 \setminus \{z - axis\}$ axis given by $\Phi(\rho, \theta, z) = (\rho \cos\theta, \rho \sin\theta, z)$, the solid of revolution

$$\Omega = \left\{(x, y, z) \in \mathbb{R}^3 : \left(\sqrt{x^2 + y^2}, z\right) \in S\right\}$$

is given by

$$\Phi^{-1}(\Omega) = \{(\rho, \theta, z) : \theta \in [0, 2\pi), (\rho, z) \in S\}.$$

Therefore, if $f \in R(\Omega)$, then

$$\iiint_\Omega f = \int_0^{2\pi} \left(\iint_S f(\rho\cos\theta, \rho\sin\theta, z)\rho d\rho dz\right) d\theta.$$

In particular, upon taking $f \equiv 1$, the function identically 1 on Ω,

Fig. 5.5 In grey the region S contained in the yz-plane and the solid of revolution Ω obtained by rotating S around the z-axis; the dot is the barycenter of S, with horizontal coordinate y_B

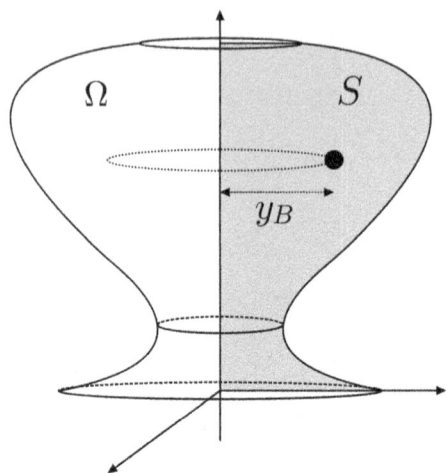

$$|\Omega| = \int_0^{2\pi} \left(\iint_S \rho d\rho dz \right) d\theta = 2\pi \left(\frac{1}{|S|} \iint_S y dy dz \right) |S| = 2\pi y_B |S|,$$

whereby $B = (y_B, z_B)$ is the barycenter of S in the yz-plane (Fig. 5.5).

The above formula is actually very old and is also known as Pappus'theorem.

Theorem 5.6 (Pappus' theorem, or Guldino's first theorem) *The volume of a solid of revolution is the product of the area of its meridian section times the length of the circumference traced by its barycenter.*

The above theorem takes an easy form when the meridian section is normal to the z-axiz, or if it is given as the hypograph of a function $\phi : [a, b] \to \mathbb{R}$, of z, namely

$$S = \{(y, z) : a \le z \le b, 0 \le y \le \varphi(z)\}.$$

Then clearly

$$|\Omega| = \pi \int_a^b \varphi^2(z) dz.$$

5.5 Guided Exercises

5.1 Compute $\iint_R f$ where $R = [0, 1] \times [2, 4]$ and $f(x, y) = \log(1 + x + 2y)$.

Answer. The rectangle R is inside the domain $\{(x, y) \in \mathbb{R}^2 : y > -(1+x)/2\}$, of the function f, where f is continuous and hence integrable, see Fig. 5.6.

Fig. 5.6 The domain of $f(x, y) = \log(1 + x + 2y)$ contains the rectangle $R = [0, 1] \times [2, 4]$

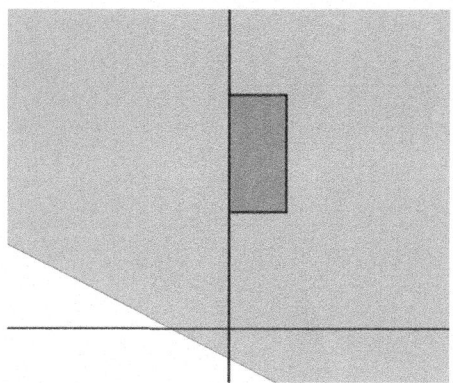

Using the reduction formulae, and recalling that

$$\int \log t \, dt = t \log t - t + c$$

the required integral becomes

$$\iint_R f = \int_2^4 \left(\int_0^1 \log(1 + x + 2y) dx \right) dy$$
$$= \int_2^4 \left[(1 + x + 2y) \log(1 + x + 2y) - x \right]_0^1 dy$$
$$= \int_2^4 ((2 + 2y) \log(2 + 2y) - (1 + 2y) \log(1 + 2y) - 1) dy.$$

A simple integration by parts reveals that

$$\int t \log t \, dt = \frac{1}{4} t^2 (2 \log t - 1) + c.$$

It follows that

$$\iint_R f = \int_2^4 (2 + 2y) \log(2 + 2y) dy - \int_2^4 (1 + 2y) \log(1 + 2y) dy - 2$$
$$= \frac{1}{2} \int_6^{10} t \log t \, dt - \frac{1}{2} \int_5^9 s \log s \, ds - 2$$
$$= \frac{125}{4} \log 5 + 16 \log 2 - \frac{99}{2} \log 3 - 3.$$

This is a straightforward application of the reduction formula for a double integral of a continuous function over a rectangle, plus a reminder of basic one-variable integration techniques.

5.2 Compute $\iint_R f$ where $R = [1, 2] \times [1, 3]$ and $f(x, y) = x^3 e^{yx^2}$.

Answer. Using the reduction formulae

$$\iint_R f = \int_1^2 \left(\int_1^3 x^3 e^{yx^2} dy \right) dx$$
$$= \int_1^2 x \left(\int_1^3 x^2 e^{yx^2} dy \right) dx$$
$$= \int_1^2 x \left[e^{yx^2} \right]_1^3 dx$$
$$= \int_1^2 x \left(e^{3x^2} - e^{x^2} \right) dx.$$

From the formula

$$\int x e^{ax^2} dx = \frac{1}{2a} e^{ax^2} + c$$

valid for any $a \neq 0$, the conclusion is that

$$\iint_R f = \frac{1}{6}(e^{12} - e^3) - \frac{1}{2}(e^4 - e).$$

This seemingly innocent integral hides a trap: if one attempts to use the reduction formula in the reverse order, the integration with respect to x is doable, but yields a very hard integral with respect to y. Indeed,

$$\int_1^2 x^3 e^{yx^2} dx = e^{yx^2} \left(\frac{x^2}{2y} - \frac{1}{2y^2} \right) \Big|_1^2$$

and the resulting function of y has no elementary primitive.

5.3 Compute $\iint_R f$ where $R = [0, 1] \times [1, 2]$ and $f(x, y) = \dfrac{x}{\sqrt{x^2 + y^2}}$.

Answer. The square R has positive distance from the origin, hence it is contained in the set where f is continuous, hence integrable. By reduction,

$$\iint_R f = \int_1^2 \left(\int_0^1 \frac{x}{\sqrt{x^2 + y^2}} dx \right) dy = \int_1^2 \left[\sqrt{x^2 + y^2} \right]_0^1 dy = \int_1^2 \left(\sqrt{1 + y^2} - y \right) dy.$$

5.5 Guided Exercises

The integral $\int \sqrt{1+y^2}\,dy$ may be computed classically in two different ways. The first way is to put

$$\sqrt{1+y^2} = t - y$$

which yields the rational integral

$$\frac{1}{4}\int \frac{(t^2+1)^2}{t^3}\,dt = \frac{t^4-1}{8t^2} + \frac{1}{2}\log t + c$$

that in the end produces

$$\int \sqrt{1+y^2}\,dy = \frac{1}{2}y\sqrt{1+y^2} + \frac{1}{2}\log\left(y+\sqrt{1+y^2}\right) + c.$$

The second way is to put

$$y = \sinh t$$

which yields the integral

$$\int \cosh^2 t\,dt = \frac{1}{2}(t + \sinh t \cosh t) + c.$$

Now, the inverse settsinh(y) of sinh t has the expression

$$\text{settsinh}(y) = \log\left(y + \sqrt{1+y^2}\right)$$

so that one recovers the formula obtained by the first method. The conlcusion of the exercise is therefore

$$\iint_R f = \left[\frac{1}{2}y\sqrt{1+y^2} + \frac{1}{2}\log\left(y+\sqrt{1+y^2}\right) - \frac{1}{2}y^2\right]_1^2$$

$$= \sqrt{5} - \frac{\sqrt{2}}{2} + \frac{1}{2}\log\left(\frac{2+\sqrt{5}}{1+\sqrt{2}}\right) - \frac{3}{2}.$$

This exercise presents no difficulties in terms of understanding the domain of integration, and is meant to recall how to integrate a classical function, and the basics on hyperbolic functions.

5.4 Compute $\iint_\Omega f$ where $f(x,y) = \dfrac{x}{x^2+y^2}$ and where

$$\Omega = \left\{ (x,y) \in \mathbb{R}^2 : \frac{x^2}{2} \le y \le x^2, 1 \le x \le 2 \right\}.$$

Answer. The domain of integration is normal with respect to the *x*-axis (see Fig. 5.8), so that the reduction formula yields (Fig. 5.7)

$$\iint_\Omega f = \int_1^2 \left(\int_{x^2/2}^{x^2} \frac{x}{x^2+y^2} dy \right) dx = \int_1^2 \left(\int_{x^2/2}^{x^2} \frac{x}{x^2(1+(y/x)^2)} dy \right) dx.$$

With the substitution $y/x = t$ the inner integral is easily computed to be

$$\int_{x^2/2}^{x^2} \frac{1}{(1+(y/x)^2)} \frac{dy}{x} = \arctan x - \arctan(x/2).$$

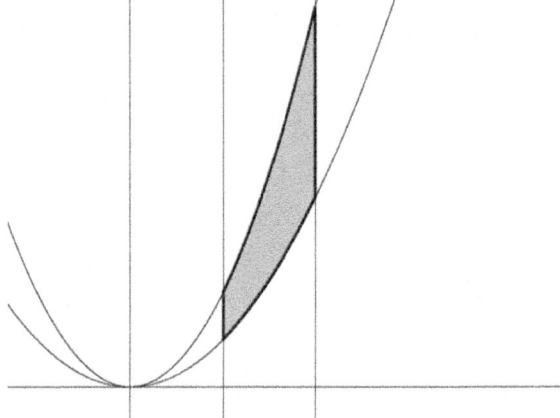

Fig. 5.7 The domain Ω in Exercise 5.4 in grey, a prototypical domain normal to the *x*-axis

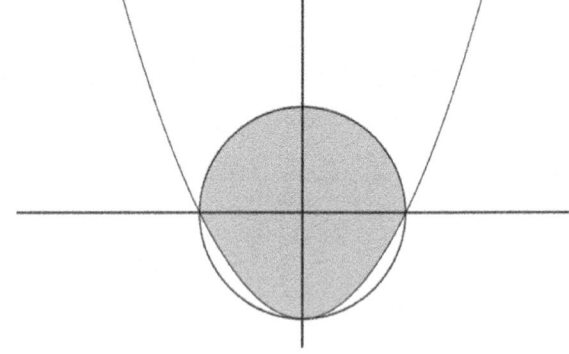

Fig. 5.8 The domain Ω in Exercise 5.5 is normal w.r.t. both axes, but the description is easier w.r.t. the *x*-axis: below a circle and above a parabola

5.5 Guided Exercises

It is therefore useful to recall that a simple integration by parts gives

$$\int \arctan x \, dx = x \arctan x - \frac{1}{2}\log(1+x^2) + c$$

which produces

$$\iint_\Omega f = \int_1^2 (\arctan x - \arctan(x/2)) dx$$

$$= \left[x(\arctan x - \arctan(x/2)) + \frac{1}{2}\left(2\log\left(1+\frac{x^2}{4}\right) - \log(1+x^2)\right) \right]_1^2$$

$$= -\frac{1}{4}\pi + \arctan 2 + \frac{7}{2}\log 2 - \frac{3}{2}\log 5.$$

In this exercise the integration region Ω is a prototypical normal domain. Once the reduction formulae are applied, it becomes a standard exercise in integration.

5.5 Compute $\iint_\Omega f$ where $f(x,y) = xy$ and where

$$\Omega = \left\{ (x,y) \in \mathbb{R}^2 : x^2 - 1 \leq y \leq \sqrt{1-x^2}, -1 \leq x \leq 1 \right\}.$$

Answer. Clearly, Ω is normal with respect to the x-axis, as drawn in Fig. 5.8 and

$$\iint_\Omega f = \int_{-1}^1 x \left[\int_{x^2-1}^{\sqrt{1-x^2}} y \, dy \right] dx$$

$$= \frac{1}{2} \int_{-1}^1 x [y^2]_{x^2-1}^{\sqrt{1-x^2}} dx$$

$$= \frac{1}{2} \int_{-1}^1 \left[x(1-x^2) - x(x^2-1)^2 \right] dx$$

$$= \frac{1}{2} \int_{-1}^1 (x^3 - x^5) dx$$

$$= 0.$$

This is a very simple exercise which is meant to show how symmetries can be useful: the domain of integration is actually normal with respect to both axes but when seen as normal with respect to the x-axis it is symmetric, and the function to be integrated with respect to x turns out to be odd. This fact can be seen from the very first line of the calculation, which can be written in the form

$$\iint_\Omega f = \int_{-1}^1 x \psi(x^2) dx.$$

Fig. 5.9 The domain Ω in Exercise 5.6

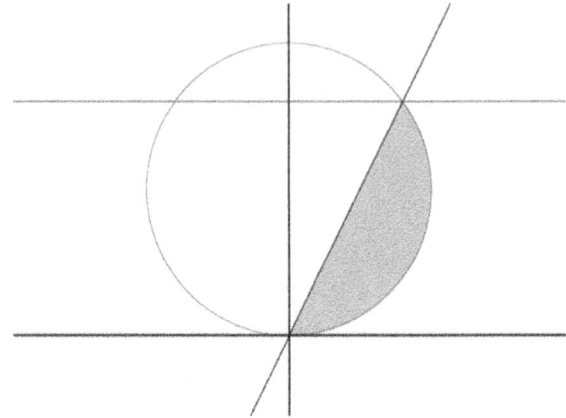

5.6 Compute $\iint_\Omega f$ where $f(x, y) = \dfrac{x \sin y}{y}$ and where

$$\Omega = \{(x, y) \in \mathbb{R}^2 : 0 \le y \le 2x,\, x^2 + (y-1)^2 \le 1\}.$$

Answer. The domain Ω is clearly normal to the y-axis, as illustrated in Fig. 5.9. The intersection points of the two loci given by the equations $y = 2x$ and $x^2 + (y-1)^2 = 1$ are found by solving

$$x^2 + (2x - 1)^2 = 1 \iff x^2 + 4x^2 - 4x + 1 - 1 = 0 \iff x(5x - 4) = 0.$$

Hence the two points are $(0, 0)$ and $(4/5, 8/5)$. Furthermore, the portion of the circle $x^2 + (y-1)^2 = 1$ where $0 \le x \le 4/5$ defines x as a funcion of y, and precisely

$$x^2 + (y-1)^2 = 1 \iff 2y - y^2 = x^2 \iff x = \sqrt{2y - y^2}.$$

Therefore

$$\Omega = \left\{(x, y) \in \mathbb{R}^2 : 0 \le y \le \frac{8}{5},\, \frac{y}{2} \le x \le \sqrt{2y - y^2}\right\}.$$

It follows that

$$\iint_\Omega f = \int_0^{8/5} \left[\int_{y/2}^{\sqrt{2y-y^2}} x\,dx\right] \frac{\sin y}{y} dy$$

$$= \int_0^{8/5} \frac{1}{2}\left[2y - y^2 - \frac{y^2}{4}\right] \frac{\sin y}{y} dy$$

5.5 Guided Exercises

$$= \frac{1}{8} \int_0^{8/5} (8-5y)\sin y\, dy$$

$$= [-\cos y]_0^{8/5} - \frac{5}{8} \int_0^{8/5} y\sin y\, dy.$$

An integration by parts gives

$$\int_0^{8/5} y\sin y\, dy = [-y\cos y + \sin y]_0^{8/5} = -\frac{8}{5}\cos(8/5) + \sin(8/5)$$

and so

$$\iint_\Omega f = 1 - \frac{5}{8}\sin(8/5).$$

The domain Ω of integration is best seen as a domain normal to the y-axis. The reduction formulae then give rise to a very simple iterated integral.

5.7 Compute the area of the region E described by the inequality $\dfrac{x^2}{a^2} + \dfrac{y^2}{b^2} \le 1$, where a,b > 0.

Answer. The region is clearly an elliptic disc (see Fig. 5.10). A suitable change of coordinates is given by

$$\begin{cases} x = au \\ y = bv \end{cases}$$

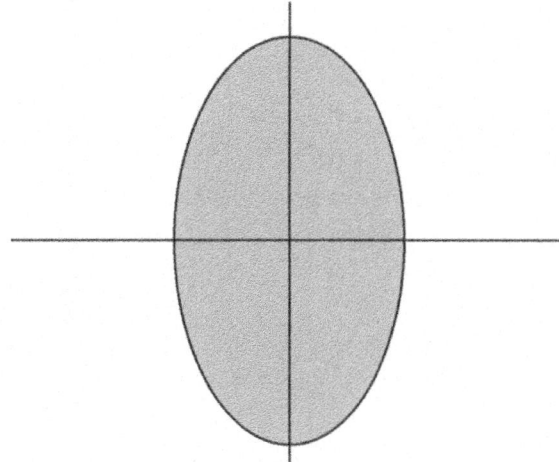

Fig. 5.10 An elliptic disc like that of Exercise 5.7; here $a < b$

Indeed, the map $\Phi : \mathbb{R}^2 \to \mathbb{R}^2$ given by $\Phi(u, v) = (au, bv)$ is obviously bijective and regular, with Jacobian matrix

$$J\Phi(x, y) = \begin{bmatrix} a & 0 \\ 0 & b \end{bmatrix},$$

which is manifestly nonsingular with determinant $ab > 0$. Further, it maps the unit disc $D = \{(u, v) \in \mathbb{R}^2 : u^2 + v^2 < 1\}$ onto the interior of E (and its boundary to the boundary of E). It follows that, by the change of variable formula

$$|E| = \iint_E dxdy = \iint_D abdudv = ab|D| = ab\pi.$$

This is an application of the change of variable formula. The ellipitic disc is just a deformation of the unit circular disc obtained by dilating or contracting each axis by the appropriate factor. This suggests the change of variable that has been chosen.

5.8 Compute the area of $\Omega = \{(x, y) \in \mathbb{R}^2 : x^2 \leq y \leq 2x^2, y^2 \leq x \leq 3y^2\} \setminus \{(0, 0)\}$.

Answer. A convenient change of coordinates is

$$\begin{cases} u = y/x^2 \\ v = x/y^2. \end{cases}$$

Indeed, in the coordinates (u, v) the region Ω corresponds to the region given by

$$\Omega' = \{(u, v) \in \mathbb{R}^2 : 1 \leq u \leq 2, 1 \leq v \leq 3\}.$$

The assignment $(x, y) \mapsto (y/x^2, x/y^2)$ defines a bijective map Ψ from the positive quadrant $\{x > 0, y > 0\}$ onto itself. Clearly, $\Omega' = \Psi(\Omega)$. Furthermore, the map Ψ has Jacobian matrix

$$J\Psi(x, y) = \begin{bmatrix} u_x & u_y \\ v_x & v_y \end{bmatrix} = \begin{bmatrix} -2y/x^3 & 1/x^2 \\ 1/y^2 & -2x/y^3 \end{bmatrix}$$

which is nonsingular, with determinant

$$\frac{4xy}{x^3y^3} - \frac{1}{x^2y^2} = \frac{3}{x^2y^2} = 3u^2v^2.$$

Now, in order to apply the change of variable formula, the relevant determinant is that of the inverse map, that is the mapping $\Psi^{-1} : \Omega' \to \Omega$. However,

$$\left|\det(J\Psi^{-1})\right| = \frac{1}{|\det J\Psi|} = \frac{1}{3u^2v^2}.$$

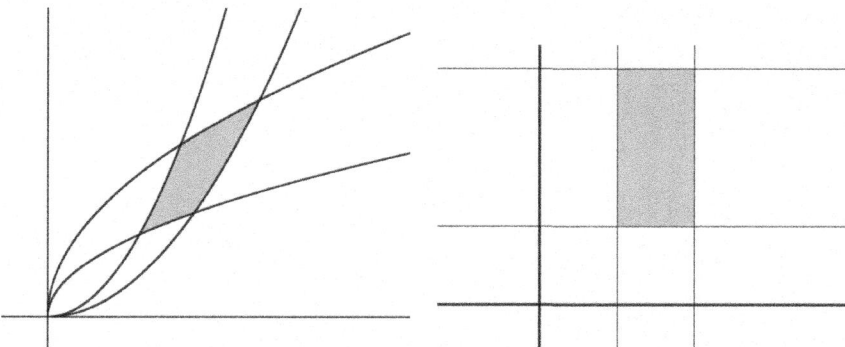

Fig. 5.11 The regions of Exercise 5.8, in the original coordinates and in the coordinates (u, v)

Therefore

$$|\Omega| = \iint_{\Omega'} \frac{1}{3u^2v^2} du dv = \frac{1}{3}\int_1^2 \frac{du}{u^2} \int_1^3 \frac{dv}{v^2} = \frac{1}{9}.$$

The change of variables is strongly suggested by the equations describing Ω, which is a deformation of a rectangle (see Fig. 5.11).

The mapping $\Psi(x, y) = (y/x^2, x/y^2)$ is explicitly invertible on the positive quadrant, and in fact $\Phi := \Psi^{-1}$ is given by

$$\Phi(u, v) = \left(\frac{1}{\sqrt[3]{u^2v}}, \frac{1}{\sqrt[3]{uv^2}}\right).$$

as the reader is invited to check. Sometimes the Jacobian determinant has an expression which does not require the full inversion procedure, as in the case at hand.

5.9 Compute $\iint_\Omega f$ where $f(x, y) = \sqrt{x^2 + y^2}$ and where

$$\Omega = \left\{(x, y) \in \mathbb{R}^2 : x^2 + y^2 - 2x \le 0\right\}.$$

Answer. The domain is the circle centered at (1, 0) of radius 1. Using the standard polar coordinates centered at the origin with $\theta \in (-\pi/2, \pi/2)$

$$(x-1)^2 + y^2 \le 1 \quad \Leftrightarrow \quad \rho^2 - 2\rho\cos\theta + 1 \le 1 \quad \Leftrightarrow \quad \rho \le 2\cos\theta.$$

Therefore

$$\iint_\Omega f = \int_{-\pi/2}^{\pi/2} \left[\int_0^{2\cos\theta} \rho^2 d\rho\right] d\theta$$

Fig. 5.12 The region Ω in Exercise 5.10: the use of polar coordinates centered at (1, 0) is natural

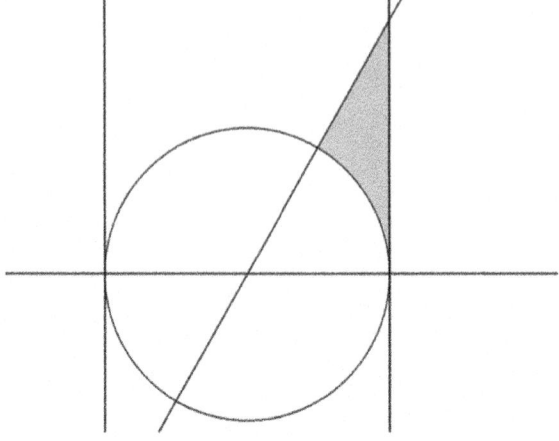

$$= \int_{-\pi/2}^{\pi/2} \left[\frac{\rho^3}{3}\right]_0^{2\cos\theta} d\theta$$

$$= \frac{8}{3} \int_{-\pi/2}^{\pi/2} \cos^3\theta \, d\theta$$

$$= \frac{8}{3} \int_{-\pi/2}^{\pi/2} \cos\theta \left(1 - \sin^2\theta\right) d\theta$$

$$= \frac{8}{3}[\sin\theta]_{-\pi/2}^{\pi/2} - \frac{8}{3}\left[\frac{\sin^3\theta}{3}\right]_{-\pi/2}^{\pi/2}$$

$$= \frac{32}{9}.$$

This simple exercise shows that in polar coordinates centered at the origin (in fact at any pole) the equation of a circle shifted to the right by the amount of the radius, with the origin (or the pole) removed, is given by the simple equation $\rho = 2\cos\theta$ (or, in general, $\rho = 2R\cos\theta$) whereby the polar angle ranges only in $(-\pi/2, \pi/2)$. Further, it is shown how to integrate the function $\cos^3\theta$.

5.10 Compute $\iint_\Omega f$ where $f(x, y) = \dfrac{x - 1}{(x-1)^2 + y^2}$ and where

$$\Omega = \left\{(x, y) \in \mathbb{R}^2 : (x-1)^2 + y^2 \geq 1, 0 \leq y \leq \sqrt{3}(x-1), 1 \leq x \leq 2\right\}.$$

Answer. Using polar coordinates centered at (1, 0), the region Ω, which is depicted in Fig. 5.12, is given by the constraints

$$0 \leq \rho\sin\theta \leq \sqrt{3}\rho\cos\theta \quad \Rightarrow \quad 0 \leq \theta \leq \frac{\pi}{3}$$

and by $\rho \geq 1$ and $1 \leq 1 + \rho\cos\theta \leq 2$, which together yield $1 \leq \rho \leq 1/\cos\theta$.

5.5 Guided Exercises

Therefore

$$\iint_\Omega \frac{x-1}{(x-1)^2+y^2}dxdy = \int_0^{\pi/3}\left(\int_1^{1/\cos\theta}\frac{\rho\cos\theta}{(\rho\cos\theta)^2+(\rho\sin\theta)^2}\rho d\rho\right)d\theta$$

$$= \int_0^{\pi/3}\cos\theta\left(\frac{1}{\cos\theta}-1\right)d\theta$$

$$= [\theta - \sin\theta]_0^{\pi/3}$$

$$= \frac{\pi}{3} - \frac{\sqrt{3}}{2}.$$

This is a more or less immediate application of the change of variables technique. All there is to understand is when $\tan\theta = 3$.

5.11 Compute $\iint_\Omega f$ where $f(x,y) = xy\sqrt[3]{x^2+y^2}$ and

$$\Omega = \{(x,y) \in \mathbb{R}^2 : (x-1)^2+y^2 \le 1, x^2+(y-1)^2 \le 1\}.$$

Answer. In polar coordinates centered at the origin:

$$(x-1)^2+y^2 \le 1 \quad \Leftrightarrow \quad \rho \le 2\cos\theta$$

$$x^2+(y-1)^2 \le 1 \quad \Leftrightarrow \quad \rho \le 2\sin\theta.$$

As depicted in Fig. 5.13 on the left, in the region Ω, which is the intersection of two discs, the angular variable θ takes values in $[0, \pi/2]$ because the points lie all in the first quadrant. When $\theta \in [0, \pi/4]$, the point ($\rho\cos\theta, \rho\sin\theta$) lies in the lower part Ω_1 of the region (darker in the figure), the portion which is bounded below by the circle centered at $(0,1)$ with polar function $\rho = 2\sin\theta$ and above by the line $\theta = \pi/4$. When $\theta \in [\pi/4, \pi/2]$, the point ($\rho\cos\theta, \rho\sin\theta$) lies in the upper part Ω_2 of the region (lighter in the figure), the portion which is bounded above by the circle centered at $(1,0)$ with polar function $\rho = 2\cos\theta$ and below by the line $\theta = \pi/4$.

Therefore, observing that $f(\rho\cos\theta, \rho\sin\theta) = \rho^{8/3}\cos\theta\sin\theta$ and taking into account the change of variables' formula

$$\iint_\Omega f = \iint_{\Omega_1} f(x,y)dxdy + \iint_{\Omega_2} f(x,y)dxdy$$

$$= \int_0^{\pi/4}\left(\int_0^{2\sin\theta}\rho^{11/3}\cos\theta\sin\theta d\rho\right)d\theta$$

$$+ \int_{\pi/4}^{\pi/2}\left(\int_0^{2\cos\theta}\rho^{11/3}\cos\theta\sin\theta d\rho\right)d\theta$$

$$= \int_0^{\pi/4}\cos\theta\sin\theta\left[\frac{3}{14}\rho^{14/3}\right]_0^{2\sin\theta}d\theta$$

$$+ \int_{\pi/4}^{\pi/2} \cos\theta \sin\theta \left[\frac{3}{14}\rho^{14/3}\right]_0^{2\cos\theta} d\theta$$

$$= \frac{3}{14} 2^{14/3} \int_0^{\pi/4} \cos\theta (\sin\theta)^{17/3} d\theta + \frac{3}{14} 2^{14/3} \int_{\pi/4}^{\pi/2} (\cos\theta)^{17/3} \sin\theta\, d\theta$$

$$= \frac{3}{14} 2^{14/3} \left\{ \left[\frac{3(\sin\theta)^{20/3}}{20}\right]_0^{\pi/4} - \left[\frac{3(\cos\theta)^{20/3}}{20}\right]_{\pi/4}^{\pi/2} \right\}$$

$$= \frac{9}{70} \sqrt[3]{2}.$$

This exercise requires a little geometric insight when it comes to deciding where the angular variable ranges. To this end, drawing the circles is of big help because on the one hand it clarifies that $\theta \in [0, \pi/2]$ and on the other hand it suggests the natural subdivision of Ω in the two portions Ω_1 and Ω_2. As for the integration procedure, the actually simple formulae

$$\int \cos\theta (\sin\theta)^\alpha d\theta = \frac{(\sin\theta)^{\alpha+1}}{\alpha+1} + c, \quad \int (\cos\theta)^\alpha \sin\theta\, d\theta = -\frac{(\cos\theta)^{\alpha+1}}{\alpha+1} + c$$

valid for $\alpha \neq -1$ have been used. The final numerical value is computed by practising patience and endurance.

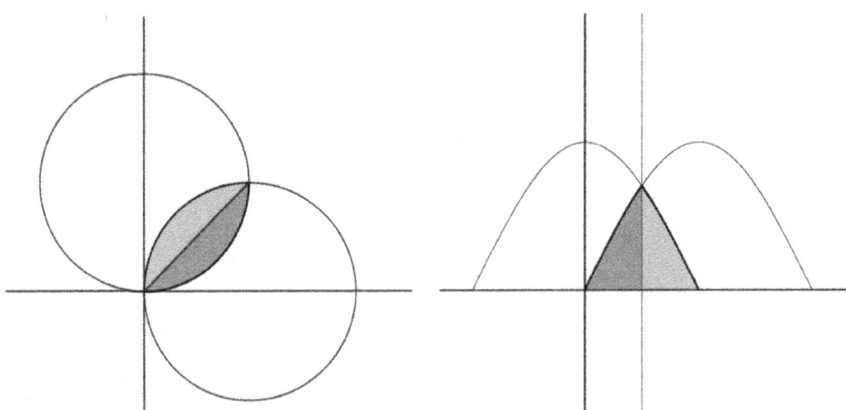

Fig. 5.13 On the left, the region Ω of Exercise 5.11 viewed as the union of two portions; on the right, the same region in polar coordinates centered at (0, 0) has an easy representation as the union of two normal domains in the (θ, ρ)-plane

5.5 Guided Exercises

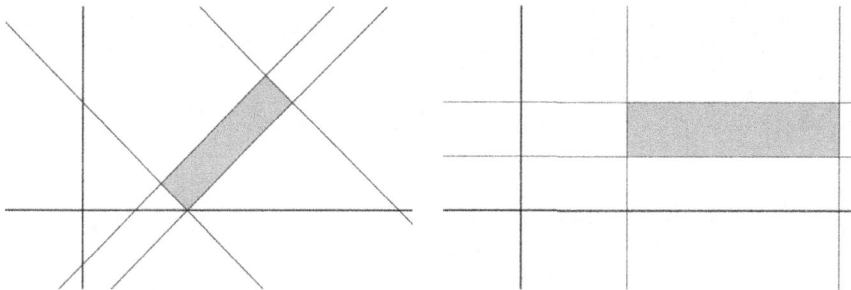

Fig. 5.14 The region Ω in Exercise 5.12 in the original coordinates and in the coordinates (u, v)

5.12 Compute $\iint_\Omega f$ where $f(x, y) = (x + y) \log(x - y)$ and

$$\Omega = \left\{ (x, y) \in \mathbb{R}^2 : 1 - x \le y \le 3 - x, x - 1 \le y \le x - \frac{1}{2} \right\}.$$

Answer. A convenient change of coordinates is

$$\begin{cases} u = x + y \\ v = x - y. \end{cases}$$

Indeed, in the coordinates (u, v) the region Ω (see Fig. 5.14) corresponds to the region $\{(u, v) \in \mathbb{R}^2 : 1 \le u \le 3, 1/2 \le v \le 1\}$. The map $(x, y) \mapsto (x+y, x-y)$ defines a linear bijection from $\mathbb{R}^2$ onto itself with linear inverse Φ given by

$$(x, y) = \Phi(u, v) = \left(\frac{u+v}{2}, \frac{u-v}{2} \right) = \begin{bmatrix} 1/2 & 1/2 \\ 1/2 & -1/2 \end{bmatrix} \begin{bmatrix} u \\ v \end{bmatrix}.$$

Being linear, the map Φ has Jacobian given by the matrix associated to it, that is

$$J\Phi(u, v) = \begin{bmatrix} x_u & x_v \\ y_u & y_v \end{bmatrix} = \begin{bmatrix} 1/2 & 1/2 \\ 1/2 & -1/2 \end{bmatrix}$$

which has determinant $-1/2$. Therefore

$$\iint_\Omega (x+y) \log(x-y) dx dy = \frac{1}{2} \int_1^3 \left(\int_{1/2}^1 u \log v \, dv \right) du$$
$$= \frac{1}{2} \left[\frac{u^2}{2} \right]_1^3 [v \log v - v]_{1/2}^1$$
$$= \log 2 - 1$$

This is just using a very simple, yet often handy, linear change of variables. The only significant observation is that a linear transformation T has Jacobian given by the matrix associated to T. This is sometimes expressed with slight abuse of language by saying that the Jabobian of a linear map is itself, an extended version of the fact that the derivative of $x \mapsto ax$ is a.

5.13 Compute $\iint_\Omega f$ where $f(x, y) = \exp\left(\frac{2x-y}{x+3y}\right)$ and where

$$\Omega = \{(x, y) \in \mathbb{R}^2 : 1 \le x + 3y \le 2, x \ge 0, y \ge 0\}.$$

Answer. A convenient change of coordinates is this time

$$\begin{cases} u = 2x - y \\ v = x + 3y. \end{cases}$$

The map $(x, y) \mapsto (2x - y, x + 3y)$ defines a linear bijection from $\mathbb{R}^2$ onto itself with linear inverse Φ given by

$$(x, y) = \Phi(u, v) = \left(\frac{3u+v}{7}, \frac{2v-u}{7}\right) = \begin{bmatrix} 3/7 & 1/7 \\ -1/7 & 2/7 \end{bmatrix}\begin{bmatrix} u \\ v \end{bmatrix}.$$

The map Φ therefore has Jacobian matrix

$$J\Phi(u, v) = \begin{bmatrix} 3/7 & 1/7 \\ -1/7 & 2/7 \end{bmatrix}$$

that has determinant $1/7$. The various constraints for (x, y) to belong to Ω become:

$$1 \le x + 3y \le 2 \Leftrightarrow 1 \le v \le 2;$$
$$x \ge 0 \Leftrightarrow 3u + v \ge 0 \Leftrightarrow u \ge -v/3;$$
$$y \ge 0 \Leftrightarrow 2v - u \ge 0 \Leftrightarrow u \le 2v.$$

Therefore, in the coordinates (u, v) the region Ω (see Fig. 5.15) corresponds to the region $\{(u, v) \in \mathbb{R}^2 : 1 \le v \le 2, -v/3 \le u \le 2v\}$, which is clearly normal to the v-axis. It follows that

$$\iint_\Omega \exp\left(\frac{2x-y}{x+3y}\right) dx dy = \frac{1}{7}\int_1^2\left[\int_{-v/3}^{2v} e^{u/v} du\right] dv$$

$$= \frac{1}{7}\int_1^2 v\left[\int_{-v/3}^{2v} e^{u/v} \frac{du}{v}\right] dv$$

$$= \frac{1}{7}\int_1^2 v\left[e^{u/v}\right]_{-v/3}^{2v} dv$$

$$= \frac{e^2 - e^{-1/3}}{7}\left[\frac{v^2}{2}\right]_1^2$$

$$= \frac{3}{14}(e^2 - e^{-1/3})$$

5.5 Guided Exercises

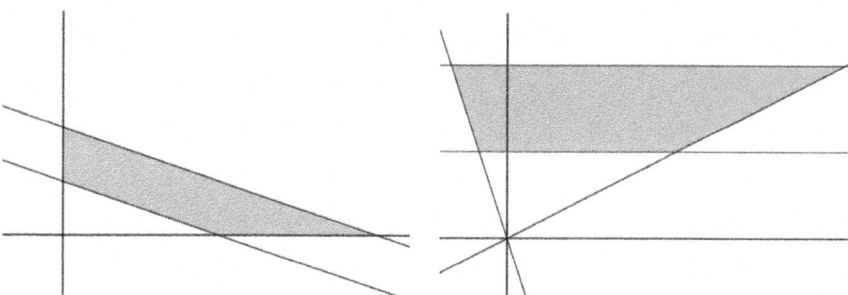

Fig. 5.15 The region Ω in Exercise 5.13 in the original coordinates and in the coordinates (u, v)

This is an application of the change of variable formula whereby the transformation is suggested by the form of the function f itself. It takes very little effort to understand the image of Ω under the transformation, because the boundary of Ω is given by lines and lines are mapped into lines by linear mappings. The calculation of the integral is essentially trivial.

5.14 Compute $\iint_\Omega f$ where $f(x, y) = x^3 \sin(xy)/y$ and

$$\Omega = \{(x, y) \in \mathbb{R}^2 : x \le y \le 2x, 1 \le xy \le 2\}.$$

Answer. A convenient change of coordinates is

$$\begin{cases} u = xy \\ v = \frac{y}{x}. \end{cases}$$

Notice that in the region Ω, both x and y are strictly positive (see also Fig. 5.16), which implies that both u and v are also srtrictly positive. The map $(x, y) \to (u, v)$ is invertible because

$$uv = y^2 \implies y = \sqrt{uv} \implies x = \frac{u}{y} = \sqrt{\frac{u^2}{uv}} = \sqrt{\frac{u}{v}}.$$

Thus, the inverse Φ is given by

$$(x, y) = \Phi(u, v) = \left(\sqrt{\frac{u}{v}}, \sqrt{uv}\right)$$

with jacobian matrix

$$J\Phi(u, v) = \begin{bmatrix} x_u & x_v \\ y_u & y_v \end{bmatrix} = \begin{bmatrix} \frac{1}{2\sqrt{uv}} & -\frac{1}{2}\sqrt{\frac{u}{v^3}} \\ \frac{1}{2}\sqrt{\frac{v}{u}} & \frac{1}{2}\sqrt{\frac{u}{v}}. \end{bmatrix}$$

that has determinant $1/2v$.

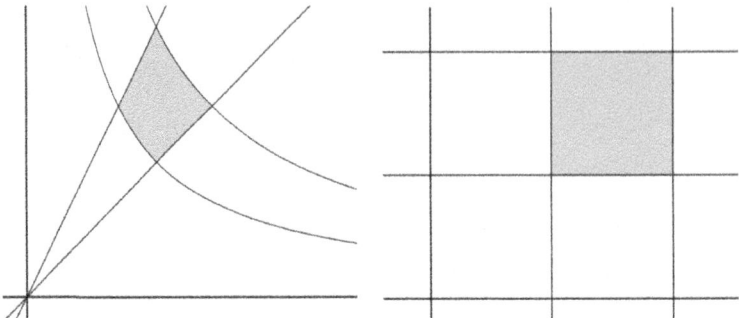

Fig. 5.16 The region Ω in Exercise 5.14 in the original coordinates and in the coordinates (u, v)

Finally, the constraints for (x, y) to belong to Ω become:

$$x \leq y \leq 2x \iff 1 \leq u \leq 2$$
$$1 \leq xy \leq 2 \iff 1 \leq v \leq 2.$$

It follows therefore that

$$\iint_\Omega \frac{x^3}{y} \sin(xy)\,dx\,dy = \frac{1}{2} \int_1^2 \left[\int_1^2 v^{-3} u \sin u\,du \right] dv$$
$$= \frac{3}{16}\left([-u\cos u]_1^2 - \int_1^2 \cos u\,du \right)$$
$$= \frac{3}{16}(\cos 1 - 2\cos 2 + \sin 2 - \sin 1).$$

The appropriate change of variables is suggested by the form of the function f, which depends on xy and x/y. In some sense, this is the multiplicative version *versus* the additive version considered in Exercise 5.12. The Jacobian needs a little care, because the components of Φ are non-trivial functions of u and v. The outcoming integral is particularly easy, beacuse in the new variables (u, v) the integration region is a square and the function factors in the product of a function of u and a function of v. This shows that the change of variables is indeed appropriate.

5.15 Compute the integral $\iint_\Omega f$, where $f(x, y) = x^2 + y^2$ and where Ω is the portion of the upper half plane consisting of the points $(x, y) = (\rho \cos\theta, \rho \sin\theta)$ that satisfy the following conditions:

(a) $0 < \theta < \frac{3}{4}\pi$,
(b) $\rho < \theta$,
(c) $\left(x + \frac{\pi}{2}\right)^2 + y^2 > \frac{\pi^2}{4}$,

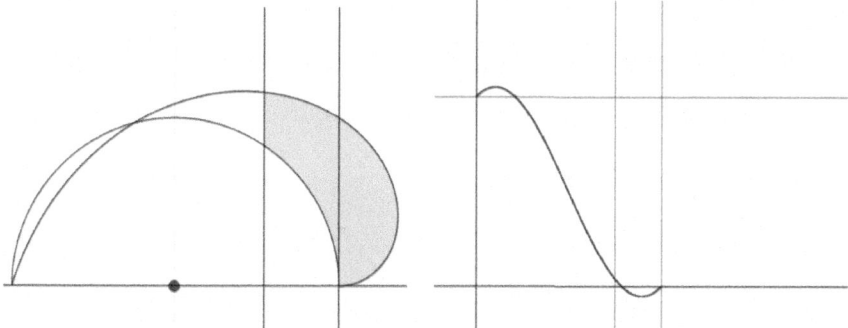

Fig. 5.17 On th left the region of Exercise 5.15, and on the right the graph of $g(\theta)$; in red the line $x = 3\pi/4$ where g is positive

Answer. The region is depicted in Fig. 5.17. It is clear that the portion of circle centered at $(-\pi/2, 0)$ and of radius $\pi/2$ that lies in the upper half plane is entirely to the left of the y-axis, and that Ω is in its complement. It is also clear that for $\theta \in (0, \pi/2)$ it holds $0 < \cos\theta < 1$ and also $0 < \sin\theta < 1$. Thus, for these values of θ the point $(x, y) = (\rho \cos\theta, \rho \sin\theta)$ lies to the right of the y-axis and to the left of the locus $\rho = \theta$ if $\rho < \theta$.

Consider now $\theta \in [\pi/2, 3\pi/4)$. Clearly, the point $(0, \pi/2)$ lies on the locus $\rho = \theta$ and above the circle. The question arise whether for any $\theta \in [\pi/2, 3\pi/4)$ the loci $\rho = \theta$ and $\left(x + \frac{\pi}{2}\right)^2 + y^2 = \frac{\pi^2}{4}$ meet. This happens if and only if the following equation has solutions for $\theta \in [\pi/2, 3\pi/4)$:

$$\left(\theta \cos\theta + \frac{\pi}{2}\right)^2 + \theta^2 \sin^2\theta = \frac{\pi^2}{4} \quad \Leftrightarrow \quad \theta + \pi \cos\theta = 0.$$

Put then $g(\theta) = \theta + \pi \cos\theta$, which we study in the larger interval $[\pi/2, \pi]$. Clearly, $g(0) = \pi$. Furthermore, since $g'(\theta) = 1 - \pi \sin\theta < g'(3\pi/4) = 1 - \pi\sqrt{2}/2 < 0$ for every $\theta \in [\pi/2, 3\pi/4]$, g is decreasing in that interval. Observe that,

$$g\left(\frac{3}{4}\pi\right) = \frac{3}{4}\pi - \pi\frac{\sqrt{2}}{2} > 0,$$

and hence the points on the locus $\rho = \theta$ lie above the circle for $\theta \in (\pi/2, 3\pi/4)$. Finally, observe that the points on the circle satisfy

$$\left(\rho \cos\theta + \frac{\pi}{2}\right)^2 + \rho^2 \sin^2\theta = \frac{\pi^2}{4} \quad \Leftrightarrow \quad \rho + \pi \cos\theta = 0.$$

The domain Ω can thus be written as $\Omega = \Omega_1 \cup \Omega_2$ where

Fig. 5.18 In light grey the region Ω_1, in darker grey the region Ω_2 of Exercise 5.15 represented in the plane (θ, ρ)

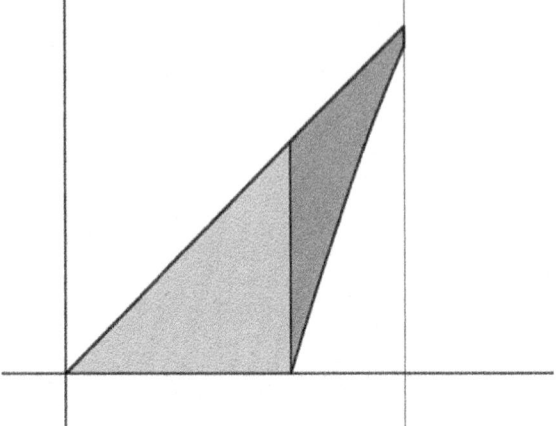

$$\Omega_1 = \left\{ (\theta, \rho) : 0 < \theta \le \frac{\pi}{2}, 0 < \rho < \theta \right\}$$

$$\Omega_2 = \left\{ (\theta, \rho) : \frac{\pi}{2} < \theta \le \frac{3}{4}\pi, -\pi \cos\theta < \rho < \theta \right\}$$

as depicted in Fig. 5.18. Both regions are normal with respect to the θ-axis.
As for the integrals,

$$\iint_{\Omega_1} (x^2 + y^2) dx dy = \int_0^{\pi/2} \int_0^\theta \rho^2 \rho \, d\rho \, d\theta = \frac{1}{4} \int_0^{\pi/2} \theta^4 \, d\theta = \frac{\pi^5}{640}.$$

Furthermore

$$\iint_{\Omega_2} (x^2 + y^2) dx dy = \int_{\pi/2}^{3\pi/4} \int_{-\pi\cos\theta}^{\theta} \rho^2 \rho \, d\rho \, d\theta$$

$$= \frac{1}{4} \int_{\pi/2}^{3\pi/4} (\theta^4 - \pi^4 \cos^4\theta) d\theta$$

$$= \frac{211}{20480} \pi^5 - \frac{\pi^4}{4} \int_{\pi/2}^{3\pi/4} \cos^4\theta \, d\theta.$$

Now, observe that

$$\cos^4\theta = (\cos^2\theta)^2$$
$$= \left(\frac{1 + \cos 2\theta}{2} \right)^2$$
$$= \frac{1}{4} + \frac{1}{2} \cos 2\theta + \frac{1}{4} \cos^2 2\theta$$

5.5 Guided Exercises

$$= \frac{1}{4} + \frac{1}{2}\cos 2\theta + \frac{1}{4}\left(\frac{1+\cos 4\theta}{2}\right)$$

$$= \frac{3}{8} + \frac{1}{2}\cos 2\theta + \frac{1}{8}\cos 4\theta$$

so that

$$\int_{\pi/2}^{3\pi/4} \cos^4\theta\, d\theta = \frac{3}{8}\left[\frac{3}{4}\pi - \frac{1}{2}\pi\right] + \frac{1}{2}\int_{\pi/2}^{3\pi/4} \cos 2\theta\, d\theta + \frac{1}{8}\int_{\pi/2}^{3\pi/4} \cos 4\theta\, d\theta$$

$$= \frac{3}{32}\pi + \frac{1}{4}\int_{\pi}^{3\pi/2} \cos\varphi\, d\varphi + \frac{1}{32}\int_{2\pi}^{3\pi} \cos\psi\, d\psi$$

$$= \frac{3}{32}\pi - \frac{1}{4}.$$

In conclusion,

$$\iint_\Omega f = \frac{\pi^5}{640} + \frac{211}{20480}\pi^5 - \frac{\pi^4}{4}\left(\frac{3}{32}\pi - \frac{1}{4}\right) = -\frac{237}{20480}\pi^5 + \frac{1}{16}\pi^4.$$

5.16 Compute the barycenter of the region Ω bounded by the four lines $y = x/2$, $y = 2x$, $y = -x + 1$ and $y = -x + 3$, with respect to the density function $\mu(x, y) = 1$.

Answer. The appropriate change of variables is suggested by the shape of Ω (see Fig. 5.19), namely

$$\begin{cases} u = \dfrac{y}{x} \\ v = x + y. \end{cases}$$

The inverse Φ of the mapping $(x, y) \mapsto (u(x, y), v(x, y))$ is given by

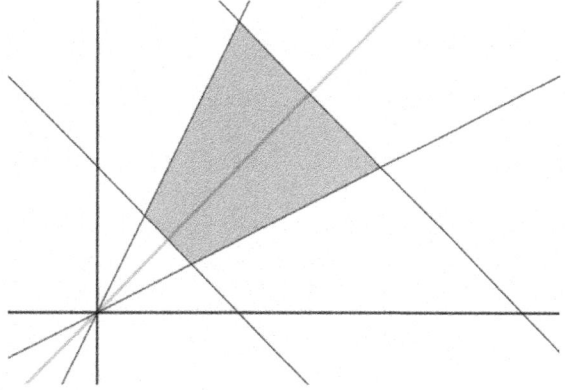

Fig. 5.19 The region Ω of Exercise 5.16 is symmetric with respect to the line $y = x$

$$(x, y) = \Phi(u, v) = \left(\frac{v}{u+1}, \frac{uv}{u+1}\right)$$

with absolute value of the Jacobian determinant $v/(u+1)^2$. The constraints on v and u are $1 \leq v \leq 3$ and $1/2 \leq u \leq 2$. Therefore the mass of Ω is

$$m(\Omega) = \iint_\Omega dxdy = \int_1^3 \left(\int_{1/2}^2 \frac{v}{(u+1)^2} du\right) dv = \left[\frac{v^2}{2}\right]_1^3 \left[-\frac{1}{u+1}\right]_{1/2}^2 = \frac{4}{3},$$

so that the barycenter has coordinates

$$\begin{aligned} x_B &= \frac{1}{m(\Omega)} \iint_\Omega xdxdy \\ &= \frac{3}{4} \int_1^3 \left(\int_{1/2}^2 \frac{v}{(u+1)} \frac{v}{(u+1)^2} du\right) dv \\ &= \frac{3}{4} \left[\frac{v^3}{3}\right]_1^3 \left[-\frac{1}{2(u+1)^2}\right]_{1/2}^2 \\ &= \frac{13}{12}, \end{aligned}$$

$$\begin{aligned} y_B &= \frac{1}{m(\Omega)} \iint_\Omega ydxdy \\ &= \frac{3}{4} \int_1^3 \left(\int_{1/2}^2 \frac{uv}{(u+1)} \frac{v}{(u+1)^2} du\right) dv \\ &= \frac{3}{4} \left[\frac{v^3}{3}\right]_1^3 \int_{1/2}^2 \left(\frac{1}{(u+1)^2} - \frac{1}{(u+1)^3}\right) du \\ &= \frac{13}{2} \left[-\frac{1}{u+1} + \frac{1}{2(u+1)^2}\right]_{1/2}^2 \\ &= \frac{13}{12}. \end{aligned}$$

In conclusion, $B = (13/12, 13/12)$.

The choice of the new coordinates is suggested by the observation that Ω is bounded by two lines of the form $x + y = c$ (with $c = 1, 3$) and two lines of the form $y/x = d$ (with $d = 1/2, 2$). Also, the region exhibits symmetry with respect to the line $y = x$ and has constant density, so that the barycenter is expected to lie on the symmetry line, that is, it is natural to obtain $x_B = y_B$.

5.17 Compute the barycenter of the region

$$\Omega = \{(x, y) \in \mathbb{R}^2 : 2x^2 \leq y \leq x^2 + 1, -1 \leq x \leq 1\}$$

with respect to the density function $\mu(x, y) = 4 - x$.

Answer. The mass of Ω, which is normal to the x-axis, is

$$m(\Omega) = \int_{-1}^{1} (4-x) \left(\int_{2x^2}^{x^2+1} dy \right) dx$$

$$= \int_{-1}^{1} (4-x)(1-x^2) dx$$

$$= \int_{-1}^{1} (x^3 - 4x^2 - x + 4) dx$$

$$= \frac{16}{3}.$$

Hence the coordinates of the barycenter are (Fig. 5.20)

$$x_B = \frac{1}{m(\Omega)} \int_{-1}^{1} (4-x) \left(\int_{2x^2}^{x^2+1} dy \right) dx$$

$$= \frac{1}{m(\Omega)} \int_{-1}^{1} (x^4 - 4x^3 - x^2 + 4x) dx$$

$$= \frac{-4/15}{16/3}$$

$$= -\frac{1}{20}.$$

$$y_B = \frac{1}{m(\Omega)} \int_{-1}^{1} x(4-x) \left(\int_{2x^2}^{x^2+1} y \, dy \right) dx$$

$$= \frac{1}{m(\Omega)} \int_{-1}^{1} (4-x) \frac{1}{2} \left((x^2+1)^2 - 4x^4 \right) dx$$

$$= \frac{64/15}{16/3}$$

$$= \frac{4}{5}.$$

All the calculations are essentially trivial. The point of this exercise is to show that the density of mass does play a role: despite the symmetry of Ω, the density weighs much more the points to the left of the y-axis than those to its right, and this in the end produces a negative value of x_B.

5.18 Consider the region $\Omega = \{(x, y) \in \mathbb{R}^2 : x^2 \leq y \leq 1, 0 \leq x \leq 1\}$ with density function $\mu(x, y) = 2xy$.

(a) Compute the barycenter of Ω.
(b) Compute the moments of inertia of Ω with respect to the coordinate axes and with respect to the origin.

Fig. 5.20 The region Ω of Exercise 5.17

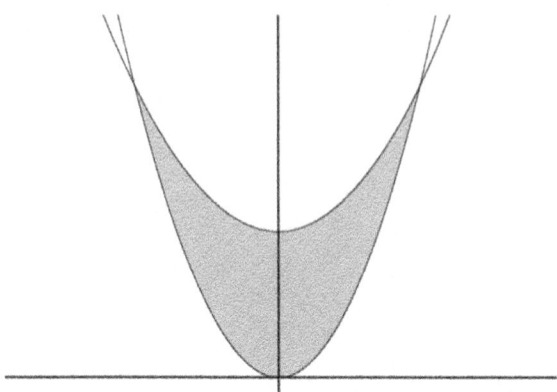

Answer.

(a) Evidently,

$$m(\Omega) = \int_0^1 \left(\int_{x^2}^1 2xy\,dy \right) dx = \int_0^1 2x \left[\frac{y^2}{2} \right]_{x^2}^1 dx = \int_0^1 (x - x^5)\,dx = \frac{1}{3}$$

so that

$$x_B = 3 \iint_\Omega 2x^2 y\,dxdy = 3 \int_0^1 (x^2 - x^6)\,dx = \frac{4}{7}$$

and

$$y_B = 3 \iint_\Omega 2xy^2\,dxdy = 3 \int_0^1 2x \left[\frac{y^3}{3} \right]_{x^2}^1 dx = 2 \int_0^1 (x - x^7)\,dx = \frac{3}{4},$$

that is $B = (4/7,\ 3/4)$.

(b) From the defining formulae,

$$I_x = \iint_\Omega y^2 \mu(x,y)\,dxdy = 2 \iint_\Omega xy^3\,dxdy = \frac{1}{2} \int_0^1 (x - x^9)\,dx = \frac{1}{5}$$

$$I_y = \iint_\Omega x^2 \mu(x,y)\,dxdy = 2 \iint_\Omega x^3 y\,dxdy = \int_0^1 (x^3 - x^7)\,dx = \frac{1}{8}$$

and hence $I_o = I_x + I_y = 13/40$.

Just apply the various formulae. It is amusing to see that the final integral is always of the form $c \int_0^1 (x^p - x^q)\,dx$ for some constant c and positive integers p and q. This is no coincidence, because $\int_{x^2}^1 y^{n-1}\,dy = (1 - x^{2n})/n$.

5.5 Guided Exercises

5.19 Consider the region $\Omega \subset \mathbb{R}^3$ bounded by the cylinder $x^2 + z^2 = 1$, by the coordinate planes and by the plane $y = 2$.

(a) Compute the integral $\iiint_\Omega y \, dx \, dy \, dz$ by segments perpendicular to the xz-plane.
(b) Compute the integral $\iiint_\Omega y \, dx \, dy \, dz$ by sections parallel to the xy-plane.

Answer.

(a) Evidently, $\Omega = \bigcup_{(x,z) \in D} F(x, z)$ where

$$F(x, z) = \{(x, y, z) \in \mathbb{R}^3 : 0 \leq 2\}$$
$$D = \{(x, z) \in \mathbb{R}^2 : x^2 + z^2 \leq 1, \, x > 0, \, z > 0\}.$$

Therefore (Fig. 5.21)

$$\iiint_\Omega y \, dx \, dy \, dz = \iint_D \left(\int_0^2 y \, dy \right) dx \, dz = 2 \iint_D dx \, dz = 2|D| = \frac{\pi}{2}.$$

(b) Since $\Omega = \bigcup_{z \in [0,1]} \Omega_z$ where

$$\Omega_z = \left\{ (x, y, z) \in \mathbb{R}^3 : 0 \leq y \leq 2, \, 0 \leq x \leq \sqrt{1 - z^2} \right\}$$

it follows that, upon setting $S_z = \{(x, y) \in \mathbb{R}^2 : (x, y, z) \in \Omega_z\}$

$$\iiint_\Omega y \, dx \, dy \, dz = \int_0^1 \left(\iint_{S_z} y \, dx \, dy \right) dz$$
$$= \int_0^1 \left[\int_0^2 y \left(\int_0^{\sqrt{1-z^2}} dx \right) dy \right] dz$$
$$= 2 \int_0^1 \sqrt{1 - z^2} \, dz.$$

Upon putting $z = \sin t$ the integral becomes

Fig. 5.21 The region Ω of Exercise 5.19; in grey the region D in the plane xz, a string based on D, and, also in grey, the section Ω_z parallel to the xy-plane

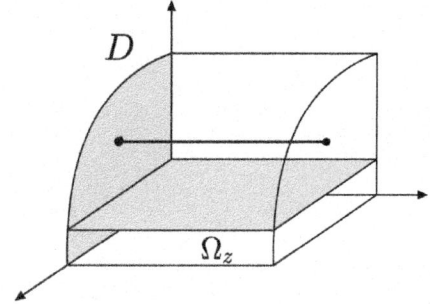

$$2\int_0^{\pi/2}\sqrt{1-\sin^2 t}\cos t\,dt = 2\int_0^{\pi/2}\cos^2 t\,dt = [t+\sin t\cos t]_0^{\pi/2} = \frac{\pi}{2}.$$

The point of this exercise is to practice with the representation of a domain either as a union of segments or as a union of sections. It is actually clear that the first choice leads to a much faster computation than the second.

5.20 Consider the tetrahedron Ω with vertices $(0, 0, 0)$, $(1, 0, 0)$, $(0, 1, 0)$ and $(0, 0, 1)$.

(a) Compute the integral $\iiint_\Omega y\,dxdydz$ by segments perpendicular to the xy-plane.
(b) Compute the integral $\iiint_\Omega y\,dxdydz$ by sections parallel to the xy-plane.

Answer. The equation of the plane passing through the points $(1, 0, 0)$, $(0, 1, 0)$ and $(0, 0, 1)$ is $x+y+z=1$, so that

$$\Omega = \{(x,y,z)\in\mathbb{R}^3 : x+y+z \le 1, x \ge 0, y \ge 0, z \ge 0\}.$$

(a) The segments issue from the planar region D contained in the positive quadrant of the xy-plane bounded by the axes and by the line of equation $x+y=1$. Hence (Fig. 5.22)

$$\iiint_\Omega y\,dxdydz = \iint_D \left(\int_0^{1-(x+y)} y\,dz\right)dxdy$$
$$= \iint_D y(1-x-y)\,dxdy$$
$$= \int_0^1\left(\int_0^{1-x}[y(1-x)-y^2]dy\right)dx$$
$$= \int_0^1\left((1-x)\left[\frac{y^2}{2}\right]_0^{1-x} - \left[\frac{y^3}{3}\right]_0^{1-x}\right)dx$$
$$= \frac{1}{6}\int_0^1 (1-x)^3\,dx$$
$$= \frac{1}{24}.$$

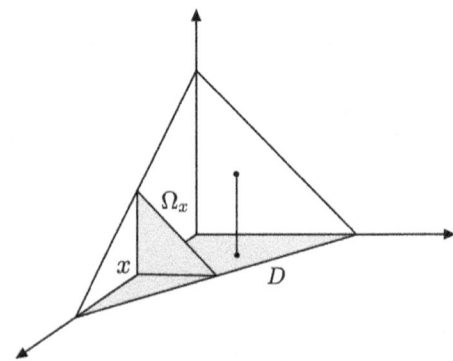

Fig. 5.22 The region Ω of Exercise 5.20; in grey the region D in the plane xy, a string based on D, and, also in grey, the section Ω_x parallel to the yz-plane

(b) For fixed $x \in [0, 1]$ the intersection Ω_x of Ω with the plane parallel to the yz-plane meeting the x-axis at x is parametrized by the subset S_x of $\mathbb{R}^2$ given by

$$S_x = \{(y, z) \in \mathbb{R}^2 : 0 \leq y \leq 1 - x, 0 \leq z \leq 1 - (x+y)\}$$

so that, as x ranges in $[0, 1]$,

$$\iiint_\Omega y\,dx\,dy\,dz = \int_0^1 \left(\iint_{S_x} y\,dy\,dz \right) dx$$
$$= \int_0^1 \left(\int_0^{1-x} \left[y \int_0^{1-(x+y)} dz \right] dy \right) dx$$
$$= \int_0^1 \left(\int_0^{1-x} y(1 - x - y)\,dy \right) dx$$
$$= \frac{1}{24},$$

whereby the last equality was obtained in part (a).

As in the previous exercise, here the task is to properly describe the domain Ω either as a union of segments issuing from a planar section or as a union of planar sections parametrized by a transversal coordinate. It is worthwhile observing that this particular tetrahedron is called a *simplex*, the convex hull of the origin and of the standard basis vectors. Interestingly, this kind of region can be considered in all dimensions, and has volume $1/(n + 1)!$ in $\mathbb{R}^n$.

5.21 Compute the integral $\iiint_\Omega xy\,dx\,dy\,dz$ where Ω is the region contained in the positive octant, i.e. where $x \geq 0$, $y \geq 0$ and $z \geq 0$, bounded by the sphere $x^2 + y^2 + z^2 = 4$, and by the planes with equations $2y + z = 2$ and $y + z = 2$.

Answer. Integrating by sections parallel to the xy-plane, the domain Ω is the union of the domains

$$\Omega_z = \left\{ (x, y, z) \in \mathbb{R}^3 : 0 \leq x \leq \sqrt{4 - y^2 - z^2}, 1 - \frac{z}{2} \leq y \leq 2 - z \right\}$$

with $z \in [0, 2]$. Thus, writing $\Omega_z = \{(x, y, z) \in \mathbb{R}^3 : (x, y) \in S_z\}$,

$$\iiint_\Omega y\,dx\,dy\,dz = \int_0^2 \left[\iint_{S_z} xy\,dx\,dy \right] dz$$
$$= \int_0^2 \left[\int_{1-z/2}^{2-z} \left(\int_0^{\sqrt{4-y^2-z^2}} x\,dx \right) y\,dy \right] dz$$
$$= \int_0^2 \left[\int_{1-z/2}^{2-z} y\,\frac{(4 - y^2 - z^2)}{2}\,dy \right] dz$$
$$= \frac{21}{20},$$

whereby the last equality does require a few lines of tedious but straightforward computations.

This exercise requires some geometric intuition. The integration region is the portion of one octant of a ball between two planes. Figure 5.23 represents Ω in three different ways.

5.22 Compute the integral $\iiint_\Omega xyz\,dxdydz$ where

$$\Omega = \left\{(x, y, z) \in \mathbb{R}^3 : x \geq 0, y \geq 0, \sqrt{x} + \sqrt{y} \leq 2, , 0 \leq z \leq \sqrt{xy}\right\}.$$

Answer. The inequality $\sqrt{x} + \sqrt{y} \leq 2$ describes a domain $D \subset \mathbb{R}^2$ normal to both axes and contained in the square $[0, 4] \times [0, 4]$, for if any of the two variables were larger than 4 than its square root would be larger than 2. Thus,

$$D = \left\{(x, y) \in \mathbb{R}^2 : 0 \leq x \leq 4, 0 \leq y \leq (2 - \sqrt{x})^2\right\}.$$

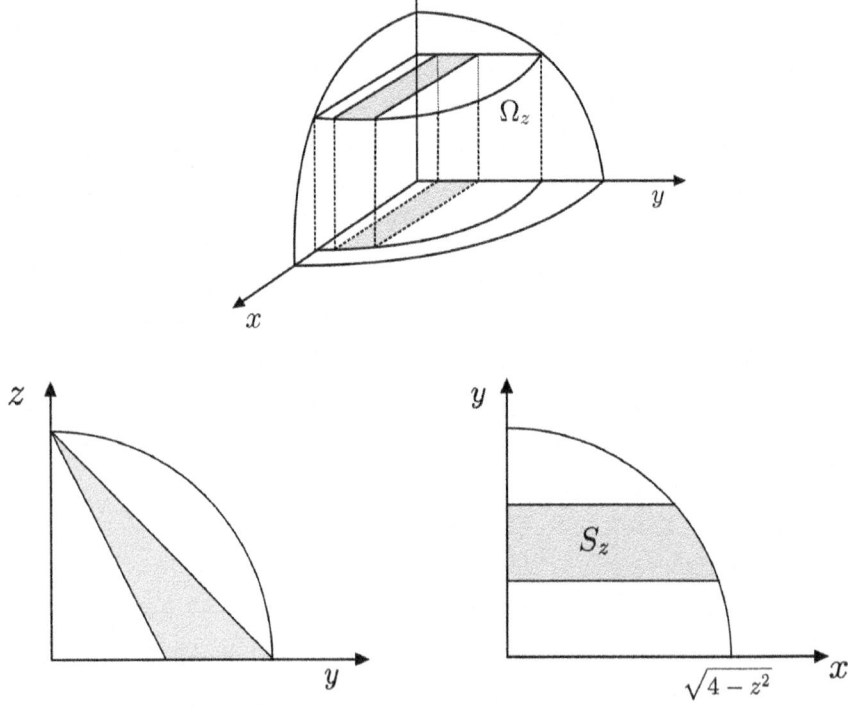

Fig. 5.23 The region Ω of Exercise 5.21; in the third figure the projection S_z of Ω_z onto the xy-plane

Hence, integrating by segments

$$\iiint_\Omega xyz\,dxdydz = \iint_D \left(\int_0^{\sqrt{xy}} xyz\,dz\right)dxdy$$

$$= \int_0^4 \left[\int_0^{(2-\sqrt{x})^2} \left(\int_0^{\sqrt{xy}} xyz\,dz\right)dy\right]dx$$

$$= \int_0^4 \left[\int_0^{(2-\sqrt{x})^2} xy\frac{z^2}{2}\bigg|_0^{\sqrt{xy}} dy\right]dx$$

$$= \frac{1}{2}\int_0^4 x^2\left[\int_0^{(2-\sqrt{x})^2} y^2\,dy\right]dx$$

$$= \frac{1}{6}\int_0^4 x^2(2-\sqrt{x})^6\,dx.$$

Now, expanding explicitely $(2-\sqrt{x})^6$ one gets

$$(2-\sqrt{x})^6 = 64 - 192\sqrt{x} + 240x - 160x\sqrt{x} + 60x^2 - 12x^2\sqrt{x} + x^3$$

and integrating out the various powers the final result is 512/2079.

This is a standard integral, whereby the domain clearly admits a representation by segments. The only slight difficulty is the realization that, in turn, the domain $D \subset \mathbb{R}^2$ is a normal domain. The integration is then essentially trivial, and in the final part it is left to the reader.

5.23 Compute the volume of

$$\Omega = \{(x,y,z) \in \mathbb{R}^3 : z \leq 1, x^2 + y^2 \leq z\}.$$

Answer. In the cylindrical coordinates

$$\begin{cases} x = \rho\cos\theta \\ y = \rho\sin\theta \\ z = z \end{cases}$$

the domain Ω is described by the conditions $0 \leq z \leq 1$, $0 \leq \theta \leq 2\pi$ and, for any $z \in [0,1]$, $0 \leq \rho \leq \sqrt{z}$ Thus, the domain admits a representation by sections and upon setting $S_z = \{(\rho,\theta) \in (0,+\infty) \times [0,2\pi) : 0 \leq \rho \leq \sqrt{z}\}$, the volume of Ω is

$$|\Omega| = \int_0^1 \left(\iint_{S_z} \rho\,d\rho d\theta\right)dz = 2\pi\int_0^1 \left[\frac{\rho^2}{2}\right]_0^{\sqrt{z}} dz = \frac{\pi}{2}.$$

Here the idea is to combine change of variables and, in the new variables (ρ, θ, z), integration by sections. One could also reverse the point of view, and say that integration by sections is used first, and then a change of coordinates in each section, namely polar coordinates.

5.24 Compute the integral $\iiint_\Omega (x^2 + y^2) dxdydz$, where Ω is the region bounded by the xy-plane and by the paraboloid with Eq. $9 - z = x^2 + y^2$.

Answer. In the cylindrical coordinates as in Exercise 5.22, the domain Ω becomes

$$\Omega' = \{(\rho, \theta, z) : 0 \leq \rho \leq 3, 0 \leq \theta < 2\pi, 0 \leq z \leq 9 - \rho^2\}.$$

Hence, integrating by sections

$$\iiint_\Omega (x^2 + y^2) dxdydz = \int_0^3 \left[\int_0^{2\pi} \int_0^{9-\rho^2} \rho^2 \rho dz d\theta \right] d\rho$$

$$= 2\pi \int_0^3 \rho^3 (9 - \rho^2) d\rho$$

$$= 2\pi \left[9 \frac{\rho^4}{4} - \frac{\rho^6}{6} \right]_0^3$$

$$= \frac{243}{2}\pi.$$

Essentially the same comments as those for Exercise 5.22 apply to this Exercise.

5.25 Compute the integral $\iiint_\Omega z dxdydz$, where

$$\Omega = \left\{ (x, y, z) \in \mathbb{R}^3 : x^2 + z^2 - 4 \leq y, z \geq \sqrt{x^2 + 2y^2} \right\}.$$

Answer. The domain is the region inside the upper sheet of the cone of equation $z^2 = x^2 + 2y^2$ and contained in the convex region determined by the paraboloid of equation $y = x^2 + z^2 - 4$. It admits the representation by segments

$$\Omega = \left\{ (x, y, z) \in \mathbb{R}^3 : (x, y) \in D, \sqrt{x^2 + 2y^2} \leq z \leq \sqrt{y + 4 - x^2} \right\},$$

where D is the region in $\mathbb{R}^2$ where both inequalities are meaningful, namely where

$$\sqrt{x^2 + 2y^2} \leq \sqrt{y + 4 - x^2} \Leftrightarrow x^2 + 2y^2 \leq y + 4 - x^2 \Leftrightarrow x^2 + \left(y - \frac{1}{4}\right)^2 \leq \frac{33}{16}.$$

Hence D is the circle of radius $\sqrt{33}/4$ centered at $(0, 1/4)$. Thus,

5.5 Guided Exercises

$$\iiint_\Omega z\,dx\,dy\,dz = \iint_D \left[\int_{\sqrt{x^2+2y^2}}^{\sqrt{y+4-x^2}} z\,dz \right] dx\,dy$$

$$= \iint_D \left[\frac{z^2}{2} \right]_{\sqrt{x^2+2y^2}}^{\sqrt{y+4-x^2}} dx\,dy$$

$$= \frac{1}{2} \iint_D (y + 4 - 2x^2 - 2y^2)\,dx\,dy.$$

The latter integral is best computed in the polar coordinates

$$\begin{cases} x = \rho\cos\theta \\ y = \rho\sin\theta \end{cases}$$

under the restriction $0 < \rho \leq \sqrt{33}/4$, and obviously $0 \leq \theta < 2\pi$. Hence,

$$\iiint_\Omega z\,dx\,dy\,dz = \frac{1}{2} \int_0^{2\pi} \left[\int_0^{\sqrt{33}/4} (\rho\sin\theta + 4 - 2\rho^2)\rho\,d\rho \right] d\theta$$

$$= \frac{1}{2} \int_0^{2\pi} \left[\frac{\rho^3}{3}\sin\theta + 2\rho^2 - \frac{\rho^4}{2} \right]_0^{\sqrt{33}/4} d\theta$$

$$= \frac{1023}{512}\pi.$$

In this exercise some geometric intuition is asked for. In Fig. 5.24 there is a represen-taion of Ω in the sense that, as explained above, the region has a natural description as a union of vertical segments issuing from the circle D, whereby the lower extreme of each segment lies on the cone and the upper extreme lies on the paraboloid. Thus, if one slices the domain with a vertical plane based on a line parallel to the x-axis, one sees two planar regions that intersect, whereby the intersection is a union of segments. The intersection of the cone with a plane clearly gives a conic, and the intersection of the paraboloid with a plane gives a parabola: these are the two conics that bound the planar "slices".

Since the region D is a circle, it is completely natural to introduce the polar coordinates for the final integral. Again, one could switch point of view and introduce cylindrical coordinates from the start and then integrate by segments in the transformed region. Notice that, in the cylindrical coordinates, the planar region R that corresponds to the circle D is of course a rectangle, namely $R = [0, 2\pi) \times (0, \sqrt{33}/4]$.

5.26 Compute the volume of the region bounded by the surfaces $x^2 + y^2 + z^2 = a^2$, $x^2 + y^2 + z^2 = b^2$, where $0 < a < b$, and by $z = \sqrt{x^2 + y^2}$.

Answer. In the spherical coordinates

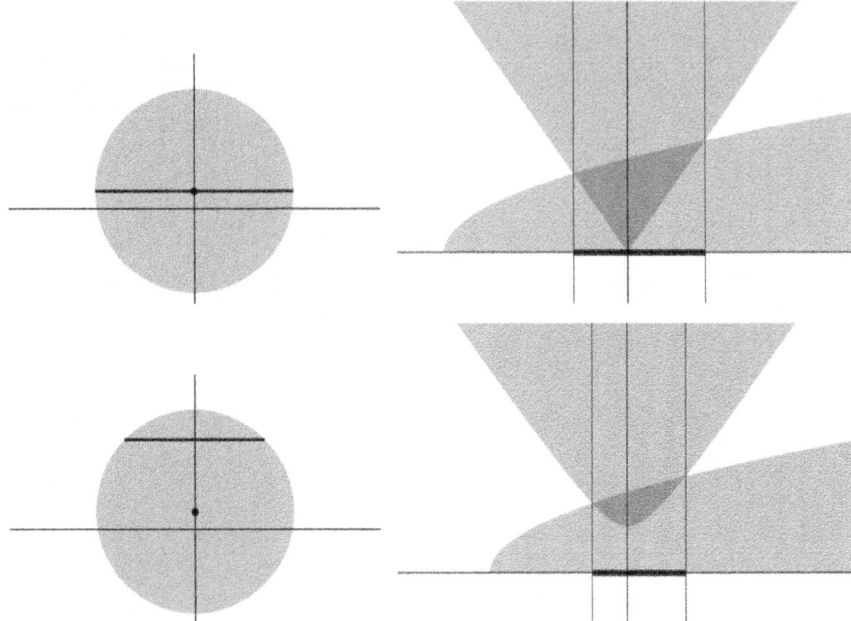

Fig. 5.24 On the left the circle D of Exercise 5.25; upon cutting the region Ω with a vertical plane based on the line containing the black segments, one gets the slice of Ω that is depicted on the right

$$\begin{cases} x = \rho \sin\theta \cos\varphi \\ y = \rho \sin\theta \sin\varphi \\ z = \rho \cos\theta \end{cases}$$

where $(\theta, \phi, \rho) \in [0, \pi] \times [0, 2\pi) \times (0, +\infty)$ the region is

$$\Omega = \{(\theta, \varphi, \rho) : 0 \leq \theta \leq \pi/4, 0 \leq \varphi \leq 2\pi, a \leq \rho \leq b\}.$$

Recalling that the Jacobian determinant has absolute value $\rho^2 \sin\theta$, the required volume is (Fig. 5.25)

$$|\Omega| = \int_0^{2\pi} \left[\int_a^b \left(\int_0^{\pi/4} \sin\theta\, d\theta \right) \rho^2 d\rho \right] d\varphi = \pi \left(\frac{b^3 - a^3}{3} \right) (2 - \sqrt{2}).$$

Given the spherical symmetry of the problem, it is natural to introduce spherical coordinates. As a matter of fact, the region Ω in the spherical coordinates is a regular parallelepiped, namely $[0, \pi/4] \times [0, 2\pi) \times [a, b]$, and the function to be integrated is of course $\rho^2 \sin\theta$. The integral factors in the product of three elementary one-dimensional integrals.

Fig. 5.25 The region of Exercise 5.26 is obtained by rotating around the vertical axis the darker region represented here, which is the piece of circular corona which lies inside the cone

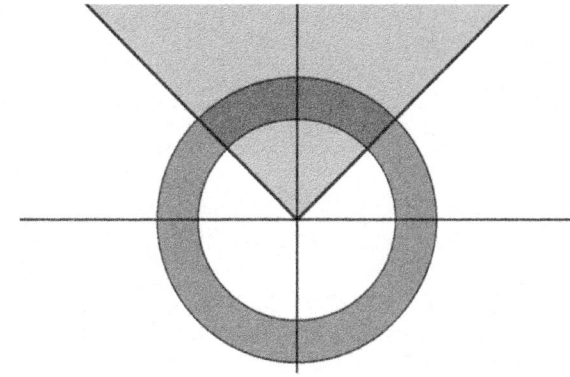

5.27 Compute the integral $\iiint_\Omega \dfrac{2x+1}{\sqrt{y^2+z^2}}\,dxdydz$, where

$$\Omega = \left\{(x,y,z) \in \mathbb{R}^3 : 1 \leq y^2 + z^2 \leq 9,\, 0 \leq x \leq 4 - \sqrt{16 - y^2 - z^2}\right\}.$$

Answer. In the cylindrical coordinates

$$\begin{cases} x = x \\ y = \rho \cos\theta \\ z = \rho \sin\theta \end{cases}$$

the domain Ω is described by $1 \leq \rho \leq 3$, $0 \leq \theta \leq 2\pi$ and $0 \leq x \leq 4 - \sqrt{16 - \rho^2}$. Therefore

$$\iiint_\Omega \frac{2x+1}{\sqrt{y^2+z^2}}\,dxdydz = \int_0^{2\pi}\left[\int_1^3\left(\int_0^{4-\sqrt{16-\rho^2}}\frac{2x+1}{\rho}\,dx\right)\rho\,d\rho\right]d\theta$$

$$= 2\pi \int_1^3 \left[(x^2 + x)\right]_0^{4-\sqrt{16-\rho^2}}\,d\rho$$

$$= 2\pi \int_1^3 \left(36 - 9\sqrt{16-\rho^2} - \rho^2\right)d\rho$$

$$= 2\pi\left[36\rho - 9\left(\frac{\rho}{2}\sqrt{16-\rho^2} + 8\arcsin(\rho/4)\right) - \frac{\rho^3}{3}\right]_1^3.$$

Here it was used that:

$$\int \sqrt{16-\rho^2}\,d\rho = \frac{\rho}{2}\sqrt{16-\rho^2} + 8\arcsin(\rho/4) + c$$

which may be derived with the substitution $\rho = 4\cos t$. A tedious computation gives then the numerical result

$$2\pi \left[\frac{190}{3} - \frac{27}{2}\sqrt{7} + \frac{9}{2}\sqrt{15} + 72(\arcsin(1/4) - \arcsin(3/4)) \right].$$

The geometry of the problem is not entirely trivial, as Ω is the region contained between the two cylinders with symmetry axis the x-axis with bases $y^2 + z^2 = 1$ and $y^2 + z^2 = 9$, respectively, region that also lies inside the hemisphere tangent to the yz-plane with radius 4 and centered at $(4, 0, 0)$. All this, however, can be safely ignored and one can simply observe that the "important" variable is $\sqrt{y^2 + z^2}$, so that in the appropriate cylindrical coordinates everything becomes quite easy. In other words, the algebra is easier than the geometry, so one exploits this.

5.28 Compute the volume of the so-called Viviani's window, namely

$$\Omega = \left\{ (x, y, z) \in \mathbb{R}^3 : x^2 + y^2 + z^2 \le 4, x^2 + (y-1)^2 \le 1 \right\}.$$

Answer. The conditions on z read $-\sqrt{4 - (x^2 + y^2)} \le z \le \sqrt{4 - (x^2 + y^2)}$, whereby $x^2 + (y-1)^2 \le 1$. Using the cylindrical coordinates as in Exercise 5.22 but with $\theta \in [-\pi, \pi)$, the second condition becomes

$$\rho^2 \cos^2 \theta + (\rho \sin \theta - 1)^2 \le 1 \Leftrightarrow \rho(\rho - 2\sin\theta) \le 0 \Leftrightarrow 0 \le \rho \le 2\sin\theta$$

and hence, putting

$$D = \left\{ (\rho, \theta) \in (0, +\infty) \times [-\pi, \pi) : 0 \le \rho \le 2\sin\theta, -\frac{\pi}{2} \le \theta \le \frac{\pi}{2} \right\},$$

one gets

$$|\Omega| = \iint_D \left(\int_{-\sqrt{4-\rho^2}}^{\sqrt{4-\rho^2}} dz \right) \rho \, d\rho \, d\theta$$

$$= 2 \iint_D \sqrt{4 - \rho^2} \, \rho \, d\rho \, d\theta$$

$$= - \int_{-\pi/2}^{\pi/2} \left[\int_0^{2\sin\theta} (-2\rho)\sqrt{4-\rho^2} \, d\rho \right] d\theta$$

$$= - \int_{-\pi/2}^{\pi/2} \left[\frac{(4-\rho^2)^{3/2}}{3/2} \right]_0^{2\sin\theta} d\theta$$

$$= \frac{2}{3} \int_{-\pi/2}^{\pi/2} \left(8 - 8(1 - \sin^2\theta)^{3/2} \right) d\theta$$

$$= \frac{16}{3} \int_{-\pi/2}^{\pi/2} (1 - \cos^3\theta) d\theta$$

$$= \frac{16}{3}\pi - \frac{16}{3} \int_{-\pi/2}^{\pi/2} \cos\theta (1 - \sin^2\theta) d\theta$$

$$= \frac{16}{3}\pi - \frac{64}{9}.$$

Viviani's window is the portion of the sphere that lies inside the vertical cylinder with basis the circle D in the xy-plane that has center at $(0, 1)$ and has radius 1. The intersection of the sphere and the cylinder is an eight-shaped curve on the sphere, called Viviani's curve. Thus, one integrates by vertical segments based on the region D, and each segment has one extreme inside one of the two loops of Viviani's curve, and the other extreme in the opposite loop of Viviani's curve. The standard cylindrical coordinates are easier to work with than those centered at $(0, 1)$ in the xy-plane.

5.29 Compute the integral $\iiint_\Omega \sqrt{4 - x^2} dx dy dz$, where

$$\Omega = \left\{ (x, y, z) \in \mathbb{R}^3 : x^2 + 2|y| \le 4,\ 0 \le z \le \sqrt{\frac{4 - x^2 - 2|y|}{3}} \right\}.$$

Answer. Observe that the function $f(x, y, z) = \sqrt{4 - x^2}$ is defined (and continuous) in the region

$$A = \{(x, y, z) \in \mathbb{R}^3 : |x| \le 2\}$$

which strictly contains Ω. Integrating by segments,

$$\iiint_\Omega \sqrt{4 - x^2} dx dy dz = \iint_D \left(\int_0^{\sqrt{\frac{4-x^2-2|y|}{3}}} \sqrt{4 - x^2} dz \right) dx dy,$$

where

$$D = \{(x, y) \in \mathbb{R}^2 : x^2 + 2|y| \le 4\}.$$

The region D is normal to the x-axis, for if $x \in [-2, 2]$ and $|y| \le -(x^2/2) + 2$, then

$$D = \left\{ (x, y) \in \mathbb{R}^2 : x \in [-2, 2],\ -2 + \frac{x^2}{2} \le y \le 2 - \frac{x^2}{2} \right\}.$$

Therefore, using the parity of $y \mapsto \sqrt{4 - x^2 - 2|y|}$,

$$\iiint_\Omega \sqrt{4-x^2} dx dy dz = \int_{-2}^{2} \left[\int_{-2+\frac{x^2}{2}}^{2-\frac{x^2}{2}} \sqrt{4-x^2} \frac{1}{\sqrt{3}} \sqrt{4-x^2-2|y|} dy \right] dx$$

$$= \frac{2}{\sqrt{3}} \int_{-2}^{2} \left[\int_{0}^{2-\frac{x^2}{2}} \sqrt{4-x^2-2y} dy \right] \sqrt{4-x^2} dx$$

$$= \frac{2}{\sqrt{3}} \int_{-2}^{2} \left[-\frac{1}{3}(4-x^2-2y)^{3/2} \right]_{0}^{2-\frac{x^2}{2}} \sqrt{4-x^2} dx$$

$$= \frac{2}{3\sqrt{3}} \int_{-2}^{2} (4-x^2)^2 dx$$

$$= \frac{1024}{135} \sqrt{3}.$$

The shape of Ω and that of its "basis" D are represented in Fig. 5.26. Clearly, there is some symmetry due to the parity of the function $x^2 + 2|y|$. Further, it is clear that the solution of the equation $x^2 + 2|y| = 4$ splits into two branches: when $y > 0$ it amounts to $y = 2 - \frac{1}{2}x^2$ and when $y < 0$ it amounts to $y = -2 + \frac{1}{2}x^2$, which explains the picture. The parabolic rather than circular nature of the symmetries suggestes to stick with Cartesian coordinates rather than passing to cylindrical ones.

5.30 Compute the integral $\iiint_\Omega \dfrac{dxdydz}{x^2+y^2+z^2+1}$ where

$$\Omega = \left\{ (x,y,z) \in \mathbb{R}^3 : x^2+y^2+(z+3)^2 \leq 9, x^2+y^2+(z+1)^2 \geq 1 \right\}.$$

Answer. The region Ω is the ball B_1 of radius 3 and centered at $(0, 0, -3)$ from which the ball B_2 of radius 1 centered at $(0, 0, -1)$ has been removed. Both balls are tangent to the origin and lie below the plane $z = 0$. Using the spherical coordinates as in Exercise 5.26 the inequalities defining Ω become

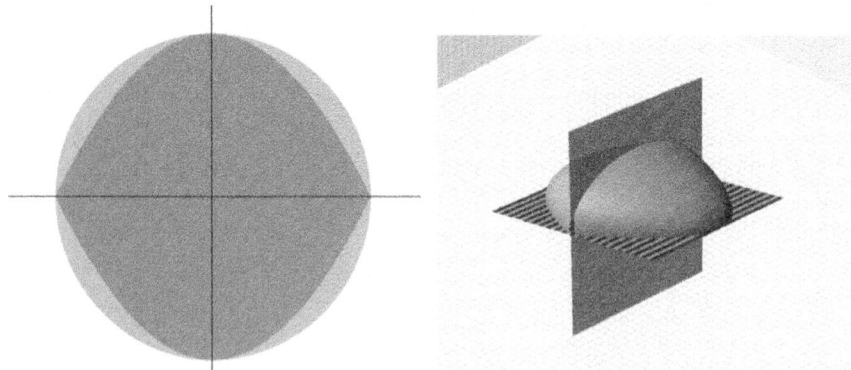

Fig. 5.26 On the left the circle of radius 2 containing the region D, and on the right the region Ω of Exercise 5.29

5.5 Guided Exercises

$$\begin{cases} x^2+y^2+z^2+6z \le 0 \\ x^2+y^2+z^2+2z \ge 0 \end{cases} \Leftrightarrow \begin{cases} \rho^2+6\rho\cos\theta \le 0 \\ \rho^2+2\rho\cos\theta \ge 0 \end{cases}$$

and hence $-2\cos\theta \le \rho \le -6\cos\theta$. Since both balls lies below the plane $z=0$, $\pi/2 \le \theta \le \pi$. Recalling that the Jacobian determinant has absolute value $\rho^2\sin\theta$ and denoting by I the required integral,

$$I = \int_0^{2\pi}\left[\int_{\pi/2}^{\pi}\left(\int_{-2\cos\theta}^{-6\cos\theta}\frac{\rho^2\sin\theta}{\rho^2+1}d\rho\right)d\theta\right]d\varphi$$

$$= 2\pi\int_{\pi/2}^{\pi}\left(\int_{-2\cos\theta}^{-6\cos\theta}\frac{\rho^2}{\rho^2+1}d\rho\right)\sin\theta d\theta$$

$$= 2\pi\int_{\pi/2}^{\pi}[\rho]_{-2\cos\theta}^{-6\cos\theta}\sin\theta d\theta - 2\pi\int_{\pi/2}^{\pi}[\arctan\rho]_{-2\cos\theta}^{-6\cos\theta}\sin\theta d\theta$$

$$= -8\pi\int_{\pi/2}^{\pi}\sin\theta\cos\theta d\theta$$

$$- 2\pi\int_{\pi/2}^{\pi}[-\arctan(6\cos\theta)+\arctan(2\cos\theta)]\sin\theta d\theta.$$

The first integral is simply 4π, while for the second it is convenient to make the change of variable $u=\cos\varphi$, which leads to

$$2\pi\int_{-1}^{0}(\arctan(2u)-\arctan(6u))du = 2\pi[u(\arctan(2u)-\arctan(6u))]_{-1}^{0}$$

$$-2\pi\int_{-1}^{0}\left(\frac{6u}{1+36u^2}-\frac{2u}{1+4u^2}\right)du$$

$$= 2\pi(\arctan 2 - \arctan 6)$$

$$-\frac{\pi}{6}[\log(1+36u^2)]_{-1}^{0} + \frac{\pi}{2}[\log(1+4u^2)]_{-1}^{0}.$$

The final result is tediously checked to be

$$\iiint_\Omega \frac{dxdydz}{x^2+y^2+z^2+1} = 2\pi(2+\arctan 2-\arctan 6) + \frac{\pi}{6}\log 37 - \frac{\pi}{2}\log 5.$$

In Fig. 5.27 it is represented a planar section of Ω that clarifies the geometry. It is useful to introduce various polar coordinates in this plane and work out the inequalities that describe the region in lighter grey. It might be a little surprising that the the best description is achieved with polar coordinates centered at the origin and not with either of the alternative systems centered at the centers of the two circles. The computation of the integral is then a standard exercise in integration.

Fig. 5.27 The intersection of the region Ω of Exercise 5.30 with the plane $y = 0$

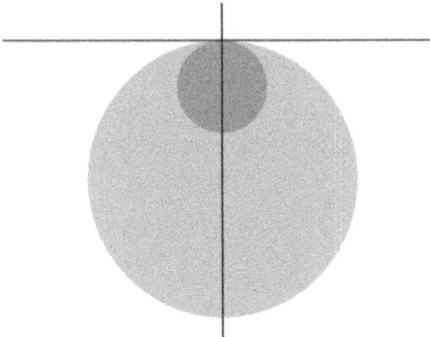

5.31 Compute the volume of the solid of revolution Ω obtained by rotating around the x-axis the planar section

$$S = \{(x, y) \in \mathbb{R}^2 : 0 \leq y \leq \min(1 - x^2, 1 - x)\}.$$

Answer. Observe that the section S is normal to the y-axis, with $y \in [0, 1]$ and $-\sqrt{1-y} \leq x \leq 1 - y$, see Fig. 5.28 on the left. Using Guldino's theorem,

$$|\Omega| = 2\pi \iint_S y \, dx \, dy$$
$$= 2\pi \int_0^1 \left(\int_{-\sqrt{1-y}}^{1-y} y \, dx \right) dy$$
$$= 2\pi \int_0^1 \left(y - y^2 + y\sqrt{1-y} \right) dy.$$

It is useful to observe that with the change of variables $u = 1 - y$

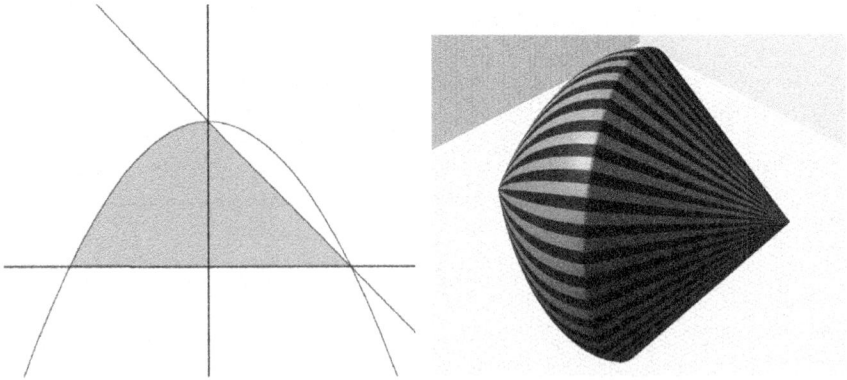

Fig. 5.28 On the left the section S, and on the right the solid of revolution Ω of Exercise 5.31

5.5 Guided Exercises

$$\int y\sqrt{1-y}\,dy = \int (u^{3/2} - u^{1/2})\,du = \frac{2}{5}(1-y)^{5/2} - \frac{2}{3}(1-y)^{3/2} + c,$$

which is easily checked to yield $|\Omega| = \frac{13}{15}\pi$.

Here it is just required to understand S and to view it as a region normal to the y-axis. Then one plainly applies Guldino's theorem.

5.32 Consider the iterated integral

$$I = \int_{-1}^{1}\left[\int_{0}^{\sqrt{1-x^2}}\left(\int_{-2}^{y+1} f(x,y,z)\,dz\right)dy\right]dx.$$

(a) Express I as a triple integral over a region Ω, and describe Ω.
(b) Express I in two other ways as iterated integrals.
(c) Compute I when $f(x, y, z) = xe^{z-1}$.

Answer. (a) Evidently, $I = \iiint_\Omega f(x, y, z)\,dxdydz$, where

$$\Omega = \left\{(x, y, z) \in \mathbb{R}^3 : -1 \le x \le 1, 0 \le y \le \sqrt{1-x^2}, -2 \le z \le y+1\right\}.$$

Geometrically, the region Ω is the portion of the cylinder $x^2 + y^2 \le 1$ where $y \ge 0$, which is bounded below by the plane $z = -2$ and above by the plane $z = y + 1$. The way I appears as an iterated integral is an integration by segments of the kind

$$I = \iint_D \left(\int_{-2}^{y+1} f(x, y, z)\,dz\right)dxdy$$

where D is the semicircle $x^2 + y^2 \le 1$ where $y \ge 0$, namely

$$D = \{(x, y) \in \mathbb{R}^2 : x^2 + y^2 \le 1, y \ge 0\}$$

and where the integral over D is then realized by viewing it as normal to the x-axis.

(b) Two possible ways to write I are either as an integration by segments viewing D as normal to the y-axis, or by sections parallel to the xy-plane, see Fig. 5.29. Concretely,

$$I = \int_{0}^{1}\left[\int_{-\sqrt{1-y^2}}^{\sqrt{1-y^2}}\left(\int_{-2}^{y+1} f(x, y, z)\,dz\right)dx\right]dy$$

$$I = \int_{-2}^{1}\left[\iint_S f(x, y, z)\,dxdx\right]dz + \int_{1}^{2}\left[\iint_{T_z} f(x, y, z)\,dxdx\right]dz,$$

Fig. 5.29 The various planes and the hemicylinder that bound the region Ω of Exercise 5.32

where

$$S = \left\{(x,y) \in \mathbb{R}^2 : -1 \le x \le 1, 0 \le y \le \sqrt{1-x^2}\right\}$$
$$T_z = \left\{(x,y) \in \mathbb{R}^2 : -1 \le x \le 1, z-1 \le y \le \sqrt{1-x^2}\right\}.$$

(c) Using the first expression of I,

$$I = \int_{-1}^{1} \left[\int_{0}^{\sqrt{1-x^2}} \left(\int_{-2}^{y+1} xe^{z-1} dz \right) dy \right] dx$$

$$= \int_{-1}^{1} x \left[\int_{0}^{\sqrt{1-x^2}} (e^y - e^{-3}) dy \right] dx$$

$$= \int_{-1}^{1} x \left[e^y - ye^{-3} \right]_{0}^{\sqrt{1-x^2}} dx$$

$$= \int_{-1}^{1} x \left(e^{\sqrt{1-x^2}} - e^{-3}\sqrt{1-x^2} - 1 \right) dx = 0$$

because the function $x \mapsto x\left(e^{\sqrt{1-x^2}} - e^{-3}\sqrt{1-x^2} - 1\right)$ is odd and is symmetric with respect to the origin.

The first task consists in understanding the geometry of Ω, which is not very hard because the equations are commanded by the iterated integral I. Since the innermost integral appearing in I is $\int_{-2}^{y+1} f(x,y,z) dz$, it is natural to guess that I is the result of an integration by segments over some region D, the shape of which is then easily found to be a hemicylinder. Since D is actually normal both to the y-axis and to the x-axis, the integration can be carried out decomposing D as a union of segments in two possible ways. This leads to the second expression of I. Alternatively, one can integrate by sections orthogonal to the z-axis. In this latter case, however, the sections have a constant semicircular shape for $-2 \leq z \leq 1$, namely $S_z = S$, and for $1 \leq z \leq 2$ they have the varying shapes T_z. Finally, it is quite clear that since xe^{z-1} is a product of functions of different variables that does not depend on y, the best way to compute I is to start by z and end by x. Symmetry considerations allow to conclude with no further calculations.

5.33 Compute the moment of inertia of the parallelepiped Ω (with constant density of mass equal to 1) bounded by the planes $y = x + 3$, $y = x + 4$, $z = -1$, $z = 2$, $x = 1$ and $x = 2$, relative to the line of equation $x = y = z$.

Answer. A parametric expression of the line l of equation $x = y = z$ is given by $\{(t, t, t) : t \in \mathbb{R}\}$, so that for $P = (x, y, z) \in \mathbb{R}^3$

$$d_\ell(P) = \frac{\|(x, y, z) \wedge (1, 1, 1)\|}{\sqrt{3}} = \frac{\sqrt{(y-z)^2 + (z-x)^2 + (x-y)^2}}{\sqrt{3}}.$$

Therefore

$$I_\ell = \iiint_\Omega d_\ell^2(x, y, z)\,dxdydz = \iiint_\Omega \frac{(y-z)^2 + (z-x)^2 + (x-y)^2}{3}\,dxdydz.$$

Now, since $\Omega = \{(x, y, z) \in \mathbb{R}^3 : (x, z) \in [-1, 2] \times [-1, 2], x + 3 \leq y \leq x + 4\}$, it is natural to integrate by segments normal to $D = [-1, 2] \times [-1, 2]$, that is

$$\begin{aligned}
I_\ell &= \frac{1}{3} \iint_D \left[\int_{x+3}^{x+4} (y-z)^2 + (z-x)^2 + (x-y)^2\,dy \right] dxdz \\
&= \frac{1}{3} \iint_D \left[\frac{(y-z)^3}{3} + y(z-x)^2 - \frac{(x-y)^3}{3} \right]_{x+3}^{x+4} dxdz \\
&= \frac{1}{9} \int_{-1}^{2} \left[\int_{-1}^{2} \left((x+4-z)^3 - (x+3-z)^3 + 37 + 3(x-z)^2 \right) dx \right] dz \\
&= \frac{1}{9} \int_{-1}^{2} \left[\frac{(x+4-z)^4}{4} - \frac{(x+3-z)^4}{4} + 37x + (x-z)^3 \right]_{-1}^{2} dz \\
&= \frac{1259}{15}
\end{aligned}$$

where in the last line a number of tedious but essentially trivial computations have been omitted Fig. 5.30.

Apart from applying the appropriate formulae for the computation of the moment of inertia, here it is useful to observe that the parallelepiped is best viewed by taking as reference plane the xz-plane, in the sense that Ω is a union of segments issuing from

Fig. 5.30 A qualitative drawing of Ω in Exercise 5.33: the horizontal plane is the xz-plane

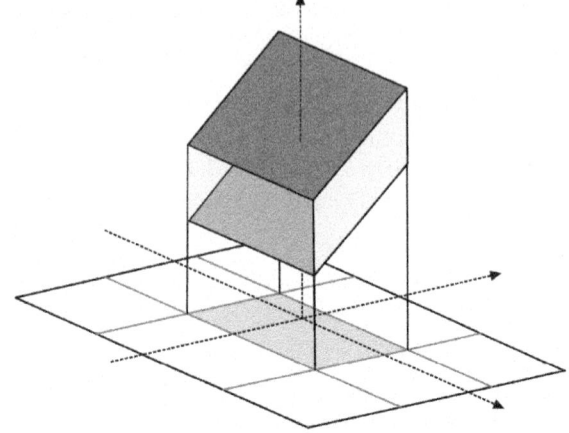

the square $D = [-1, 2] \times [-1, 2]$ in that plane (see Fig. 5.30). The actual integral is then a rather tedious but ultimately elementary manipulation of monomials.

5.6 Problems

5.34 Compute $\iint_\Omega f(x, y)\,dxdy$, where:

(a) $f(x, y) = e^{y^3}$,
$$\Omega = \left\{(x, y) \in \mathbb{R}^2 : 0 \leq x \leq 1, \sqrt{x} \leq y \leq 1\right\}$$

(b) $f(x, y) = x(1 + y)$,
$$\Omega = \left\{(x, y) \in \mathbb{R}^2 : 0 \leq x \leq 2, -2 \leq y \leq 0, xy \geq -2\right\}$$

(c) $f(x, y) = \dfrac{y}{x^2 + y^2}$
$$\Omega = \left\{(x, y) \in \mathbb{R}^2 : 0 \leq x \leq 3, \sqrt{1 - x^2/9} \leq y \leq \sqrt{9 - x^2}\right\}$$

(d) $f(x, y) = x^2 y^2$,
$$\Omega = \left\{(x, y) \in \mathbb{R}^2 : |x| + |y| \geq 4, \max(|x|, |y|) \leq 4\right\}$$

(e) $f(x, y) = \sqrt{x + y} + \sin(x - y)$,

$$\Omega = \{(x,y) \in \mathbb{R}^2 : |x-1| + |y-1| \le 1\}$$

(f) $f(x,y) = \dfrac{x^3 + y}{x + y}$,
$$\Omega = \{(x,y) \in \mathbb{R}^2 : 1 \le x + y \le 2, x \ge 0, y \ge 0\}$$

(g) $f(x,y) = x^2 + y^2 + xy + x - y - 1/2$,
$$\Omega = \{(x,y) \in \mathbb{R}^2 : x^2 + y^2 \le 1\}$$

(h) $f(x,y) = \dfrac{xy}{x^2 + y^2}$,
$$\Omega = \left\{(x,y) \in \mathbb{R}^2 : x^2 + y^2 \ge 4, 0 \le y \le -\sqrt{3}x/3, -2 \le x \le 0\right\}$$

(i) $f(x,y) = x(y+1)$,
$$\Omega = \left\{(x,y) \in \mathbb{R}^2 : x^2 + y^2 \le 1, y \ge -x - 1, y \ge \sqrt{3}x - 1\right\}$$

(j) $f(x,y) = \sin(x^2 + y^2) - \dfrac{y}{x}$,
$$\Omega = \left\{(x,y) \in \mathbb{R}^2 : 1 \le x^2 + y^2 \le 4, -x \le y \le \sqrt{3}x\right\}$$

(k) $f(x,y) = x\sqrt{x^2 + y^2}$,
$$\Omega = \{(x,y) \in \mathbb{R}^2 : x^2 + (y-2)^2 \le 4, (x+2)^2 + y^2 \le 4\}$$

(l) $f(x,y) = xy^2$,
$$\Omega = \left\{(x,y) \in \mathbb{R}^2 : 1/\sqrt{x} \le y \le 2/\sqrt{x}, \sqrt{x} \le y \le \sqrt{2x}\right\}$$

(m) $f(x,y) = \dfrac{x^2}{y^2}$,
$$\Omega = \{(x,y) \in \mathbb{R}^2 : x^2 + y^2 \ge 9, x^2 + (y-2)^2 \le 4\}.$$

5.35 Compute the moment of inertia with respect to the x-axis of the planar region $A = \left\{(x,y) \in \mathbb{R}^2 : -1 \le x \le 2, |y| \le \sqrt[3]{1 + x^3}\right\}$ constant mass density $\mu(x,y) = 1$.

5.36 Let $\Omega = \{(x,y) \in \mathbb{R}^2 : (x-1)^2 + (y-1)^2 \le 2, (x+1)^2 + (y-1)^2 \ge 2\}$ and

$$f(x, y) = \begin{cases} \dfrac{x+y}{\sqrt{x^2+y^2}} & (x, y) \neq (0, 0) \\ 3 & (x, y) = (0, 0). \end{cases}$$

(a) Show that $f \in \mathbf{R}(\Omega)$.
(b) Compute the integral $\iint_\Omega f(x, y) dx dy$.

5.37 Compute $\iint_\Omega f(\theta, \rho) d\theta d\rho$, where $f(\theta, \rho) = \rho^2 \cos \theta$ and where

$$\Omega = \{(\theta, \rho) \in [0, 2\pi) \times (0, +\infty) : 0 < \rho \leq 1,$$
$$\rho^2 \sin \theta \cos \theta + 2\rho (\sin \theta + \cos \theta) \leq 2.\}$$

5.38 Compute the area of the region between the curves whose expressions in polar coordinates are $\rho = \theta$, $\rho = \theta^2$, $\theta = 1$ and $\theta = 5\pi/4$.

5.39 Compute the area of the region inside the curve $\left(\dfrac{x^2}{9} + \dfrac{y^2}{25} \right)^2 + \dfrac{x^2}{9} - \dfrac{y^2}{25} = 0$.

5.40 Compute the coordinates of the barycenter of the region

$$\Omega = \{(x, y) \in \mathbb{R}^2 : |x| \leq y \leq 2|x|, x^2 + y^2 \leq 4\}$$

with respect to any constant mass density function μ.

5.41 Compute the coordinates of the barycenter of the region

$$\Omega = \{(x, y) \in \mathbb{R}^2 : 4x^2 + 9y^2 \leq 36, y \geq 0\}$$

with respect to the mass density function $\mu(x, y) = x^2 y^2$.

5.42 Compute the coordinates of the barycenter of the region

$$\Omega = \{(x, y) \in \mathbb{R}^2 : x \geq y^2 - 2y, y \geq -x, 2y \geq x\}$$

with respect to any constant mass density function μ.

5.43 Compute $\iiint_\Omega f(x, y) dx dy$, where

(a) $f(x, y, z) = xyz$,

$$\Omega = \{(x, y, z) \in \mathbb{R}^3 : x^2 + y^2 + z^2 \leq 1, x \geq 0, y \geq 0, z \geq 0\}$$

(b) $f(x, y, z) = x^2 - y^2 + zy$,

$$\Omega = \{(x, y, z) \in \mathbb{R}^3 : x^2 + y^2 + z^2 \leq 1, x^2 + y^2 - z^2 \leq 0, x \geq 0, y \geq 0, z \geq 0\}$$

(c) $f(x, y, z) = y$,

5.6 Problems

$$\Omega = \{(x, y, z) \in \mathbb{R}^3 : x^2 + z^2 \leq y \leq x + z + 1\}$$

(d) $f(x, y, z) = z$,

$$\Omega = \{(x, y, z) \in \mathbb{R}^3 : 1 \leq x^2 + y^2 + z^2 \leq 2z\}$$

(e) $f(x, y, z) = z$,

$$\Omega = \{(x, y, z) \in \mathbb{R}^3 : 4 \leq x^2 + y^2 + z^2 \leq 9, x + y \geq 0, x \geq 0, z \geq 0\}$$

(f) $f(x, y, z) = x\sqrt{y^2 + z^2}$,

$$\Omega = \{(x, y, z) \in \mathbb{R}^3 : x^2 + y^2 + z^2 \leq 16, x \geq 2\}$$

(g) $f(x, y, z) = z$,

$$\Omega = \{(x, y, z) \in \mathbb{R}^3 : x^2 + y^2 \leq (2 + \sin z)^2, 0 \leq z \leq 2\pi\}$$

(h) $f(x, y, z) = (1 + x^2 + y^2 + z^2)^{-1}$

$$\Omega = \{(x, y, z) \in \mathbb{R}^3 : x^2 + y^2 + (z-1)^2 \geq 1, x^2 + y^2 + (z-2)^2 \leq 4\}$$

(i) $f(x, y, z) = xyz$,

$$\Omega = \{(x, y, z) \in \mathbb{R}^3 : 2y^2 - x^2 - z^2 \geq 2, x^2 + (z+2)^2 \geq 4, 1 \leq y \leq 3, x \geq 0\}.$$

5.44 Compute $\iiint_\Omega x\,dx\,dy\,dz$ where Ω is the tetrahedron with vertices $(0, 0, 0)$, $(4, 0, 0)$, $(0, 4, 0)$ and $(0, 0, 2)$.

5.45 Compute the volume of $\{(x, y, z) \in \mathbb{R}^3 : z \geq x^2 + y^2, (z-2)^2 \geq x^2 + y^2, z \leq 2\}$.

5.46 Compute the volume of $\left\{(x, y, z) \in \mathbb{R}^3 : \frac{x^2}{16} + \frac{y^2}{9} + \frac{z^2}{4} \leq 1\right\}$.

5.47 Compute the volume of the sphere centered at the origin and radius 3, using spherical, cylindrical and Cartesian coordinates.

5.48 Consider the region $\Omega(h) = \left\{(x, y, z) \in \mathbb{R}^3 : x^2 + y^2 \leq 1/z^2, \sqrt[4]{8} \leq z \leq h\right\}$.

(a) Compute the volume $|\Omega(h)|$ of $\Omega(h)$.
(b) Compute, if existing, $\lim_{h \to +\infty} |\Omega(h)|$.

5.49 Compute the volume of $\{(x, y, z) \in \mathbb{R}^3 : x^2 + (z-1)^2 - 2 \le y \le 1 - x^2 - (z+1)^2\}$.

5.50 Compute the integral $\iiint_\Omega z\,dxdydz$ where Ω is the revolution solid around the z-axis of the planar region $\{(y, z) \in \mathbb{R}^2 : 1 \le z \le \sqrt{2}, 0 \le y \le z^2\}$.

5.51 Compute the barycenter of the region

$$\Omega = \left\{(x, y, z) \in \mathbb{R}^3 : 0 \le z \le 8 - \sqrt{x^2 + y^2}, x^2 + y^2 \le 4, x^2 + y^2 + z^2 \ge 1\right\}$$

for any constant density of mass.

5.52 Compute the moment of inertia with respect to the z-axis of the region

$$\Omega = \left\{(x, y, z) \in \mathbb{R}^3 : x^2 + z^2 \le (y-2)^2, 1 \le y \le 4\right\}$$

with constant density of mass $\mu(x, y, z) = 1$.

Reference

1. Baronti, M., De Mari, F., van der Putten, R., Venturi, I.: Calculus problems, Unitext, 101, La Matematica per il 3+2. Springer (2016)

Chapter 6
Sequences and Series of Functions

Sequences of functions are pervasive in Mathematics and arise naturally in the process of approximating a given function by simpler ones, at least locally. This is for example the case when one looks at the Taylor polynomials of a given function f which is very regular in an open neighborhood I of a given point x_0, say for simplicity $f \in C^\infty(I)$. Upon computing higher and higher order derivatives of f at x_0 one builds the sequence $(T_{n,x_0} f)_{n \geq 0}$ of the Taylor polynomials centered at x_0 that are actually defined on the whole $\mathbb{R}$. It is natural to ask under what circumstances the numerical sequence obtained by evaluating each polynomial at one at the same point, say $x \in I$, converges to $f(x)$. This example explains also how the idea of series of functions comes about, for each of the polynomials $T_{n,x_0} f$ is constructed by summing powers so that again it is natural to ask what happens by fixing a point and considering the numerical series that arises. Both these examples will be discussed below but are just instances of a much more general picture.

In this chapter, I denotes a non empty subset of $\mathbb{R}$. Typically, I is an interval but this is actually not always necessary.

6.1 Pointwise and Uniform Convergence

6.1.1 Sequences

A *sequence* of functions defined on I, denoted $(f_n)_{n \geq 0}$, is a countable collection of functions $f_n : I \to \mathbb{R}$, each of which is defined on I. Here n is any natural number, though sometimes $n \geq n_0$ for some fixed positive integer n_0. The sequence $(f_n)_{n \geq 0}$ can also be thought of as a collection of numerical sequences, one for each $x \in I$, namely the sequences $(f_n(x))_{n \geq 0}$ obtained by evaluating the functions f_n at $x \in I$. It is worth observing that possibly some of the f_n is defined on a larger set, but what is crucial is that a non empty set I is contained in all the domains of the functions of the sequence, at least for n large enough.

Definition 6.1 (*Pointwise convergence of sequences*) The sequence $(f_n)_{n \geq 0}$ defined on I *converges pointwise* on the subset J of I to the function $f : J \to \mathbb{R}$ if for every $x \in J$
$$\lim_n f_n(x) = f(x).$$

If $J = I$ then $(f_n)_{n \geq 0}$ is said to converge pointwise to f. It is common to write $f_n \to f$ on J, or simply $f_n \to f$ whenever $J = I$.

A sequence may converge pointwise on a proper subset of its domain. For example, the sequence defined on $I = [0, +\infty)$ by $f_n(x) = x^n$ clearly converges pointwise on $J = [0, 1]$ to the function
$$f(x) = \begin{cases} 0 & 0 \leq x < 1 \\ 1 & x = 1. \end{cases}$$

However, x^n diverges to $+\infty$ whenever $x > 1$, so that on no larger subset of I does the sequence converge pointwise to some well-defined function.

Definition 6.2 (*Uniform convergence of sequences*) The sequence $(f_n)_{n \geq 0}$ defined on I *converges uniformly* on the subset J of I to the function $f : J \to \mathbb{R}$ if
$$\lim_n \sup_{x \in J} |f_n(x) - f(x)| = 0.$$

If $J = I$ then $(f_n)_{n \geq 0}$ is said to converge uniformly to f. It is common to write $f_n \rightrightarrows f$ on J, or simply $f_n \rightrightarrows f$ whenever $J = I$.

An example of a sequence converging uniformly on $[0, 1]$ is $f_n(x) = x^{1+1/n}$. As $n \to \infty$, $f_n(x) \to x$ pointwise and furthermore
$$\sup_{0 \leq x \leq 1} |f_n(x) - f(x)| = \max_{0 \leq x \leq 1} \left(x - x^{1+1/n}\right) = \frac{1}{\left(1 + \frac{1}{n}\right)^n} \frac{1}{n+1}$$

as the reader may verify by computing the maximum of $x - x^{1+1/n}$ on $[0, 1]$. This obviously tends to 0 and hence the convergence is uniform.

It is easy to see that a sequence that converges uniformly on J also converges pointwise on J. The converse is not true. For example, as shown above, the sequence $f_n(x) = x^n$ converges pointwise on $[0, 1)$ to the function $f(x) = 0$, but the convergence is not uniform because
$$\sup_{0 \leq x < 1} |f_n(x) - f(x)| = \sup_{0 \leq x < 1} x^n = 1$$

which of course does not tend to 0 as $n \to \infty$.

6.1 Pointwise and Uniform Convergence

It is useful to observe that $f_n \rightrightarrows f$ on I means that for any fixed $\varepsilon > 0$ the graph of f_n for n large enough is entirely contained in the strip

$$S_\varepsilon(f) = \{(x, y) \in I \times \mathbb{R} : |y - f(x)| < \varepsilon\}$$

of the plane (Fig. 6.1).

Proposition 6.1 (Cauchy criterion for uniform convergence) *The sequence $(f_n)_{n \geq 0}$ defined on I converges uniformly on the subset J of I to the function $f : J \to \mathbb{R}$ if and only if for every $\varepsilon > 0$ there exists n_ε such that if $n, m > n_\varepsilon$, then*

$$\sup_{x \in J} |f_n(x) - f_m(x)| < \varepsilon.$$

Uniform convergence is important because many properties of the individual functions of the sequence $(f_n)_{n \geq 0}$ are enjoyed also by the limit function f provided that $f_n \rightrightarrows f$, as the following results show.

Theorem 6.1 (Uniform convergence and continuity) *The uniform limit of a sequence of continuous functions is continuous. Thus, if $f_n \rightrightarrows f$ on I and all the f_n are continuous on I then also f is continuous on I.*

Theorem 6.2 (Uniform convergence and differentiability) *Let I be an interval and suppose that the functions $f_n : I \to \mathbb{R}$ are differentiable on I; suppose further that $f_n \to f$ and that $f_n' \rightrightarrows g$ on I. Then f is differentiable on I and $f' = g$.*

Theorem 6.3 (Uniform convergence and integration) *Let I be an interval and suppose that the functions $f_n : I \to \mathbb{R}$ are continuous on I and that $f_n \rightrightarrows f$ on I. Then*

$$\lim_n \int_a^b f_n(x)\,dx = \int_a^b f(x)\,dx$$

for any $a, b \in I$.

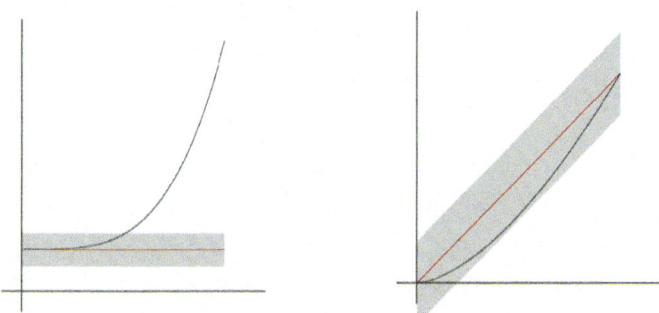

Fig. 6.1 The graph of $x^n + a$ is not entirely contained in the strip $S_\varepsilon(a)$ if ε is small (left); conversely, for any $\varepsilon > 0$ the graph of $x^{1+1/n}$ is entirely contained in $S_\varepsilon(x)$ for n large enough (right)

Theorem 6.4 (Dominated convergence) *Suppose that $(f_n)_{n\geq 0}$ are continuous on $[a, +\infty)$ and suppose that there is a function $f : [a, +\infty) \to \mathbb{R}_+$ such that;*

(i) $|f_n(x)| \leq f(x)$ for every $x \in [a, +\infty)$;
(ii) the integral $\int_a^{+\infty} f(x)\,dx$ converges.

Then $(f_n)_{n\geq 0}$ converges uniformly on $[a, +\infty)$ and

$$\lim_n \int_a^{+\infty} f_n(x)\,dx = \int_a^{+\infty} \lim_n f_n(x)\,dx.$$

Finally, some remarks on the nature of the set of uniform convergence of a sequence are in order. It is often the case that a sequence $(f_n)_{n\geq 1}$ converges uniformly on the compact intervals $[a, b]$ contained in the convergence set I but not in I itself. This is the case for example when $f_n(x) = x^n$, which converges in $I = [0, 1)$ and uniformly in every interval $[a, b]$ contained in I but not in I. In many circumstances, the uniform convergence on every interval $[a, b]$ can be inferred from that on nicer intervals. For example, if $I = \mathbb{R}$ and one shows uniform convergence on every interval of the form $[-\alpha, \alpha]$ for any positive α, then it also takes place on every interval $[a, b]$ because if $\alpha = \max\{|a|, |b|\}$, then obviously $[a, b] \subset [-\alpha, \alpha]$ and uniform convergence on a set I implies uniform convergence on any subset of I. Conversely, of course, uniform convergence on every interval $[a, b] \subset \mathbb{R}$ implies that on those of the form $[-\alpha, \alpha]$. Similarly, uniform convergence on a finite number of sets implies uniform convergence on their union, though this conclusion is in general false for infinite unions. Hence, for example, uniform convergence on every interval $[-\alpha, \alpha]$ does not in general imply uniform convergence on $\mathbb{R}$.

The above remarks have a particular relevance when it comes to proving either continuity or differentiability on a set I of the pointwise limit f of a sequence $(f_n)_{n\geq 1}$. For example, continuity of f at any $x_0 \in I$, will follow if $f_n \rightrightarrows f$ in some small interval $I_\varepsilon = [x_0 - \varepsilon, x_0 + \varepsilon] \subset I$ provided that all the f_n are continuous on I_ε. Similarly, if $(f_n)_{n\geq 1}$ converges and $(f'_n)_{n\geq 1}$ converges uniformly on I_ε then f is differentiable at x_0. Thus, since both properties are local, uniform convergence on the compact intervals contained in I allows to infer regularity properties of f on I.

6.1.2 Series

If a sequence $(f_n)_{n\geq 0}$ defined on I is given, it is natural to ask whether there is a natural process of summing its terms at each point $x \in I$.

Definition 6.3 (*Pointwise convergence of a series of functions*) Given the sequence $(f_n)_{n\geq 0}$ defined on I, the associated *sequence of partial sums* $(S_n)_{n\geq 0}$ is the sequence of functions on I defined by

6.1 Pointwise and Uniform Convergence

$$S_n(x) = \sum_{k=0}^{n} f_k(x), \quad x \in I.$$

The *series* $\sum_{k=0}^{\infty} f_k$ is said to converge pointwise on the subset J of I to the *sum* $S: J \to \mathbb{R}$ if for every $x \in J$

$$\lim_n S_n(x) = S(x).$$

If $J = I$ then the series $\sum_{k=0}^{\infty} f_k$ is said to converge pointwise to S. It is common to write $\sum_{k=0}^{\infty} f_k = S$.

Clearly, a necessary condition for the series $\sum_{k=0}^{\infty} f_k$ to converge pointwise on J is that $\lim_k f_k(x) = 0$ for every $x \in J$.

An example of converging series, for example for $x \in I = (-1, 1)$, is the so-called *geometric series* $\sum_{n=0}^{\infty} x^n$. More precisely, from the well known formula

$$1 + x + x^2 + \cdots + x^n = \frac{1 - x^{n+1}}{1 - x}$$

valid for $x \neq 1$ and every positive integer n, it follows that for $x \in (-1, 1)$

$$S_n(x) = \sum_{k=0}^{n} x^k = \frac{1 - x^{n+1}}{1 - x} \tag{6.1}$$

so that

$$\sum_{k=0}^{\infty} x^k = \lim_n \frac{1 - x^{n+1}}{1 - x} = \frac{1}{1 - x}, \quad x \in (-1, 1).$$

Thus the geometric series converges in $(-1, 1)$. The very same formulæ, in particular formula (6.1), show that the geometric series diverges positively in $(1, +\infty)$. A minute thought reveals that in fact it diverges positively in $[1, +\infty)$, and that $\lim_n S_n(x)$ does not exist for $x \leq -1$.

Particularly simple series are the so-called *telescopic series*, namely those of the form

$$\sum_{n=0}^{\infty} (f_{n+1} - f_n)$$

where $(f_n)_{n \geq 0}$ is a given sequence of functions. In complete analogy with the case of numerical series, the associated partial sum is

$$\sum_{k=0}^{n} (f_{k+1} - f_k) = f_{n+1} - f_0$$

and thus the convergence of the series reduces to that of the sequence.

There is a slight ambiguity in the notation $\sum_{k=0}^{\infty} f_k$ as well as in the term "series". What is really meant by either one is that starting from $(f_n)_{n\geq 0}$ one builds $(S_n)_{n\geq 0}$ and considers its limit, which may or may not exist. Formally, therefore,

$$\sum_{k=0}^{\infty} f_k(x) := \lim_n \sum_{k=0}^{n} f_k(x)$$

independently of the existence of the indicated limit. It should be clear that for any positive integers k_1 and k_2 the convergence of each of the series

$$\sum_{k=k_1}^{\infty} f_k(x), \quad \sum_{k=k_2}^{\infty} f_k(x)$$

implies and is implied by the convergence of the other, and that if one diverges so does the other and if one of the two limits does not exist neither does the other. For this reason, both for simplicity and in order to stress that the index set is irrelevant in the sense just explained, it is often best to write $\sum f_k$ without specifying the range of summation. Likewise, the label of the summation index is irrelevant, so $\sum f_k$ and $\sum f_n$ are identical expressions. If convergent, the actual value of the sum is obviously affected by the choice of the initial value of the summation index.

Definition 6.4 (*Uniform, absolute and total convergence of a series*) Let $(f_n)_{n\geq 0}$ be defined on I and let $J \subset I$ be non empty. The series $\sum f_n$ is said to converge:

(i) *uniformly* on J if $(S_n)_{n\geq 0}$ converges uniformly on J; in this case the sequence $(f_n)_{n\geq 0}$ converges uniformly to 0 on J;

(ii) *absolutely* on J if the series $\sum_{n=0}^{\infty} |f_n|$ converges pointwise on J;

(iii) *totally* on J if the numerical series $\sum_{n=0}^{\infty} \sup_{x \in J} |f_n(x)|$ converges.

Theorem 6.5 *Consider the series $\sum f_n$, where $(f_n)_{n\geq 0}$ are defined on I and suppose that $J \subset I$.*

(i) *If $\sum f_n$ converges totally on J, then it converges absolutely and uniformly on J.*
(ii) *If $\sum f_n$ converges absolutely on J, then it converges pointwise on J.*
(iii) *If $\sum f_n$ converges uniformly on J, then it converges pointwise on J.*

None of the converse implications to (i), (ii) or (iii) in Theorem 6.5 is true and neither uniform convergence implies absolute convergence, nor does absolute convergence imply uniform convergence. Thus, total convergence is the strongest notion and pointwise convergence is the weakest one; absolute and uniform convergence are in some sense intermediate notions. Sometimes a series converging pointwise on J is simply said to converge on J.

6.1 Pointwise and Uniform Convergence

The geometric series converges absolutely in $(-1, 1)$ but not uniformly, because

$$\sup_{x \in (-1,1)} \left| S_n(x) - \frac{1}{1-x} \right| = \sup_{x \in (-1,1)} \left| \frac{x^{n+1}}{1-x} \right| = +\infty,$$

as the reader is urged to check. The series does not converge totally in $(-1, 1)$ for otherwise it would converge uniformly. This can also be checked directly because

$$\sup_{x \in (-1,1)} |x^n| = 1$$

and evidently $\sum 1$ diverges. It is interesting to observe that the very same series does converge totally in any closed interval of the form, say, $[-a, a]$ contained in $(-1, 1)$, because if $0 < a < 1$

$$\sup_{x \in [-a,a]} |x^n| = a^n$$

and $\sum_{n=0}^{\infty} a^n = 1/(1-a)$.

As for sequences, much can be said about the properties of the sum of a series using uniform convergence, as the next three results illustrate.

Theorem 6.6 (Continuity of a series) *If $(f_n)_{n \geq 0}$ are continuous on the interval I and $\sum f_n \rightrightarrows F$ on I, then $f_n \rightrightarrows 0$ on I and F is continuous on I.*

Theorem 6.7 (Differentiability of a series) *Suppose that $(f_n)_{n \geq 0}$ are differentiable on the interval I and assume that:*

(i) $\sum f_n$ *converges in at least one point $x_0 \in I$;*
(ii) $\sum f'_n \rightrightarrows G$ *on I.*

Then $\sum f_n$ converges uniformly in I to a differentiable sum F and $F' = G$, that is:

$$\sum \frac{d}{dx} f_n(x) = \frac{d}{dx} \sum f_n(x). \tag{6.2}$$

It is customary to call $\sum f'_n$ the *formal derivative* of the series $\sum f_n$. If conclusion (ii) of Theorem 6.7 holds, then formula (6.2) is referred to as *term by term differentiation*.

Theorem 6.8 (Integrability of a series) *Suppose that $(f_n)_{n \geq 0}$ are bounded and integrable on the interval $[a, b]$. If $\sum f_n \rightrightarrows F$ on $[a, b]$, then F is bounded and integrable on $[a, b]$ and*

$$\sum \int_a^b f_n(x)\, dx = \int_a^b \sum f_n(x)\, dx. \tag{6.3}$$

It is customary to call $\sum \int_a^b f_n(x)\,dx$ the *formal integral* of the series $\sum f_n$. If the conclusion of Theorem 6.8 holds, then formula (6.3) is referred to as *term by term integration*.

6.2 Power Series

Definition 6.5 A *power series centered at* $x_0 \in \mathbb{R}$ is a series of the form

$$\sum_{n=0}^{\infty} a_n (x - x_0)^n,$$

where $(a_n)_{n \geq 0}$ is a fixed sequence of real numbers, called the *coefficients* of the power series.

Notice that a power series always converges at its center x_0, with sum a_0. A simple example of power series is the geometric series, which is centered at $x_0 = 0$ and has coefficients $a_n = 1$ for every n.

The convergence properties of a power series are much stronger than those of arbitrary series, as illustrated in the next result.

Theorem 6.9 *For the power series $\sum_{n=0}^{\infty} a_n (x - x_0)^n$ one and only one of the following holds true:*

(i) *the series converges only at x_0;*
(ii) *there exists $r > 0$ such that the series converges pointwise in $(x_0 - r, x_0 + r)$ and does not converge in $\mathbb{R} \setminus [x_0 - r, x_0 + r]$; in this case, the series converges totally in every closed interval contained in $(x_0 - r, x_0 + r)$;*
(iii) *the series converges in $\mathbb{R}$ and totally in every closed and bounded interval.*

As a consequence of Theorem 6.9, with each power series there is an associated interval, called the *convergence interval*. In case (i) it degenerates to the point x_0, and in case (iii) it is the whole real line, while in case (ii) it is the bounded open interval $(x_0 - r, x_0 + r)$. It is thus customary to call *convergence radius* the number r appearing in case (ii) and to say that $r = 0$ in case (i) and $r = +\infty$ in case (iii).

Observe that nothing can be said a priori about the convergence of the series at the boundary points of the convergence interval in case (ii), that is, at $x_0 \pm r$. Indeed, the geometric series converges exactly in $(-1, 1)$ (it does not converge at $x = \pm 1$), whereas the series

$$\sum_{n=0}^{\infty} (-1)^n \frac{x^{n+1}}{n+1}$$

converges in $(-1, 1]$. In both cases $r = 1$ but the boundary behaviour is different.

It is clear that the convergence radius tells almost the whole story about a power series and that its determination is therefore of fundamental importance. It is possible

6.2 Power Series

to prove that, including the cases $r = 0$ or $r = +\infty$ explained above

$$r = \sup\{y \in \mathbb{R}^+ : \sum_{n=0}^{\infty} |a_n| y^n \text{ converges}\}. \tag{6.4}$$

In the context of power series, the notation $\overline{\mathbb{R}}$ refers to $\mathbb{R} \cup \{+\infty\}$.

Theorem 6.10 (Root and ratio tests) *Consider the power series $\sum a_n(x - x_0)^n$. Suppose that one of the following limits*

$$\lim_n \sqrt[n]{|a_n|}, \quad \lim_n \left|\frac{a_{n+1}}{a_n}\right| \tag{6.5}$$

exists in $\overline{\mathbb{R}}$, and denote it by ℓ. Then the radius of convergence of the series is

$$r = \begin{cases} 0 & \text{if } \ell = +\infty \\ 1/\ell & \text{if } 0 < \ell < +\infty \\ +\infty & \text{if } \ell = 0. \end{cases}$$

Theorem 6.10 implicitly asserts that if both limits in (6.5) exist, then they necessarily coincide. Also, formula (6.4) and formulae (6.5) show that the radius of convergence is independent of the summation index. In other words, both series

$$\sum_{n=n_1}^{\infty} a_n(x - x_0)^n, \quad \sum_{n=n_2}^{\infty} a_n(x - x_0)^n$$

have the same convergence interval and radius, for all $n_1, n_2 \geq 0$. This is why when it comes to convergence properties it is sometimes preferable to write $\sum a_n(x - x_0)^n$ without specifying the summation range.

In the case in which a power series converges at one of the boundary points of the interval of convergence, more can be said about the uniform convergence.

Theorem 6.11 (Abel's theorem) *Suppose that the power series $\sum a_n(x - x_0)^n$ has a finite and positive radius of convergence r. If the series converges at $x_0 + r$, then it converges uniformly in $[a, x_0 + r]$ for every $a \in (x_0 - r, x_0 + r)$. Similarly, if it converges at $x_0 - r$, then it converges uniformly in $[x_0 - r, a]$ for every $a \in (x_0 - r, x_0 + r)$. If it converges at $x_0 \pm r$, then it converges uniformly in $[x_0 - r, x_0 + r]$.*

Given the power series $\sum_{n=0}^{\infty} a_n(x - x_0)^n$ it is natural to consider the series which is obtained by differentiating it term by term, namely the power series

$$\sum_{n=1}^{\infty} n a_n(x - x_0)^{n-1} = \sum_{n=0}^{\infty} (n+1) a_{n+1}(x - x_0)^n. \tag{6.6}$$

Theorem 6.12 (Derivative of power series) *Suppose that a power series has radius of convergence* $r \in (0, +\infty]$ *and denote by* f *the sum of the series in the convergence interval, that is:*

$$f(x) = \sum_{n=0}^{\infty} a_n (x - x_0)^n, \qquad x \in (x_0 - r, x_0 + r). \tag{6.7}$$

Then f *is differentiable in* $(x_0 - r, x_0 + r)$, *the convergence radius of the series of derivatives (6.6) is also* r *and*

$$f'(x) = \sum_{n=1}^{\infty} n a_n (x - x_0)^{n-1}, \qquad x \in (x_0 - r, x_0 + r).$$

By successively applying Theorem 6.12 it is possible to obtain the following important result.

Theorem 6.13 *Suppose that a power series has radius of convergence* $r \in (0, +\infty]$ *and denote by* f *the sum of the series in the convergence interval* $I = (x_0 - r, x_0 + r)$, *as in (6.7). Then* $f \in C^{\infty}(I)$ *and*

$$a_n = \frac{f^{(n)}(x_0)}{n!}. \tag{6.8}$$

It follows from Theorem 6.13 that the coefficients of a power series are uniquely determined by the sum of the series. This implies in particular the following fact.

Theorem 6.14 (Identity principle of power series) *Suppose that for some* $\delta > 0$ *the identity*

$$\sum_{n=0}^{\infty} a_n (x - x_0)^n = \sum_{n=0}^{\infty} b_n (x - x_0)^n, \qquad |x - x_0| < \delta$$

holds in the sense of converging power series. Then $a_n = b_n$ *for every* $n \geq 0$.

6.3 Taylor Series

Inspired by formula (6.8), and by the theory of Taylor polynomials, it is natural to consider power series associated with smooth functions, as in the definition that follows.

Definition 6.6 (*Taylor series*) Let I be an open interval, take $x_0 \in I$ and suppose that $f \in C^{\infty}(I)$. The series

6.3 Taylor Series

$$\sum_{n=0}^{\infty} \frac{f^{(n)}(x_0)}{n!}(x - x_0)^n \qquad (6.9)$$

is called the *Taylor series* of f centered at x_0. The function f is said to *admit a Taylor series expansion in I centered at x_0* if the series (6.9) converges to $f(x)$ for every $x \in I$. If $x_0 = 0$, the Taylor series is also known as *McLaurin series*.

Theorem 6.13 shows that the sum f of a convergent power series admits a Taylor series expansion in the interval of convergence centered at its center x_0. The Taylor series of f coincides with the original power series thanks to formula (6.8). Therefore, Taylor series are not special power series: every power series is the Taylor series of its sum. The enphasis in the terminology of Taylor series is in the fact that the function f is the given datum.

Not every function $f \in C^\infty(I)$ admits a Taylor series expansion in I. Consider for example

$$f(x) = \begin{cases} e^{-1/x^2} & x \neq 0 \\ 0 & x = 0. \end{cases}$$

Since $f^{(n)}(0) = 0$ for every $n \geq 0$ the McLaurin series of f vanishes identically and hence converges to f only at $x = 0$.

Theorem 6.15 *Let I be an open interval, take $x_0 \in I$ and suppose that $f \in C^\infty(I)$. If there exist $L, M > 0$ such that for every $n \in \mathbb{N}$*

$$|f^{(n)}(x)| \leq L \cdot M^n, \qquad x \in I$$

then f admits a Taylor series expansion in I centered at x_0.

Definition 6.7 (*Analytic function*) Let I be an open interval. A function $f \in C^\infty(I)$ is called *analytic in I* if for every point $x \in I$ there is an interval $I_{x,\delta} = (x - \delta, x + \delta) \subset I$ such that f admits a Taylor series expansion in $I_{x,\delta}$ centered at x.

Theorem 6.16 *Let I be an open interval, and suppose that $f, g \in C^\infty(I)$ are analytic in I. Then*

(i) *$f + g$ is analytic in I;*
(ii) *fg is analytic in I;*
(iii) *if $g(x) \neq 0$ for every $x \in I$, then f/g is analytic in I.*

Furthermore, if $h \in C^\infty(J)$ is analytic in the open interval J and $h(J) \subset I$, then the composition $f \circ h$ is analytic in J.

6.4 Fourier Series

The theory of *Fourier series* involves *periodic functions* and natural series expansions associated with them, the terms of which are given by trigonometric functions of the form

$$a \cos\left(\frac{2\pi x}{T}\right) + b \sin\left(\frac{2\pi x}{T}\right), \qquad a, b \in \mathbb{R}, \ T > 0. \tag{6.10}$$

Definition 6.8 Let $I \subset \mathbb{R}$ be such that there exists $T > 0$ such that $x + T \in I$ whenever $x \in I$. A function $f : I \to \mathbb{R}$ is said to be periodic of *period* $T > 0$ (or also T-periodic) if $f(x + T) = f(x)$ for every $x \in I$. The space of T-periodic functions on I is denoted $\mathscr{P}_T(I)$.

An example of periodic function is the mantissa function $f(x) = x - [x]$, that has period 1. Constant functions are periodic of any period T, but no minimum positive value of T exists. If a function is defined on $\mathbb{R}$ and is T-periodic, then it is also kT periodic for every positive integer k, and the function $x \mapsto f(kx)$ is T/k periodic.

The most important class of periodic functions is of course the class of trigonometric functions of the form (6.10). which are periodic of period T.

Definition 6.9 A *trigonometric polynomial* of degree n is a linear combination of the form

$$\frac{a_0}{2} + \sum_{k=1}^{n}\left[a_k \cos\left(\frac{2\pi kx}{T}\right) + b_k \sin\left(\frac{2\pi kx}{T}\right)\right] \tag{6.11}$$

where $a_k, b_k \in \mathbb{R}$, $a_n^2 + b_n^2 \neq 0$ and $T > 0$. A series of the form

$$\frac{a_0}{2} + \sum_{k=1}^{\infty}\left[a_k \cos\left(\frac{2\pi kx}{T}\right) + b_k \sin\left(\frac{2\pi kx}{T}\right)\right] \tag{6.12}$$

is called a *trigonometric series*, provided that $a_k^2 + b_k^2 \neq 0$ for infinitely many k; in this case, the sequences $(a_k)_{k \geq 0}$ and $(b_k)_{k \geq 1}$ are its coefficients.

As already observed, the kth summand in (6.11) is T/k periodic, hence T periodic; it follows that the polynomial itself is T-periodic. When $T = 2\pi$, (6.11) becomes

$$\frac{a_0}{2} + \sum_{k=1}^{n}\left[a_k \cos(kx) + b_k \sin(kx)\right]. \tag{6.13}$$

Observe that

$$\begin{aligned} a \cos t + b \sin t &= \sqrt{a^2 + b^2}\left(\frac{a}{\sqrt{a^2 + b^2}} \cos t + \frac{b}{\sqrt{a^2 + b^2}} \sin t\right) \\ &= \sqrt{a^2 + b^2}\left(\cos \phi \cos t + \sin \phi \sin t\right) \\ &= \sqrt{a^2 + b^2} \cos(t - \phi), \end{aligned}$$

6.4 Fourier Series

where ϕ is an angle, hence a number, such that

$$(\cos\phi, \sin\phi) = \left(\frac{a}{\sqrt{a^2+b^2}}, \frac{b}{\sqrt{a^2+b^2}}\right).$$

This proves that (6.13) can also be written as

$$\frac{A_0}{2} + \sum_{k=1}^{n} A_k \cos(kx - \phi_k), \tag{6.14}$$

which exhibits a trigonometric polynomial as a linear combination of cosine functions with different amplitudes A_k, frequencies k and phases ϕ_k. Formula (6.14) is sometimes called the physical version of the trigonometric polynomial (6.13) (Fig. 6.2).

Definition 6.10 (*Piecewise continuous functions*) A function f defined on the bounded interval $I \subset \mathbb{R}$ is said to be *piecewise continuous* on I if it is continuous on I except possibly for a finite number of points $x_1, \ldots, x_n \in I$ where it is required that all the limits

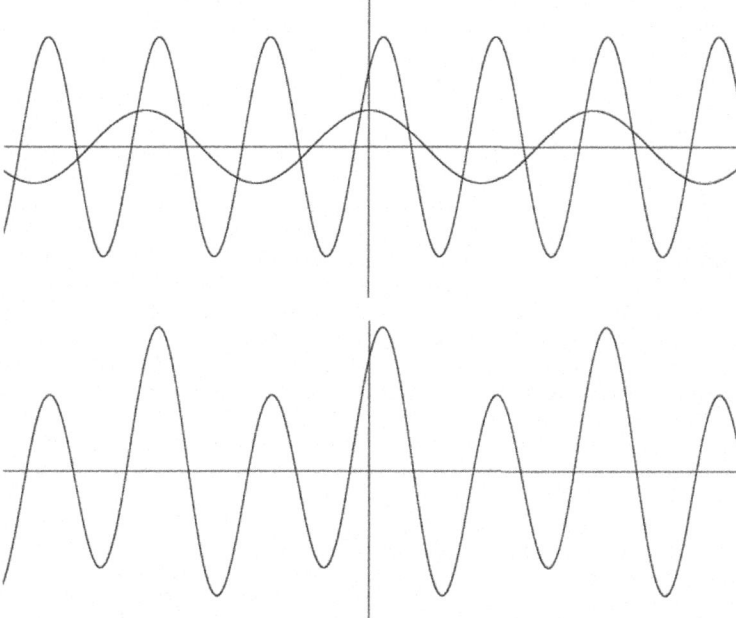

Fig. 6.2 Above are the graphs of $\cos x$ and $3\cos(2x - \pi/4)$; below is the trigonometric polynomial obtained by summing them, namely $\cos x + 3\cos(2x - \pi/4)$

$$f(x_j^-) := \lim_{x \to x_j^-} f(x), \qquad f(x_j^+) := \lim_{x \to x_j^+} f(x), \qquad j = 1, \ldots, n \qquad (6.15)$$

exist and are finite. The existence and finiteness of the limits as in (6.15) is also required at the boundary points a and b of I, in the sense that

$$f(a^+) := \lim_{x \to a^+} f(x) \in \mathbb{R}, \qquad f(b^-) := \lim_{x \to b^-} f(x) \in \mathbb{R}.$$

A function is said to be piecewise continuous on the unbounded interval I if it is piecewise continuous on every bounded interval contained in I. The space of piecewise continuous functions on I is denoted by $PC(I)$.

A few comments are in order. For any bounded interval I, a function in $PC(I)$ is necessarily bounded. This is not true for unbounded I because any continuous function on I is clearly piecewise continuous on every bounded interval, so it belongs to $PC(I)$, but it may well be unbounded.

Most functions that are of interest in the basic theory of Fourier series are both periodic and piecewise continuous. In particular, it is natural to consider the space

$$PC_T := \mathscr{P}_T(\mathbb{R}) \cap PC(\mathbb{R})$$

which consists of the T-periodic functions on $\mathbb{R}$ which are piecewise continuous on bounded intervals. A function in PC_T, being T-periodic, is completely known when restricted to any interval of the form $[a, a+T)$, where it is piecewise continuous. Any such function is bounded on $[0, T)$ and hence, by periodicity, on $\mathbb{R}$. The map

$$PC_T \to PC([0, T)), \qquad f \mapsto f\big|_{[0,T)}$$

which to each piecewise continuous and T-periodic function associates its restriction to $[0, T)$ is therefore a bijection, and the same holds for any interval of the form $[a, a+T)$. Indeed, given $f \in PC([0, T))$, it is possible to define its T-periodic extension, by setting $\tilde{f}(x) = f(\tilde{x})$ if $x = \tilde{x} + nT$ where $\tilde{x} \in [0, T)$ and $n \in \mathbb{Z}$, thereby obtaining an element in PC_T whose restriction to $[0, T)$ is of course f.

Definition 6.11 The *Fourier coefficients* of a function $f \in PC_T$ are the real numbers

$$a_n := \frac{2}{T} \int_0^T f(x) \cos\left(\frac{2\pi}{T} nx\right) dx, \qquad n = 0, 1, 2, \ldots \qquad (6.16)$$

$$b_n := \frac{2}{T} \int_0^T f(x) \sin\left(\frac{2\pi}{T} nx\right) dx, \qquad n = 1, 2, \ldots \qquad (6.17)$$

The Fourier series of $f \in PC_T$ is the series

$$\frac{a_0}{2} + \sum_{n=1}^{\infty} \left[a_n \cos\left(\frac{2\pi}{T} nx\right) + b_n \sin\left(\frac{2\pi}{T} nx\right) \right]. \qquad (6.18)$$

6.4 Fourier Series

The integration in (6.16) and (6.17) is meaningful due to the piecewise continuity of f, and can be carried out on any other interval of length T; they are thus often written with $\int_{-T/2}^{T/2}$ in place of $\int_0^T$. In the case $T = 2\pi$ they take the simpler form

$$a_n = \frac{1}{\pi} \int_0^{2\pi} f(x) \cos(nx)\, dx, \quad n = 0, 1, 2, \ldots$$

$$b_n = \frac{1}{\pi} \int_0^{2\pi} f(x) \sin(nx)\, dx, \quad n = 1, 2, \ldots$$

and the Fourier series becomes

$$\frac{a_0}{2} + \sum_{n=1}^{\infty} \left[a_n \cos(nx) + b_n \sin(nx) \right].$$

In order to formulate satisfactory convergence results for Fourier series other function spaces need to be introduced.

Definition 6.12 (*Piecewise smooth functions*) A function f defined on the bounded interval I is said to be *piecewise smooth* on I if:

(i) $f \in \mathrm{PC}(I)$;
(ii) the derivative of f exists and is continuous everywhere in I except possibly at finitely many points $y_1, \ldots, y_m \in I$ (including the points $x_1, \ldots, x_n$ at which f is not continuous, if existing) where it may fail either to exist or to be continuous, but where it is required that all the limits

$$f'(y_j^-) := \lim_{x \to y_j^-} f'(x), \qquad f'(y_j^+) := \lim_{x \to y_j^+} f'(x), \qquad j = 1, \ldots, m \quad (6.19)$$

exist and are finite. The existence and finiteness of the limits as in (6.19) is also required at the boundary points a and b of I.

A function is said to be piecewise smooth on the unbounded interval I if it is piecewise smooth on every bounded interval contained in I. The space of piecewise smooth functions on I is denoted by $\mathrm{PS}(I)$.

The function drawn in Fig. 6.3 is piecewise smooth on a bounded interval. A shorter, though slightly imprecise way, to define $\mathrm{PS}(I)$ is to say that both f and f' are piecewise continuous. Many comments similar to those that apply to piecewise continuous functions are also appropriate for piecewise smooth functions. In particular, it is useful to consider the space

$$\mathrm{PS}_T := \mathscr{P}_T(\mathbb{R}) \cap \mathrm{PS}(\mathbb{R})$$

which consists of the T-periodic functions on $\mathbb{R}$ which are piecewise smooth on every bounded interval. The map

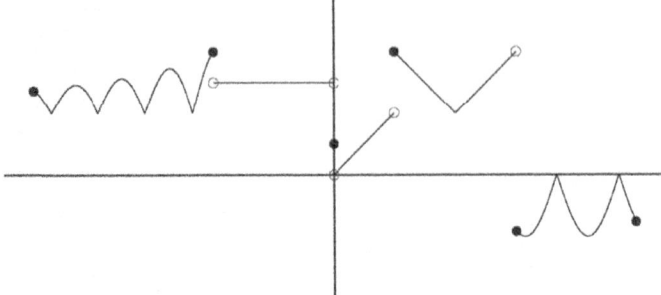

Fig. 6.3 A possible graph of a piecewise continuous function

$$\mathrm{PS}_T \to \mathrm{PS}([0, T)), \qquad f \mapsto f\big|_{[0,T)}$$

is also a bijection, and the same holds for any other interval of the form $[a, a + T)$. In Fig. 6.4 it is given an example of a function which is not piecewise smooth.

It is now possible to state the main result concerning the pointwise convergence of Fourier series.

Theorem 6.17 (Dirichlet) *The Fourier series of* $f \in \mathrm{PS}_T$ *converges to*

$$\frac{f(x^+) + f(x^-)}{2}$$

at all points; in particular, the Fourier series of f *converges to* f *at all points at which* f *is continuous.*

Below, it will be tacitly assumed that the sequences $(a_n)_{n\geq 0}$ and $(b_n)_{n\geq 1}$ that appear in a statement are the Fourier coefficients of the function that is under consideration in the same statement.

Since the cosine function is even and the sine function is odd, the Fourier coefficients $(a_n)_{n\geq 0}$ of an odd function vanish and the Fourier coefficients $(b_n)_{n\geq 1}$ of an even function vanish. Consequently, the Fourier series becomes a series of sine

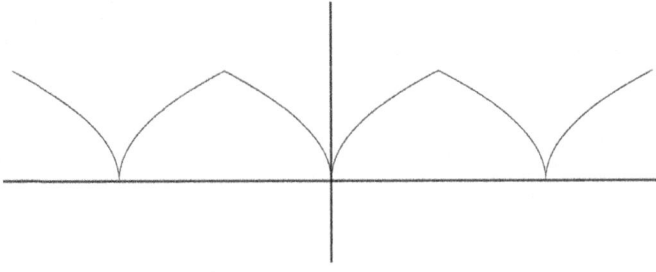

Fig. 6.4 A continuous function which is not piecewise smooth, due to unbounded derivative

6.4 Fourier Series

functions in the former case and of cosine functions in the latter. Furthermore, the coefficients that do appear may be computed on half the interval, for symmetry reasons. This leads to set

$$\text{PC}_T^e := \{ f \in \text{PC}_T : f \text{ is even} \}$$
$$\text{PC}_T^o := \{ f \in \text{PC}_T : f \text{ is odd} \},$$

both of which can be identified with $\text{PC}((0, T/2))$, and to observe that:

(i) if $f \in \text{PC}_T^e$, then $b_n = 0$ for all $n \geq 1$ and

$$a_n = \frac{4}{T} \int_0^{T/2} f(x) \cos\left(\frac{2\pi}{T} n x\right) dx, \quad n \geq 0, \tag{6.20}$$

so that the Fourier series of f is the *cosine series expansion*

$$\frac{a_0}{2} + \sum_{n=1}^{\infty} a_n \cos\left(\frac{2\pi}{T} n x\right); \tag{6.21}$$

(ii) if $f \in \text{PC}_T^o$, then $a_n = 0$ for all $n \geq 0$ and

$$b_n = \frac{4}{T} \int_0^{T/2} f(x) \sin\left(\frac{2\pi}{T} n x\right) dx, \quad n \geq 1, \tag{6.22}$$

so that the Fourier series of f is is the *sine series expansion*

$$\sum_{n=1}^{\infty} b_n \sin\left(\frac{2\pi}{T} n x\right). \tag{6.23}$$

Theorem 6.18 (Parseval) *For any $f \in \text{PS}_T$ Parseval's identity*

$$\frac{2}{T} \int_0^T |f(x)|^2 \, dx = \frac{|a_0|^2}{2} + \sum_{n=1}^{\infty} (|a_n|^2 + |b_n|^2) \tag{6.24}$$

holds true.

Clearly, the absolute values in (6.24) are not necessary because f is real valued and also $a_n, b_n \in \mathbb{R}$. However, the very same equality holds for complex-valued functions, for which this is the correct (and classical) formulation.

The next three theorems are formulated for functions in PC_T or in PS_T. It is important to observe that if instead $f \in \text{PC}(I)$ or if $f \in \text{PS}(I)$ for some bounded interval I of length T, the conclusions hold for the T-periodic extension of f.

Theorem 6.19 (Uniform convergence) *Suppose that $f \in \mathrm{PC}_T$.*

(i) *If the numerical series $\sum |a_n|$ and $\sum |b_n|$ both converge, then the Fourier series of f converges totally in $\mathbb{R}$, hence absolutely and uniformly.*

(ii) *If $f \in \mathrm{PS}_T$, then its Fourier series converges totally hence absolutely and uniformly to f in every closed interval of length less than T in which f is continuous.*

(iii) *If $f \in \mathrm{PS}_T$ and is also continuous, then its Fourier series of converges totally hence absolutely and uniformly to f in $\mathbb{R}$.*

(iv) *If $f \in C^1(\mathbb{R})$, then its Fourier series converges totally hence absolutely and uniformly to f in $\mathbb{R}$.*

Theorem 6.20 (Integration of a Fourier series) *Suppose that $f \in \mathrm{PC}_T$. Then for every $[a,b]$ with $b-a < T$*

$$\int_a^b f(x)\,dx = \frac{a_0}{2}(b-a) + \sum_{n=1}^{\infty} \int_a^b \left[a_n \cos\left(\frac{2\pi}{T}nx\right) + b_n \sin\left(\frac{2\pi}{T}nx\right) \right] dx.$$

Theorem 6.21 (Derivation of a Fourier series) *Suppose that $f \in \mathrm{PS}_T$ is continuous on $\mathbb{R}$. Then the series of derivatives*

$$\frac{2\pi}{T} \sum_{n=1}^{\infty} \left[nb_n \cos\left(\frac{2\pi}{T}nx\right) - na_n \sin\left(\frac{2\pi}{T}nx\right) \right] \tag{6.25}$$

is the Fourier series of f' and converges pointwise to $\dfrac{f'(x^+) + f'(x^-)}{2}$.

Theorem 6.22 (Riemann-Lebesgue lemma) *The Fourier coefficients of any function in PS_T satisfy $\lim_n a_n = \lim_n b_n = 0$.*

Theorem 6.23 (Smoothness and decay)

(i) *For any function in PS_T there exists a number $M \geq 0$ such that its Fourier coefficients satisfy $|a_n| \leq M/n$ and $|b_n| \leq M/n$ for every positive integer n.*

(ii) *If $f \in \mathscr{P}_T(\mathbb{R}) \cap C^{k-1}(\mathbb{R})$ for some integer $k \geq 1$ and $f^{(k)} \in \mathrm{PS}_T$, then there exists a number $M \geq 0$ such that its Fourier coefficients satisfy*

$$|a_n| \leq \frac{M}{n^{k+1}}, \qquad |b_n| \leq \frac{M}{n^{k+1}}$$

for every positive integer n.

6.5 Guided Exercises

6.1 Consider the sequence $(f_n)_{n\geq 1}$ where $f_n(x) = \dfrac{|x|^n\sqrt{1-x^{2n}}}{n+1}$.

(a) Discuss the pointwise and the uniform convergence of $(f_n)_{n\geq 1}$ on the largest possible domain.

(b) Compute, if existing, $\displaystyle\lim_n \int_0^{1/2} f_n(x)\,\mathrm{d}x$.

Answer. (a) Clearly, the largest domain is $I = [-1, 1]$, and on this set

$$0 \leq f_n(x) \leq \frac{1}{n+1}$$

so that $f_n \rightrightarrows 0$ on I, hence pointwise.

(b) Since $f_n \rightrightarrows 0$ on I and since $[0, 1/2] \subset I$, it follows that

$$\lim_n \int_0^{1/2} f_n(x)\,\mathrm{d}x = \int_0^{1/2} \lim_n f_n(x)\,\mathrm{d}x = 0.$$

This is a strightforward warming-up exercise on uniform convergence. The easiest possible situation is when an estimate of the form $0 \leq f_n(x) \leq c_n$ with $c_n \to 0$ is available, as in this case. All theorems on uniform convergence apply, as the interchange of limit and integral signs.

6.2 Consider the sequence $(f_n)_{n\geq 1}$ defined by $f_n(x) = \dfrac{\sin(n\sqrt{x})}{n+x}$ on $[0, +\infty)$.

(a) Find the pointwise limit of $(f_n)_{n\geq 1}$, if existing;

(b) Compute, if existing, $\displaystyle\lim_n \int_0^a f_n(x)\,\mathrm{d}x$ as a varies in $a \in (0, +\infty)$.

Answer. (a) Clearly, $f_n(x) \to 0$ for any fixed $x \in \mathbb{R}$. Hence the pointwise limit of $(f_n)_{n\geq 1}$ is the null function on $[0, +\infty)$.

(b) Observe that

$$\sup_{x\geq 0} |f_n(x)| = \sup_{x\geq 0}\left|\frac{\sin(n\sqrt{x})}{n+x}\right| \leq \frac{1}{n} \to 0$$

so that actually $f_n \rightrightarrows 0$, that is, $(f_n)_{n\geq 1}$ converges uniformly to the null function on $[0, +\infty)$. An application of Theorem 6.3 yields

$$\lim_n \int_0^a f_n(x)\,\mathrm{d}x = \int_0^a \lim_n f_n(x)\,\mathrm{d}x = 0$$

for every positive a. This is an elementary exercise on pointwise and uniform convergence. The oscillating numerator is bounded in modulus by 1, and the denominator is larger than n. Hence the sequence is uniformly convergent (to the null function) so that it is legitimate to exchange limit and integral signs.

6.3 Consider the sequence $(f_n)_{n\geq 1}$ where $f_n: [0, +\infty) \to \mathbb{R}$ is defined by

$$f_n(x) = \begin{cases} \dfrac{1}{n} & 0 \leq x \leq n \\ \dfrac{2n-x}{n^2} & n < x \leq 2n \\ 0 & x > 2n. \end{cases}$$

(a) Find the pointwise limit of $(f_n)_{n\geq 1}$, where existing.
(b) Discuss if $(f_n)_{n\geq 1}$ converges uniformly to its pointwise limit.
(c) Is it true that $\displaystyle\lim_n \int_0^{+\infty} f_n(x)\,dx = \int_0^{+\infty} \lim_n f_n(x)\,dx$?

Answer. (a) Clearly, $f_n(x) \to 0$ for any fixed $x \in \mathbb{R}$. Hence the pointwise limit of $(f_n)_{n\geq 1}$ is the null function on $[0, +\infty)$.

(b) Further,
$$\sup_{x \geq 0} |f_n(x)| = \frac{1}{n} \to 0$$
so that $f_n \rightrightarrows 0$, that is, $(f_n)_{n\geq 1}$ converges uniformly to the null function on $[0, +\infty)$.

(c) Obviously, $\int_0^{+\infty} \lim_n f_n(x)\,dx = 0$. However,
$$\int_0^{+\infty} f_n(x)\,dx = \int_0^n f_n(x)\,dx + \int_n^{2n} f_n(x)\,dx = 1 + \frac{1}{2} = \frac{3}{2},$$
so that $\displaystyle\lim_n \int_0^{+\infty} f_n(x)\,dx = 3/2$. Therefore the answer is no.

The graph of f_n is a trapezoid that becomes thinner and longer as n grows but has constant area $3/2$. As the height decreases precisely at the rate $1/n$, the sequence flattens out to the null function, which is therefore the uniform limit of $(f_n)_{n\geq 1}$. The tricky point of this exercise is to understand if it is possible to exchange limit and integral sign but in the improper sense, that is, the integral in really over the interval $[0, +\infty)$ so that Theorem 6.3 cannot be applied. In fact, not only the result cannot be applied but the conclusion can be false if applied to improper integrals. This exercise actually provides an example where things go wrong.

6.4 Consider the sequence $(f_n)_{n\geq 1}$ where $f_n(x) = \sqrt{x^2 + \dfrac{1}{n}}$.

(a) Discuss the pointwise and the uniform convergence of $(f_n)_{n\geq 1}$ on $\mathbb{R}$.
(b) Discuss the uniform convergence of the sequence $(f'_n)_{n\geq 1}$ in $[-1, 1]$.

6.5 Guided Exercises

Answer. (a) Evidently, $\sqrt{x^2 + \frac{1}{n}} \to \sqrt{x^2} = |x|$ for every $x \in \mathbb{R}$, so that $(f_n)_{n\geq 1}$ converges pointwise to $f(x) = |x|$ on $\mathbb{R}$. Furthermore,

$$|f_n(x) - f(x)| = \sqrt{x^2 + \frac{1}{n}} - |x|$$

$$= \frac{\left(\sqrt{x^2 + \frac{1}{n}} - |x|\right)\left(\sqrt{x^2 + \frac{1}{n}} + |x|\right)}{\left(\sqrt{x^2 + \frac{1}{n}} + |x|\right)}$$

$$= \frac{\frac{1}{n}}{\left(\sqrt{x^2 + \frac{1}{n}} + |x|\right)}$$

$$\leq \frac{\frac{1}{n}}{\sqrt{\frac{1}{n}}} = \frac{1}{\sqrt{n}}.$$

Therefore

$$\sup_{x \in \mathbb{R}} |f_n(x) - f(x)| \leq \frac{1}{\sqrt{n}} \to 0,$$

so that $f_n(x) \rightrightarrows |x|$ on $\mathbb{R}$.

(b) Clearly, each f_n is differentiable on $\mathbb{R}$ but the sequence $(f'_n)_{n\geq 1}$ cannot converge uniformly on $[-1, 1]$ for otherwise the limit function would be differentiable in $[-1, 1]$. However, the limit function $f(x) = |x|$ is not differentiable at the origin.

This is a simple exercise on (non) uniform convergence of the sequence of derivatives of differentiable functions. Indeed, observe that

$$f'_n(x) = \frac{x}{\sqrt{x^2 + \frac{1}{n}}}.$$

and hence

$$f'_n(x) \to \begin{cases} x/|x| & x \neq 0 \\ 0 & x = 0, \end{cases}$$

which is not continuous at the origin. This shows that $(f'_n)_{n\geq 1}$ does not converge uniformly, for otherwise the uniform limit of a sequence of continuous functions would be continuous. In particular, Theorem 6.2 cannot be applied, in agreement with the conclusion that the limit function is not differentiable. Notice that differentiability in the limit is lost although $(f_n)_{n\geq 1}$ converges uniformly on $\mathbb{R}$.

6.5 Discuss pointwise and uniform convergence of $(f_n)_{n\geq 1}$ on the largest possible common domain of the functions $f_n(x) = \log(1 + x^n)$.

Answer. For even n the function $f_n(x)$ is defined on $\mathbb{R}$ but for odd n only in $(-1, +\infty)$. Thus the largest possible domain is $I = (-1, +\infty)$. Evidently

$$\lim_n f_n(x) = \begin{cases} +\infty & x > 1 \\ \log 2 & x = 1 \\ 0 & |x| < 1. \end{cases}$$

Therefore the pointwise limit is the function $f \colon (-1, 1] \to \mathbb{R}$ defined by

$$f(x) = \begin{cases} \log 2 & x = 1 \\ 0 & |x| < 1. \end{cases}$$

In order to study uniform convergence it is useful to recall (see Appendix B.3 in [1]) that for any $x > -1$

$$\frac{x}{x+1} \leq \log(1+x) \leq x.$$

Hence, for any $0 < b < 1$ and any $x \in [0, b]$ it is $\log(1+x) \leq x$, so that

$$|\log(1+x^n)| = \log(1+x^n) \leq x^n \leq b^n.$$

Similar estimates hold for any $-1 < a < 0$ and any $x \in [a, 0]$ whenever n is even, whereas for n odd it is best to observe that $|\log(1+x)| \leq |x|/(1+x)$ and that the function $x \mapsto |x|/(1+x)$ is decreasing in $(-1, 0]$. This allows to estimate $|f_n(x)|$ for any $-1 < a < 0$ and any $x \in [a, 0]$, in the sense that

$$|\log(1+x^n)| \leq \begin{cases} \dfrac{|x|^n}{1+x^n} \leq \dfrac{|a|^n}{1+a^n} & n \text{ odd} \\ x^n \leq a^n & n \text{ even}. \end{cases}$$

It follows that

$$\sup_{x \in [a,b]} |\log(1+x^n)| \to 0$$

whenever $[a, b] \subset (-1, 1)$. Therefore $f_n \rightrightarrows f$ in every $[a, b] \subset (-1, 1)$. There cannot be uniform convergence in any interval containing $x_0 = 1$ because the limit function f is discontinuous at $x_0 = 1$ while all the f_n are continuous. Finally, neither can there be uniform convergence in any interval of the form $(c, 1)$, because

$$\sup_{x \in (c,1)} |\log(1+x^n)| \geq \lim_{x \to 1^-} |\log(1+x^n)| = \log 2,$$

nor can there be uniform convergence in any interval of the form $(-1, d)$, because

$$\sup_{x \in (-1,d)} |\log(1+x^n)| = +\infty$$

whenever n is odd.

It is straightforward to understand the pointwise limit of the sequence, and it is rather manifest that being a jump function as it is, uniform convergence can only occur in the compact intervals inside $(-1, 1)$. Showing this actually requires a little thought because one must understand the behaviour of $t \mapsto \log(1+t)$ in the interval $(-1, 1)$. The crucial estimates are in Appendix B.3 [1], and it is also important to notice that for n even and $x \in (-1, 0]$ obviously $x^n \in [0, 1)$. The fact that things go badly when approaching the point 1 and also the point -1 is also quite clear. The true key, though, is to observe that for $|x| < 1$ the behaviour of x^n and of $\log(1 + x^n)$ for large n are very similar, because in fact $\log(1 + x^n) \sim x^n$, so one can use the former as a guide to understand the latter.

6.6 Consider the sequence $(f_n)_{n \geq 1}$ where $f_n(x) = \dfrac{e^{nx}}{e^{nx} + 1}$.

(a) Discuss the pointwise convergence.
(b) Is the convergence of $(f_n)_{n \geq 1}$ uniform on $\mathbb{R}$?
(c) Find a family of intervals of the form $[a, b]$ on which the convergence is uniform.

Answer. (a) Clearly

$$\lim_n \frac{e^{nx}}{e^{nx} + 1} = f(x) := \begin{cases} 0 & x < 0 \\ \frac{1}{2} & x = 0 \\ 1 & x > 0. \end{cases}$$

Hence $(f_n)_{n \geq 1}$ converges pointwise to f on $\mathbb{R}$, see Fig. 6.5.

(b) Since f is not continuous at the origin, the convergence cannot be uniform on $\mathbb{R}$.
(c) Consider first $I = [a, b]$ with $b < 0$. The limit is the null function, hence

$$\sup_{x \in I} |f_n(x) - f(x)| = \sup_{x \in I} \frac{e^{nx}}{e^{nx} + 1} \leq \frac{e^{nb}}{e^{na} + 1} \to 0$$

as $n \to +\infty$. Therefore $f_n \rightrightarrows f$ on I. Similarly, if $J = [a, b]$ with $a > 0$,

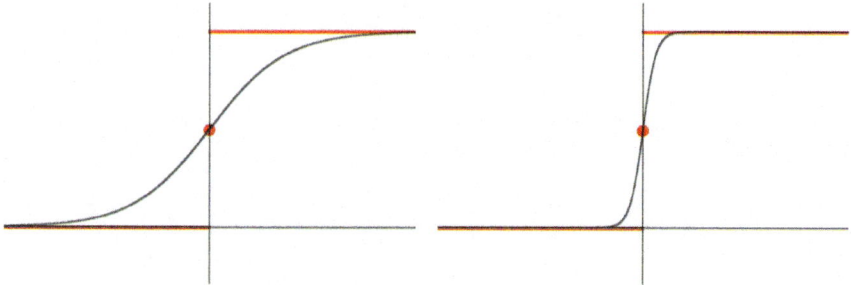

Fig. 6.5 In black two functions of the sequence in Exercise 6.6; in red the limiting jump function

$$\sup_{x\in J}|f_n(x)-f(x)| = \sup_{x\in J}\left|\frac{e^{nx}}{e^{nx}+1}-1\right| = \sup_{x\in J}\frac{1}{e^{nx}+1} = \frac{1}{e^{na}+1} \to 0$$

because $x \mapsto 1/(e^{nx}+1)$ is manifestly decreasing on $(0,+\infty)$. Again, $f_n \rightrightarrows f$ on J. In conclusion, $f_n \rightrightarrows f$ on any interval $[a,b]$ not containing the origin.

Here the pointwise limit is a prototypical jump function. Thus uniform convergence cannot occur globally, but it is expected to take place away from the discontinuity, that is, on the compact sets that do not contain the origin. It may be observed that actually $f_n \rightrightarrows f$ also on the closed half lines $(-\infty, b]$ with $b < 0$ and $[a,+\infty)$ with $a > 0$, a fact that may be proved by the very same arguments: just put $I = (-\infty, b]$ and $J = [a,+\infty)$ and follow the proof. The key point is indeed staying well away from the discontinuity.

6.7 Consider the sequence $(f_n)_{n\geq 1}$ where $x \in [0,1)$ and $f_n(x) = \dfrac{x^{2n}}{x^{2n}+1}$.

(a) What is the domain I of the pointwise limit f of the sequence?
(b) Discuss the uniform convergence.
(c) Is it true that $\lim_n f'_n\left(\left(\frac{1}{n}\right)^{1/2n}\right) = \lim_n f'\left(\left(\frac{1}{n}\right)^{1/2n}\right)$?

Answer. (a) Clearly,

$$\lim_n \frac{x^{2n}}{x^{2n}+1} = 0$$

whenever $x \in [0,1)$, so that the limiting function is the null function f defined on the whole interval $I = [0,1)$, see Fig. 6.6.

(b) Since

$$\sup_{x\in[0,1)}\left|\frac{x^{2n}}{x^{2n}+1}-0\right| = \sup_{x\in[0,1)}\frac{x^{2n}}{x^{2n}+1} \geq \lim_{x\to 1^-}\frac{x^{2n}}{x^{2n}+1} = \frac{1}{2},$$

the convergence cannot be uniform on $I = [0,1)$. Take next $b < 1$ and consider the interval $[0,b] \subset I$. It is quite clear that

$$\sup_{x\in[0,b]}\left|\frac{x^{2n}}{x^{2n}+1}\right| \leq b^{2n} \to 0.$$

Hence $f_n \rightrightarrows 0$ on $[0,b]$.

(c) First of all,

$$f'_n(x) = \frac{2nx^{2n-1}(1+x^{2n}) - 2nx^{2n-1}x^{2n}}{(1+x^{2n})^2} = \frac{2nx^{2n-1}}{(1+x^{2n})^2}.$$

Observe that upon setting for every $n \in \mathbb{N}\setminus\{0\}$

6.5 Guided Exercises

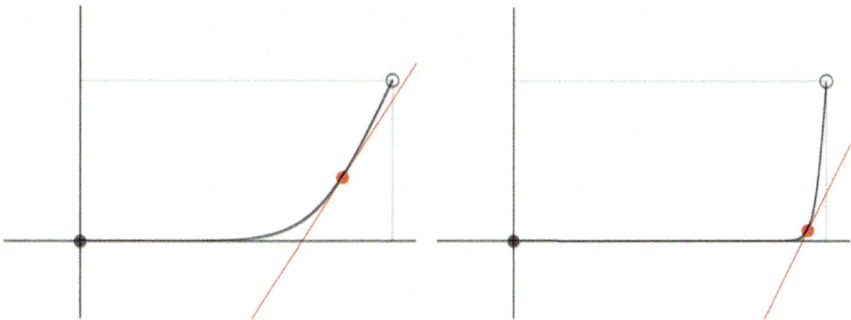

Fig. 6.6 In black two functions of the sequence in Exercise 6.7; in red the tangent lines at the points $(a_n, f(a_n))$

$$a_n := \left(\frac{1}{n}\right)^{1/2n} = \exp\left(-\frac{\log n}{2n}\right)$$

what needs to be established is whether

$$\lim_n f'_n(a_n) = \lim_n f'(a_n)$$

holds true. Observe preliminarily that $a_n \in (0, 1)$ and $a_n \to 1$ as $n \to +\infty$. Now, since $f'(x) \equiv 0$ on $(0, 1)$, it follows that $f'(a_n) = 0$ for every n and hence in particular

$$\lim_n f'(a_n) = 0.$$

Next, from the computation of f'_n and after some manipulation it turns out that

$$f'_n(a_n) = \frac{2na_n^{2n-1}}{(1+a_n^{2n})^2} = \frac{2n\left[\left(\frac{1}{n}\right)^{1/2n}\right]^{2n-1}}{\left(1+\left[\left(\frac{1}{n}\right)^{1/2n}\right]^{2n}\right)^2} = \cdots = \frac{2n^{1/2n}}{(1+\frac{1}{n})^2}$$

and evidently $f'_n(a_n) \to 2$. Therefore $\lim_n f'_n(a_n) = 2 \neq 0 = \lim_n f'(a_n)$.

Parts (a) and (b) are quite standard. The most interesting question is part (c). The point of part (c) is that the sequence $(a_n)_{n\geq 1}$ converges to 1 in a carefully chosen way, in the sense that the line tangent to the graph of f_n at $(a_n, f(a_n))$ is sufficiently steep and has a slope that tends to 2, unlike the tangent line to the graph of the limiting function which is of course horizontal at every point. This is possible because f_n becomes increasingly steep near 1 since for every n the value at $x_0 = 1$ is $f_n(1) = 1/2$ although $f_n \to 0$ in the interval $[0, 1)$. Formally this is not true because f_n is only defined in $[0, 1)$ but the analytic expression of f_n makes perfect sense at $x = 1$ and produces the value $1/2$. The computation of $f'_n(a_n)$ is not very interesting and perhaps somewhat tedious, but the geometric content is that no matter how flat has f_n become for large n, there is always room to pick a point a_n where f_n is steep enough. It would actually be possible to pick points b_n where $f'_n(b_n) = 2$ for every $n \geq 5$ because f'_n takes all values in $(0, n/2)$.

6.8 Consider the sequence $(f_n)_{n \geq 1}$ where

$$f_n(x) = \begin{cases} \dfrac{n \sin(x + \frac{1}{n})}{1 - x} & 0 \leq x < 1 \\ x/n & 1 \leq x \leq n \\ e^{n-x} & x > n. \end{cases}$$

(a) Discuss the pointwise convergence.
(b) Discuss the uniform convergence on the compact intervals contained in $[1, +\infty)$.
(c) Discuss the uniform convergence in $[1, +\infty)$.

Answer. (a) First of all,

$$\lim_n f_n(0) = \lim_n \left[n \sin(1/n)\right] = \lim_n \frac{\sin(1/n)}{1/n} = 1.$$

Next, if $0 < x < 1$, then

$$\lim_n f_n(x) = \lim_n \frac{n \sin(x + \frac{1}{n})}{1 - x} = +\infty$$

because $\sin(x + \frac{1}{n})/(1 - x) \to \sin(x)/(1 - x) > 0$. Finally, if $x \geq 1$ then $n > x$ eventually and

$$\lim_n f_n(x) = \lim_n \frac{x}{n} = 0.$$

Therefore $f_n \to f$ pointwise on the set $\{0\} \cup [1, +\infty)$, where

$$f(x) = \begin{cases} 1 & x = 0 \\ 0 & x \geq 1. \end{cases}$$

(b) Choose now $[a, b] \subset [1, +\infty)$. For n large enough, $[a, b] \subset [1, n]$ so that

$$\sup_{a \leq x \leq b} |f_n(x) - f(x)| = \sup_{a \leq x \leq b} |f_n(x)| = \sup_{a \leq x \leq b} \frac{x}{n} \leq \frac{b}{n} \to 0$$

for $n \to +\infty$. Therefore the sequence converges uniformly in every compact interval contained in $[1, +\infty)$.

(c) The simple inequality

$$\sup_{1\leq x} |f_n(x) - f(x)| = \sup_{1\leq x} |f_n(x)| \geq |f_n(n)| = 1$$

implies that there is no uniform convergence in $[1, +\infty)$.

The sequence under consideration converges only in the set $\{0\} \cup [1, +\infty)$. Thus, uniform behaviours are only meaningful to investigate in the interval $[1, +\infty)$, where the values of f_n grow from $1/n$ to 1 and then decrease exponentially from 1 to 0. The maximum value 1 is attained at $x_n = n$. Therefore, as n grows the maximum point x_n moves to the right but the maximum value is always 1. This is a prototypical situation that features uniform convergence on compacta (the maximum point leaves any compact set where the function decreases to 0) but not globally (the uniform distance from the limit is constantly 1).

6.9 Consider the sequence $(f_n)_{n\geq 1}$ where

$$f_n(x) = \begin{cases} 0 & x < 0 \\ x^{1/n} & 0 \leq x < 1 \\ n^{-(n+1)} x^n \sin(x^4) & 1 \leq x \leq n \\ 0 & x > n. \end{cases}$$

(a) Discuss the pointwise convergence.
(b) Discuss the uniform convergence in $[0, 1)$.
(c) Discuss the uniform convergence in $[1, +\infty)$.
(d) Compute, if existing, $\lim_n \int_0^{27} f_n(x)\,dx$.

Answer. (a) If $x \leq 0$ then clearly $f_n(x) \to 0$. If $0 < x < 1$, then $\lim_n x^{1/n} = 1$. Take now $x \geq 1$. Then for large enough n, namely when $n \geq 2x \geq x$, it is

$$f_n(x) = n^{-(n+1)} x^n \sin(x^4) = \frac{1}{n} \left(\frac{x}{n}\right)^n \sin(x^4)$$

and since $x/n \leq (1/2)$ for these values of n, evidently $(x/n)^n \to 0$. Since the function $x \to \sin(x^4)$ is bounded and since of course $1/n \to 0$, the conclusion for $x \geq 1$ is that $f_n(x) \to 0$. Therefore $f_n \to f$ pointwise, where

$$f(x) = \begin{cases} 0 & x \leq 0 \\ 1 & 0 < x < 1 \\ 0 & x \geq 1. \end{cases}$$

(b) Observe that f is not continuous at 0, hence it is not continuous on $[0, 1)$ whereas the functions f_n are. Hence the convergence cannot be uniform on $[0, 1)$.
(c) The expression found in item (a) implies

$$\sup_{1\leq x} |f_n(x) - f(x)| = \sup_{1\leq x} |f_n(x)| = \sup_{1\leq x\leq n} \left|\frac{1}{n}\left(\frac{x}{n}\right)^n \sin(x^4)\right| \leq \frac{1}{n} \to 0,$$

which shows that the convergence is uniform in $[1, +\infty)$.

(d) Using the uniform convergence in $[1, +\infty)$

$$\lim_n \int_0^{27} f_n(x)\,dx = \lim_n \left(\int_0^1 f_n(x)\,dx + \int_1^{27} f_n(x)\,dx\right)$$
$$= \lim_n \int_0^1 x^{1/n}\,dx + \int_1^{27} \lim_n f_n(x)\,dx$$
$$= \lim_n \int_0^1 x^{1/n}\,dx$$
$$= \lim_n \frac{1}{\frac{1}{n}+1} = 1.$$

Here the interval in which something needs to be understood is the expanding set $[1, n]$, where the sequence converges to 0. Indeed, although f_n oscillates more and more rapidly as x approaches n from the left, $|f_n|$ decreases quickly in that interval because of the multiplicative factors $(x/n)^n$ and $1/n$ which both tend to 0. The behaviour in $(0, 1)$ is that of the power $x^{1/n}$, which converges to 1. Thus the limit function is a jump function f, and more precisely the function which vanishes everywhere except in $(0, 1)$ where it is 1, otherwise known as the characteristic function of the set $(0, 1)$. Since f exhibits problems relative to uniform convergence across the point 1, when evaluating $\lim_n \int_0^{27} f_n(x)\,dx$ it is then natural to consider the splitting $(0, 27) = (0, 1) \cup [1, 27)$ and evaluate the integral separately in the two intervals. The integral over $(0, 1)$ can be determined explicitly as a function of n. For the second, uniform convergence comes to help in that the limit can be taken under the integral sign, thereby producing a vanishing summand.

6.10 Consider the sequence $(f_n)_{n\geq 1}$ where

$$f_n(x) = \frac{\log(1+e^{nx})}{\sqrt{n^2x^2+nx+1}}.$$

(a) Discuss the pointwise and uniform convergence of the sequence.
(b) Compute, if existing, $\lim_n \int_{-1}^1 f_n(x)\,dx$.

Answer. (a) First of all, observe that $y^2 + y + 1 > 0$ for any $y \in \mathbb{R}$ and hence the argument of the square root at the denomintor, namely $n^2x^2 + nx + 1$ is always strictly positive, implying that each f_n is defined on the whole real line $\mathbb{R}$. Now, it is clear that for fixed $x < 0$ the numerator tends to 0 while the denominator diverges to $+\infty$ as $n \to +\infty$. Hence $f_n(x) \to 0$ for $x < 0$. Next, writing

6.5 Guided Exercises

$$f_n(x) = \frac{\log(e^{nx}(1+e^{-nx}))}{nx\sqrt{1+\frac{1}{nx}+\frac{1}{n^2x^2}}} = \frac{nx+\log(1+e^{-nx})}{nx\sqrt{1+\frac{1}{nx}+\frac{1}{n^2x^2}}}$$

shows that $f_n(x) \to 1$ for $x > 0$. Finally, $f_n(0) = \log 2$ for every n. Therefore

$$\lim_n f_n(x) = f(x) := \begin{cases} 0 & x < 0 \\ \log 2 & x = 0 \\ 1 & x > 0. \end{cases}$$

As f is not continuous at the origin, while all the f_n are continuous everywhere, the convergence cannot be uniform in any interval containing the origin. Take then $a > 0$ and consider $[a, +\infty)$. From the inequalities

$$nx \leq \sqrt{n^2x^2+nx+1} \leq nx+1$$

which are true for any $x \geq 0$, it follows that

$$1 - \frac{1}{nx+1} = \frac{nx}{nx+1} \leq \frac{nx}{\sqrt{n^2x^2+nx+1}} \leq \frac{nx+\log(1+e^{-nx})}{\sqrt{n^2x^2+nx+1}} = f_n(x)$$

but also

$$f_n(x) = \frac{nx+\log(1+e^{-nx})}{\sqrt{n^2x^2+nx+1}} \leq \frac{nx+\log(1+e^{-nx})}{nx} = 1 + \frac{\log(1+e^{-nx})}{nx}.$$

From the achieved estimates, namely

$$-\frac{1}{nx+1} \leq f_n(x) - 1 \leq \frac{\log(1+e^{-nx})}{nx},$$

it follows that

$$|f_n(x) - 1| \leq \max\left\{\left|-\frac{1}{nx+1}\right|, \left|\frac{\log(1+e^{-nx})}{nx}\right|\right\}$$

$$\leq \frac{1}{nx+1} + \frac{\log(1+e^{-nx})}{nx}$$

$$\leq \frac{1}{na+1} + \frac{\log(1+e^{-na})}{na}$$

whenever $x \geq a$. This entails

$$\lim_n \sup_{x \in [a,+\infty)} |f_n(x) - 1| = 0,$$

so that $f_n \rightrightarrows 1$ on every interval $[a, +\infty)$ with $a > 0$. The convergence, however, is not uniform on $(0, +\infty)$ because

$$\sup_{x \in (0,+\infty)} |f_n(x) - 1| \geq \lim_{x \to 0^+} \left| \frac{nx + \log(1 + e^{-nx})}{\sqrt{n^2x^2 + nx + 1}} - 1 \right| = |\log 2 - 1|.$$

As for uniform convergence in $(-\infty, b]$ with $b < 0$, observe that

$$\sqrt{n^2x^2 + nx + 1} \geq \frac{\sqrt{3}}{2}$$

because $t^2 + t + 1 \geq 3/4$ for every $t \in \mathbb{R}$. Therefore

$$|f_n(x)| \leq \frac{\log(1 + e^{nx})}{\frac{\sqrt{3}}{2}} \leq \frac{2e^{nx}}{\sqrt{3}} \leq \frac{2e^{nb}}{\sqrt{3}}$$

whenever $x \leq b$. It follows that

$$\lim_n \sup_{x \in (-\infty, b]} |f_n(x)| = 0,$$

so that $f_n \rightrightarrows 0$ on every interval $(-\infty, b]$ with $b < 0$. Since $f_n(x) \to \log 2$ as $x \to 0^-$, the convergence is not uniform on $(-\infty, 0)$.

(b) Observe that since there is no uniform convergence on $[-1, 1]$, it is not legitimate to infer that the required limit is equal to $\int_{-1}^{1} f(t)\, dt$. By performing the change of variable $t = nx$,

$$\int_{-1}^{1} f_n(x)\, dx = \frac{1}{n} \int_{-n}^{n} \frac{\log(1 + e^t)}{\sqrt{t^2 + t + 1}}\, dt = \frac{1}{n} \left(\int_{-n}^{0} \frac{\log(1 + e^t)}{\sqrt{t^2 + t + 1}}\, dt + \int_{0}^{n} \frac{\log(1 + e^t)}{\sqrt{t^2 + t + 1}}\, dt \right).$$

Consider the first integral appearing above. Since $\log(1 + e^t) \sim e^t$ for $t \to -\infty$,

$$\lim_{t \to -\infty} \frac{\log(1 + e^t)}{\sqrt{t^2 + t + 1}} = 0$$

with order larger than any positive real number. Therefore

$$\lim_n \int_{-n}^{0} \frac{\log(1 + e^t)}{\sqrt{t^2 + t + 1}}\, dt = \int_{-\infty}^{0} \frac{\log(1 + e^t)}{\sqrt{t^2 + t + 1}}\, dt \in \mathbb{R}$$

and consequently

$$\lim_n \frac{1}{n} \int_{-n}^{0} \frac{\log(1 + e^t)}{\sqrt{t^2 + t + 1}}\, dt = 0.$$

6.5 Guided Exercises

The second integral, however, diverges with $n \to +\infty$ because $\log(1+e^t) \sim t$ for $t \to +\infty$, so that the function φ inside the integral sign converges to 1. Therefore

$$\frac{1}{n}\int_0^n \frac{\log(1+e^t)}{\sqrt{t^2+t+1}}\,dt = \frac{1}{n}\int_0^n \varphi(t)\,dt$$

is an indeterminate form as $n \to +\infty$, to evaluate which it is enough to evaluate the limit as $x \to +\infty$ of the function $g(x) = (\int_0^x \varphi(t)\,dt)/x$. The limit as $x \to +\infty$ of g can be computed using de l'Hôpital's theorem because the numerator is of class C^1 since φ is continuous. The derivative of the numerator of g is $\varphi(x)$ and that of the denominator is 1. As observed earlier, $\varphi(x) \to 1$ as $x \to +\infty$. The conclusion is therefore that

$$\lim_n \int_{-1}^1 f_n(x)\,dx = \lim_n \frac{1}{n}\int_0^n \frac{\log(1+e^t)}{\sqrt{t^2+t+1}}\,dt = \lim_{x \to +\infty} \frac{\int_0^x \varphi(t)\,dt}{x} = 1.$$

This is a rather elaborate exercise. Each of the issues to be addressed is not particularly demanding, but they all call for some attention and there are several of them. So the main difficulty is perhaps that despite the simple formulations, each of the two questions brakes down in a sequence of smaller questions. Pointwise convergence is actually a simple matter, but uniform convergence is not. Things break up at the origin, so it is natural to study $(0, +\infty)$ and $(-\infty, 0)$ separately. In a sense, the key observation is that

$$\log(1+e^t) \sim_{-\infty} e^t, \qquad \log(1+e^t) \sim_{+\infty} t$$

whereas

$$\sqrt{t^2+t+1} \sim_{\pm\infty} |t|.$$

When suitably rephrased, this says that for large values of x the convergence is uniform, that is, on intervals bounded away from the origin. The behaviour at the origin, however, forbids uniform convergence on either one of the whole rays $(0, +\infty)$ and $(-\infty, 0)$. The second question requires first the change of variable $t = nx$ which is completely natural because each function in the sequence is of the form $f_n(x) = \varphi(nx)$. Once this is done, one simply needs to understand the two improper integrals

$$\int_{-\infty}^0 \varphi(t)\,dt, \qquad \int_0^{+\infty} \varphi(t)\,dt.$$

The former converges and the latter diverges of order 1, which is all is really necessary to establish.

6.11 Consider the sequence $(f_n)_{n\geq 1}$ where

$$f_n(x) = \begin{cases} (-1)^n (x + \frac{1}{n})^2 & x \in \mathbb{Q} \\ (-1)^{n+1} (x - \frac{1}{n})^2 & x \in \mathbb{R} \setminus \mathbb{Q}. \end{cases}$$

Discuss pointwise and uniform convergence of the sequences $(f_n)_{n\geq 1}$ and $(f_n^2)_{n\geq 1}$.

Answer. First of all, observe that

$$\lim_n (x + \frac{1}{n})^2 = \lim_n (x - \frac{1}{n})^2 = x^2,$$

so that

$$\lim_n f_n(x) = \begin{cases} 0 & x = 0 \\ \nexists & x \neq 0. \end{cases}$$

Therefore, the sequence $(f_n)_{n\geq 1}$ converges only at $x_0 = 0$ to the value 0. As for the sequence $(f_n^2)_{n\geq 1}$, evidently

$$f_n^2(x) = \begin{cases} (x + \frac{1}{n})^4 & x \in \mathbb{Q} \\ (x - \frac{1}{n})^4 & x \in \mathbb{R} \setminus \mathbb{Q} \end{cases}$$

and hence $\lim_n f_n^2(x) = x^4$ for every $x \in \mathbb{R}$. In order to study the uniform convergence of $(f_n^2)_{n\geq 1}$, consider first an interval of the form $[a, b]$. From the formula $x^4 - y^4 = (x - y)(x + y)(x^2 + y^2)$ it follows

$$|f_n^2(x) - x^4| \leq \frac{1}{n}\left(2|x| + \frac{1}{n}\right)\left(x^2 + \frac{1}{n^2} + \frac{2|x|}{n} + x^2\right)$$

$$\leq \frac{1}{n}\left((2(|a| + |b|) + \frac{1}{n}\right)\left(2(|a| + |b|)^2 + \frac{1}{n^2} + \frac{2}{n}2(|a| + |b|)\right)$$

because if $x \in [a, b]$, then $|x| \leq \max\{|a|, |b|\} \leq |a| + |b|$. Therefore

$$\lim_n \sup_{x \in [a,b]} |f_n^2(x) - x^4| = 0$$

and $(f_n^2)_{n\geq 1}$ converges uniformly to $f(x) = x^4$ on every interval $[a, b]$. The convergence is not uniform on any unbounded interval. To see this consider for instance the interval $[a, +\infty)$, where

$$\sup_{x \in [a,+\infty)} |f_n^2(x) - x^4| \geq \sup_{x \in [a,+\infty) \cap \mathbb{Q}} \left|\left((x + \frac{1}{n})^2 + x^2\right)\left((x + \frac{1}{n})^2 - x^2\right)\right|$$

6.5 Guided Exercises

$$\geq \lim_{\substack{x \to \infty \\ x \in \mathbb{Q}}} \left|2x^2 + \frac{2x}{n} + \frac{1}{n^2}\right| \left|\frac{2x}{n} + \frac{1}{n^2}\right| = +\infty.$$

The argument for intervals of the kind $(-\infty, a]$ is analogous.

First of all, it is quite clear that the alternating signs play a (negative) role in the behaviour of $(f_n)_{n \geq 1}$ but not of $(f_n^2)_{n \geq 1}$, so this exercise is really about $(f_n^2)_{n \geq 1}$. The issue is uniform convergence, so the behaviour to be evaluated is that of $|f_n^2(x) - x^4|$ which for $x \in \mathbb{Q}$ is

$$\left|x^4 + 4\frac{x^3}{n} + 6\frac{x^2}{n^2} + 4\frac{x}{n^3} + \frac{1}{n^4} - x^4\right| = \left|4\frac{x^3}{n} + 6\frac{x^2}{n^2} + 4\frac{x}{n^3} + \frac{1}{n^4}\right|.$$

Quite visibly, and informally speaking, this quantity is bounded with respect to x (and tends to 0 as $n \to \infty$) if x is bounded, whereas it is not bounded if x is not bounded. The point is how to accurately express this simple observation, which in the end implies that uniform convergence occurs on compacta and does not over unbounded intervals.

6.12 Consider the series $\sum_{n=1}^{\infty} \left(\frac{x^n}{n} - \frac{x^{n+1}}{n+1}\right)$.

(a) Find the interval I where the series converges pointwise to a sum S.
(b) Prove that the convergence to S is uniform in I.
(c) Show that the series of derivatives does not converge uniformly on the set J where it converges pointwise and that it does not converge to S' on J.

Answer. (a) This is a telescopic series and the partial sum is

$$S_n(x) = \left(x - \frac{x^2}{2}\right) + \cdots + \left(\frac{x^n}{n} - \frac{x^{n+1}}{n+1}\right) = x - \frac{x^{n+1}}{n+1}.$$

Therefore, if $|x| \leq 1$, then $S_n(x) \to x$ and the series converges pointwise to $S(x) = x$ on $I = [-1, 1]$. In $\mathbb{R} \setminus I$ the partial sum S_n does not have a finite limit.

(b) For every $x \in I$

$$|S(x) - S_n(x)| = \left|x - x + \frac{x^{n+1}}{n+1}\right| \leq \frac{|x|^{n+1}}{n+1} \leq \frac{1}{n+1}.$$

Therefore

$$\limsup_n \sup_{x \in I} |S(x) - S_n(x)| = 0$$

so that the series converges uniformly on I.

(c) The series of derivatives is

$$\sum_{n=1}^{\infty}\left(n\frac{x^{n-1}}{n} - (n+1)\frac{x^n}{n+1}\right) = \sum_{n=0}^{\infty}(x^n - x^{n+1}),$$

which is also a telescopic series. The nth partial sum of the latter series is

$$T_n(x) = (1-x) + \cdots + (x^n - x^{n+1}) = 1 - x^{n+1} = S'_{n+1}(x),$$

so that

$$\lim_n T_n(x) = \begin{cases} \nexists & x = -1 \\ 1 & x \in (-1, 1) \\ 0 & x = 1. \end{cases}$$

Therefore the series of derivatives converges pointwise in $J = (-1, 1]$ to the sum

$$G(x) = \begin{cases} 1 & x \in (-1, 1) \\ 0 & x = 1. \end{cases}$$

Since the sum G is not continuous, the convergence cannot be uniform on J because of course $(x^n - x^{n+1})$ is continuous on J for every n. Finally, $G(1) = 0 \neq 1 = S'(1)$.

This exercise displays a simple example where "derivation under the series sign" cannot be performed because the series of derivates fails to converge uniformly. This is the key issue. Although the original series does converge uniformly to S on I and although the series of derivatives converges (except at -1) to a sum G, the equality $S'(x) = G(x)$ may fail. It should also be noticed that in general

$$\frac{d}{dx} S_n(x) = \frac{d}{dx} \sum_{k=1}^n f_k(x) = \sum_{k=1}^n f'_k(x).$$

Since the summation index has been changed for convenience,

$$S'_n(x) = T_{n-1}(x)$$

in the notation introduced in this exercise.

6.13 Consider the series $\sum_{n=0}^{\infty}(-1)^n \left[1 - \log\left(\frac{x}{x-1}\right)\right]^{2n}$.

(a) Determine where the series converges.
(b) Find the sum $S(x)$.
(c) Compute, if existing, $\lim_{x \to x_0^+} S(x)$ where $x_0 = \dfrac{e^2}{e^2 - 1}$.

6.5 Guided Exercises

Answer. (a) Each function $f_n(x) := \left[1 - \log\left(\frac{x}{x-1}\right)\right]^{2n}$ is defined provided that the argument of the logarithm is positive, that is, in $J = (-\infty, 0) \cup (1, +\infty)$. Furthermore,

$$(-1)^n f_n(x) = (-1)^n \left[1 - \log\left(\frac{x}{x-1}\right)\right]^{2n} = \left\{-1\left[1 - \log\left(\frac{x}{x-1}\right)\right]^2\right\}^n.$$

Hence, upon setting

$$y := y(x) = -\left[1 - \log\left(\frac{x}{x-1}\right)\right]^2$$

it is possible to rewrite the original series as the geometric series $\sum_{n=0}^{\infty} y^n$, which converges for $y \in (-1, 1)$. Since $y \leq 0$, it is enough to find $I = \{x : y(x) > -1\}$, that is, to determine for which values of x it holds

$$-\left[1 - \log\left(\frac{x}{x-1}\right)\right]^2 > -1 \quad \Longleftrightarrow \quad -1 < 1 - \log\left(\frac{x}{x-1}\right) < 1$$

which is in turn equivalent to solve the system

$$\begin{cases} \log\left(\frac{x}{x-1}\right) < 2 \\ \log\left(\frac{x}{x-1}\right) > 0 \end{cases} \Longleftrightarrow \begin{cases} \frac{x}{x-1} < e^2 \\ \frac{x}{x-1} > 1 \end{cases} \Longleftrightarrow \begin{cases} \frac{(e^2-1)x - e^2}{x-1} > 0 \\ \frac{1}{x-1} > 0. \end{cases}$$

The solution's set is $I = \{x : x > \frac{e^2}{e^2-1} = x_0\} = (x_0, +\infty)$. Since $x_0 > 1$, it is $I \subset J$, and the series converges exactly in $I = (x_0, +\infty)$.

(b) From the fact that the series may be written as a geometric series, the sum is

$$S(x) = \sum_{n=0}^{\infty} y(x)^n = \frac{1}{1 - y(x)} = \frac{1}{1 + \left[1 - \log\left(\frac{x}{x-1}\right)\right]^2}$$

and is defined in $I = (x_0, +\infty)$.

(c) From the above calculation and from the observation that

$$\lim_{x \to x_0^+} \log\left(\frac{x}{x-1}\right) = \log\left(\frac{x_0}{x_0-1}\right) = 2,$$

it follows that

$$\lim_{x \to x_0^+} S(x) = \frac{1}{1 + \left[1 - \log\left(\frac{x_0}{x_0-1}\right)\right]^2} = \frac{1}{2}.$$

The main point is to realize that the given series may be written in the form of a geometric series $\sum_{n=0}^{\infty} y(x)^n$. Once this is realized, then it is just a matter of finding where $y(x)$ is defined and when $|y(x)| < 1$. It turns out that this happens precisely in the open interval $(x_0, +\infty)$, where of course the explicit form of $S(x)$ is given by the sum of the geometric series, namely $S(x) = (1 - y(x))^{-1}$. Thus the required limit is immediate to compute because $y(x) \to -1$ as $x \to x_0$, so that $S(x) \to 1/2$. In this regard, it is worthwhile observing that the equality

$$\sum_{n=0}^{\infty} t^n = \frac{1}{1-t}$$

between a convergent series on the left hand side and a rational function on the right hand side holds exactly when $t \in (-1, 1)$, but the rational function on the right hand side is perfectly well-defined also when $t = -1$ and is equal to $1/2$. In some sense, this is the main point of the exercise.

6.14 Let $S(x) = \sum_{n=1}^{\infty} \int_0^{x^2/n^2} e^{-t^2}\, dt$.

(a) Find the domain of S.
(b) Where is S continuous?
(c) Where is S differentiable?

Answer. (a) The domain of S is the set where the indicated series converges pointwise. Observe that, since $e^{-t^2} \leq 1$,

$$f_n(x) := \int_0^{x^2/n^2} e^{-t^2}\, dt \leq \frac{x^2}{n^2}.$$

Therefore

$$0 \leq \sum_{n=1}^{\infty} f_n(x) \leq x^2 \sum_{n=1}^{\infty} \frac{1}{n^2} < +\infty,$$

and the series converges in $\mathbb{R}$, and the domain of S is thus $\mathbb{R}$.

(b) Evidently, $f_n \in C^0(\mathbb{R})$ for every n. Furthermore, for any $R > 0$ it is

$$0 \leq \sum_{n=1}^{\infty} \sup_{x \in [-R,R]} |f_n(x)| \leq R^2 \sum_{n=1}^{\infty} \frac{1}{n^2}$$

so that the series converges totally, hence uniformly, in every interval $[-R, R]$ with $R > 0$. It follows that S is continuous in any such interval, hence on $\mathbb{R}$.

(c) Observe that

$$\sum_{n=1}^{\infty} f_n'(x) = \sum_{n=1}^{\infty} \frac{2x}{n^2} e^{-\frac{x^4}{n^4}}.$$

6.5 Guided Exercises

Arguing as before, for any $R > 0$ it is

$$0 \le \sum_{n=1}^{\infty} \sup_{x \in [-R,R]} |f_n'(x)| \le 2R \sum_{n=1}^{\infty} \frac{1}{n^2}.$$

so that the series of derivatives converges totally, hence uniformly, in every interval $[-R, R]$ with $R > 0$. It follows that S is differentiable on $\mathbb{R}$ because it is such in every interval $[-R, R]$ with $R > 0$.

Here the key property to check is the uniform convergence, which is most easily handled with total convergence. Thus, a good estimate of $|f_n(x)|$ is needed. This fact is indeed manifest from the fact that the integral is taken over rapidly shrinking intervals, namely $[0, x^2/n^2]$, and the Gaussian function is bounded by 1. The derivatives f_n' are also dominated by $2x/n^2$ and this gives total convergence on compact sets of the series $\sum f_n'$, hence an easy answer to question (c).

6.15 Consider the series $\sum_{n=1}^{\infty} \dfrac{\sin nx}{n^2}$.

(a) Discuss the pointwise and uniform convergence of the series to its sum $S(x)$.
(b) Express $\int_0^{\pi} S(x)\,dx$ as the sum of a convergent numerical series.

Answer. (a) Evidently, for every $n \ge 1$

$$\sup_{x \in \mathbb{R}} \left| \frac{\sin nx}{n^2} \right| = \frac{1}{n^2}$$

and $\sum n^{-2} < +\infty$. Hence the given series converges totally, hence absolutely and uniformly, on $\mathbb{R}$ to its sum S.

(b) As every $f_n(x) = (\sin nx)/n^2$ is continuous on $[0, \pi]$, so is the sum S and

$$\int_0^{\pi} S(x)\,dx = \int_0^{\pi} \sum_{n=1}^{\infty} \frac{\sin nx}{n^2}\,dx$$

$$= \sum_{n=1}^{\infty} \int_0^{\pi} \frac{\sin nx}{n^2}\,dx$$

$$= \sum_{n=1}^{\infty} \frac{1}{n^3} \int_0^{\pi} n \sin nx\,dx$$

$$= \sum_{n=1}^{\infty} \frac{1}{n^3} [-\cos nx]_0^{\pi}$$

$$= \sum_{n=1}^{\infty} \frac{1}{n^3} [1 - (-1)^n].$$

Finally
$$1-(-1)^n = \begin{cases} 0 & n \text{ is even} \\ 2 & n \text{ is odd,} \end{cases}$$

whence
$$\int_0^\pi S(x)\,dx = \sum_{m=0}^\infty \frac{2}{(2m+1)^3}.$$

This is a standard exercise on term-by-term integration in the sense of Theorem 6.8, where all it is required to show is uniform convergence. The computation is then straightforward.

6.16 Consider the series $\sum_{n=1}^\infty \dfrac{\arctan(nx)}{n^2+x^2}$.

(a) Prove that the pointwise limit S belongs to $C^1(\mathbb{R}\setminus\{0\})$.
(b) Show that S is not differentiable at $x_0 = 0$.

Answer. (a) Since
$$\left|\frac{\arctan(nx)}{n^2+x^2}\right| \le \frac{\pi}{2}\frac{1}{n^2+x^2} \le \frac{\pi}{2}\frac{1}{n^2}$$

and since the series $\sum n^{-2}$ converges, the given series converges totally in $\mathbb{R}$, hence the pointwise limit S is defined and continuous on $\mathbb{R}$. Consider next the formal derivative, namely, after elementary manipulation,
$$\sum_{n=1}^\infty \left[\frac{n}{(1+n^2x^2)(n^2+x^2)} - \frac{2x\arctan(nx)}{(n^2+x^2)^2}\right]$$

Observe that if $|x| \ge \lambda > 0$, then
$$\frac{n}{(1+n^2x^2)(n^2+x^2)} \le \frac{n}{(1+n^2\lambda^2)(n^2+\lambda^2)} \le \frac{n}{n^4\lambda^2} = \frac{1}{n^3\lambda^2}.$$

whereas if $|x| \le \delta$, then
$$\left|\frac{2x\arctan(nx)}{(n^2+x^2)^2}\right| \le 2|x|\frac{\pi}{2}\frac{1}{n^4} \le \delta\pi\frac{1}{n^4}.$$

Since both series $\sum n^{-3}$ and $\sum n^{-4}$ converge, it follows that if $0 < \lambda < \delta$, then in every set of the form
$$I_{\lambda,\delta} = \{x \in \mathbb{R} : \delta \le |x| \le \lambda\} = [-\delta,-\lambda] \cup [\lambda,\delta]$$

6.5 Guided Exercises

the formal derivative converges totally. Since each f'_n is continuous on $I_{\lambda,\delta}$, it follows that $f \in C^1(I_{\lambda,\delta})$ for every choice $0 < \lambda < \delta$. The union over all such choices is

$$\bigcup_{0<\lambda<\delta} I_{\lambda,\delta} = \mathbb{R} \setminus \{0\}$$

and hence $f \in C^1(\mathbb{R} \setminus \{0\})$.

(b) Take next $x > 0$ and consider the difference quotient

$$\frac{f(x) - f(0)}{x} = \frac{1}{x} \sum_{n=1}^{\infty} \frac{\arctan(nx)}{n^2 + x^2}.$$

Suppose by contradiction that f is differentiable at $x = 0$, that is, that the above expression admits a finite limit as $x \to 0^+$. Since for every positive integer N

$$\frac{1}{x} \sum_{n=1}^{\infty} \frac{\arctan(nx)}{n^2 + x^2} \geq \frac{1}{x} \sum_{n=1}^{N} \frac{\arctan(nx)}{n^2 + x^2}$$

and since for any such N

$$\lim_{x \to 0^+} \frac{1}{x} \sum_{n=1}^{N} \frac{\arctan(nx)}{n^2 + x^2} = \sum_{n=1}^{N} \lim_{x \to 0^+} \frac{1}{x} \frac{\arctan(nx)}{n^2 + x^2} = \sum_{n=1}^{N} \frac{1}{n}$$

it follows that

$$\lim_{x \to 0^+} \frac{1}{x} \sum_{n=1}^{\infty} \frac{\arctan(nx)}{n^2 + x^2} \geq \lim_{x \to 0^+} \frac{1}{x} \sum_{n=1}^{N} \frac{\arctan(nx)}{n^2 + x^2} = \sum_{n=1}^{N} \frac{1}{n} \to +\infty,$$

as $N \to +\infty$, contradicting the existence of $f'(0)$.

In part (a) it is required to prove that the pointwise sum S (which is actually quite clearly defined and continuous for every $x \in \mathbb{R}$ by a very natural inquality that shows total convergence of the series) admits a continuous derivative everywhere except at the origin. This involves looking at the formal derivative in the hope that it actually converges totally in a large family of sets. The computation of the derivative f'_n produces a sum of two terms. One of them is easily estimated (in absolute value) from above when x is not too small and the other when x is not too large. Hence one achieves total convergence in $I_{\lambda,\delta}$, the union of two symmetric compact intervals bounded away from the origin. This is enough to prove that the derivative exists in any such set and, being the sum of a uniformly convergent series of continuous functions, it is continuous, a fact that follows from an application of Theorems 6.7 and 6.6. The property of being of class C^1 is local, and any point in $\mathbb{R} \setminus \{0\}$ is an interior point of one (actually infinitely many) of the sets $I_{\lambda,\delta}$. This concludes part (a). That things go wrong at the origin is suggested by the computation

$$f'_n(0) = \lim_{x \to 0^+} \frac{f_n(x) - f_n(0)}{x} = \lim_{x \to 0^+} \frac{1}{x} \frac{\arctan(nx)}{n^2 + x^2} = \lim_{x \to 0^+} \frac{\arctan(nx)}{nx} \frac{n}{n^2 + x^2} = \frac{1}{n}$$

which shows that $\sum f'_n(0)$ diverges, a strong hint of the insurgence of a problem.

6.17 Find the domain of the pointwise sum S of the series

$$\sum_{n=1}^{\infty} \frac{(2\cos x - 1)^{2n+1}}{n + 4^n},$$

study its monotonicity properties, compute the limits at the extreme points of its domain and draw a qualitative graph of S.

Answer. Write

$$f_n(x) = \frac{(2\cos x - 1)^{2n+1}}{n + 4^n}.$$

If $\cos x = 1/2$, then $f_n(x) = 0$ for every n and the series converges. Evidently, each f_n is defined on $\mathbb{R}$ and is 2π-periodic. Apply the ratio criterion to the numerical series $\sum |f_n(x)|$ for fixed $x \in \mathbb{R}$, assuming that $\cos x \neq 1/2$, namely consider the ratio

$$\frac{|f_{n+1}(x)|}{|f_n(x)|} = \frac{|2\cos x - 1|^{2n+3}}{(n+1) + 4^{n+1}} \frac{n + 4^n}{|2\cos x - 1|^{2n+1}}$$

$$= |2\cos x - 1|^2 \frac{4^n(1 + n4^{-n})}{4^{n+1}(1 + (n+1)4^{-n-1})}$$

$$= \frac{|2\cos x - 1|^2}{4} \frac{1 + n4^{-n}}{1 + (n+1)4^{-n-1}}.$$

Clearly,

$$\lim_n \frac{|f_{n+1}(x)|}{|f_n(x)|} = \frac{|2\cos x - 1|^2}{4}.$$

This is smaller than 1 when $|2\cos x - 1| < 2$, that is, when $-1/2 < \cos x < 3/2$. Therefore the series converges absolutely, hence pointwise, in the set

$$I = \bigcup_{k \in \mathbb{Z}} \left(-\frac{2}{3}\pi + 2k\pi, \frac{2}{3}\pi + 2k\pi \right) =: \bigcup_{k \in \mathbb{Z}} I_k.$$

Conversely, if $\cos x \leq -1/2$, then

$$f_n(x) \leq \frac{-2 \cdot 4^n}{n + 4^n} =: a_n$$

and the series $\sum a_n$ diverges to $-\infty$ because $a_n \to -2$. Therefore for such values of x the series $\sum f_n(x)$ diverges to $-\infty$ as well, by comparison. It follows that the domain of the pointwise sum S is exactly I. Furthermore, since each f_n is 2π-periodic, it is enough to study the behaviour of S in the interval $I_0 = \left(-\frac{2}{3}\pi, \frac{2}{3}\pi\right)$, and, in fact, since S is even, the interval $\left(0, \frac{2}{3}\pi\right)$ would suffice. Now, for fixed $a \in (\pi/2, 2\pi/3)$ it is easy to check that (see Fig. 6.7) for every $x \in [-a, a]$

$$|f_n(x)| \le |f_n(a)| = \frac{|2\cos a - 1|^{2n+1}}{n + 4^n} =: b_n$$

and, as observed earlier, the series $\sum b_n$ converges. Thus the series $\sum f_n(x)$ converges totally hence uniformly in every compact interval contained in I_0. Therefore S is continuous on its domain.

Next, observe that each $f_n \in C^1(\mathbb{R})$ with

$$f_n'(x) = \frac{(2n+1)(2\cos x - 1)^{2n}}{n + 4^n}(-2\sin x).$$

Consider now the odd C^∞ function

$$g_n(x) = (2\cos x - 1)^{2n} \sin x$$

restricted to $[0, 2\pi/3)$, where it is positive. A little analysis (see Fig. 6.8) reveals that $g_n(0) = 0 = g_n(\pi/3)$, $g_n(\pi/2) = 1$ and $g_n(2\pi/3) = 2^{2n-1}\sqrt{3}$; furthermore, g_n has a unique local maximum at a point $x_n \in (0, \pi/3)$, where $\cos x_n = (4n+1)/(4n+2)$ Actually, $x_n \to 0$ as $n \to \infty$ and $g_n(x_n) < 1$. Finally, g_n increases in $[\pi/3, 2\pi/3]$ and is therefore larger than 1 to the right of $\pi/2$ and smaller than 1 to the left of $\pi/2$.

It follows that for any $a > \pi/2$ and any $x \in [-a, a] \subset I_0$ it is

$$|f_n'(x)| = \frac{2n+1}{n+4^n} 2|g_n(x)|$$

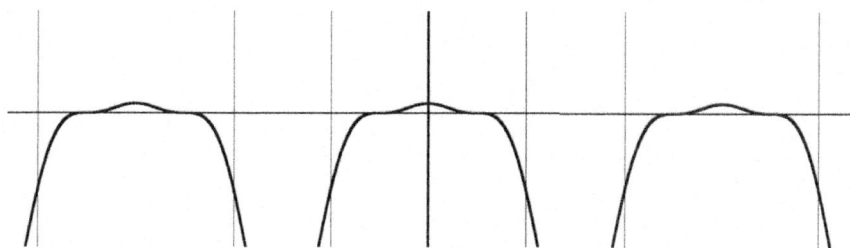

Fig. 6.7 The graph of the periodic function S in Exercise 6.17

Fig. 6.8 The bigraph of the function g_1 in Exercise 6.17, restricted to $[0, 2\pi/3]$; the local maximum is smaller than 1 and occurs to the left of $\pi/3$; then g_1 grows all the way to $2\sqrt{3}$ and is exactly 1 at $\pi/2$. The behaviour of the other functions g_n is very similar

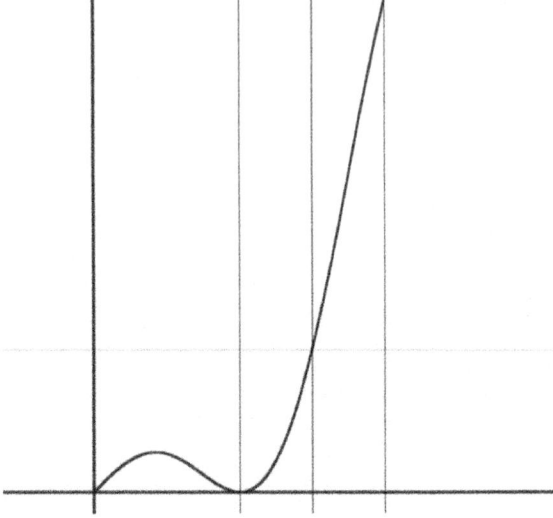

$$\leq \frac{2n+1}{n+4^n} 2|g_n(a)|$$
$$= \frac{2n+1}{n+4^n} 2(2\cos a - 1)^{2n}|\sin a|$$
$$\leq 2(2n+1)\left(\frac{(2\cos a - 1)^2}{4}\right)^n =: c_n$$

Now, since

$$\frac{(2\cos a - 1)^2}{4} < 1,$$

the series $\sum c_n$ converges so that the formal derivative converges in every compact contained in I_0, and likewise in each I_k. This implies that S is differentiable in I with

$$S'(x) = -2\sin x \sum_{n=1}^{\infty} \frac{(2n+1)(2\cos x - 1)^{2n}}{n+4^n}.$$

As argued above, it is enough to look at $x \in (-2\pi/3, 2\pi/3)$. Now, $(2\cos x - 1)^{2n} \geq 0$ for every $n \geq 1$ and equality occurs exactly at $\pm\pi/3$, for every $n \geq 1$. Therefore S' has the same sign of $-\sin x$. Thus, $S'(x) > 0$ for every $x \in (-2\pi/3, 0)$ and $S'(x) < 0$ for every $x \in (0, 2\pi/3]$. Therefore, S is strictly increasing in $(-2\pi/3, 0]$ and strictly decreasing in $[0, 2\pi/3]$.

Finally, it is possible to show that the (certainly existing) limits at $\pm 2\pi/3$ are both $-\infty$. Indeed, look at $-2\pi/3$. The function $2\cos x - 1$ is strictly negative in the interval $(-2\pi/3, -\pi/3)$ and such are also its odd powers. Hence for every $k \geq 1$

6.5 Guided Exercises

and every $x \in (-2\pi/3, -\pi/3)$ it holds

$$S(x) \le \sum_{n=1}^{k} \frac{(2\cos x - 1)^{2n+1}}{n + 4^n}.$$

Therefore

$$\lim_{x \to -\frac{2}{3}\pi^+} S(x) \le \lim_{x \to -\frac{2}{3}\pi^+} \sum_{n=1}^{k} \frac{(2\cos x - 1)^{2n+1}}{n + 4^n}$$

$$= \sum_{n=1}^{k} \lim_{x \to -\frac{2}{3}\pi^+} \frac{(2\cos x - 1)^{2n+1}}{n + 4^n}$$

$$= \sum_{n=1}^{k} \frac{-2 \cdot 4^n}{n + 4^n}.$$

Taking the limit as $k \to \infty$, it follows that $\lim_{x \to -\frac{2}{3}\pi^+} S(x) = -\infty$ because the series

$$\sum_{n=1}^{\infty} \frac{-2 \cdot 4^n}{n + 4^n}$$

diverges to $-\infty$. A similar argument handles the limit $x \to \frac{2}{3}\pi^-$. The graph of S is illustrated in Fig. 6.7.

This exercise is standard in its basic itinerary, but each step requires a little ingenuity. In order to find the set I of pointwise convergence it is wise to pause and observe that the denominator grows very quickly independently of x so that conceivably some growth of the numerator could occur without destroying the convergence. This matter may be settled precisely by applying the ratio test, which gives the condition $|2\cos x - 1| < 2$, which does indeed allow for some not negligible growth of the numerator $(2\cos x - 1)^{2n}$ when $|2\cos x - 1| > 2$. The request for convergence is of course 2π-periodic, a direct consequence of the 2π-periodicity of each f_n. Hence I may be studied by looking at its basic interval $I_0 = (-\frac{2}{3}\pi, \frac{2}{3}\pi)$. Differentiability calls to look at the formal derivative $\sum f_n'$, where $f_n'(x) = \alpha_n g_n(x)$. Some patience leads to understanding the behaviour of g_n, depicted in Fig. 6.8, hence to the estimates $|f_n'(x)| \le c_n$ in every "large" symmetric compact subinterval of I_0 with a convergent series $\sum c_n$. Hence S is differentiable in I and the derivative has the nice form $S'(x) = -2\sin x \sum \alpha_n (2\cos x - 1)^{2n}$, which reveals at once where S is either increasing or decreasing because $\sum \alpha_n (2\cos x - 1)^{2n}$ is nonnegative. Finally, the limiting behaviour at $\pm 2\pi/3$ is established with an argument similar to that used in Exercise 6.16, namely by combining two observations: near the limiting points the f_n all have the same sign, so partial sums either converge or diverge monotonically, and for partial sums the limit can be taken inside the summation thereby implying the suspected divergence.

6.18 Determine where the series $\sum_{n=1}^{\infty}(-1)^n \dfrac{\sin x + 2n}{n^2}$ converges pointwise and where it converges uniformly.

Answer. For any fixed $x \in \mathbb{R}$ the series is a Leibniz series. Indeed, for any $x \in \mathbb{R}$

$$f_n(x) := \frac{\sin x + 2n}{n^2} > 0$$

for $n \geq 1$ because $\sin x > -2 \geq -2n$. Further, $\lim_n \dfrac{\sin x + 2n}{n^2} = 0$, and finally

$$\frac{\sin x + 2n}{n^2} > \frac{\sin x + 2(n+1)}{(n+1)^2} \iff (n+1)^2(\sin x + 2n) > n^2 \sin x + 2n^2(n+1)$$
$$\iff [(n+1)^2 - n^2]\sin x > -2n^2 - 2n$$
$$\iff \sin x > -\frac{2n(n+1)}{2n+1}.$$

Observe now that for every positive integer

$$-\frac{2n(n+1)}{2n+1} < -1$$

because this is equivalent to $n^2 > 1/2$. In conclusion, for every $x \in \mathbb{R}$ it holds that $f_n(x) > f_{n+1}(x)$, namely $(f_n(x))_{n \geq 1}$ is a decreasing sequence. By Leibniz criterion, that is Theorem 12.10 in [1], the series converges. Hence the set of pointwise convergence is $\mathbb{R}$. Next, using again Theorem 12.10 in [1]

$$\left| \sum_{n=1}^{\infty}(-1)^n f_n(x) - \sum_{n=1}^{N}(-1)^n f_n(x) \right| \leq f_{N+1}(x) = \frac{\sin x + 2(N+1)}{(N+1)^2} \leq \frac{2N+3}{(N+1)^2}.$$

Since of course $(2N+3)/(N+1)^2 \to 0$, the series converges uniformly on $\mathbb{R}$.

In this exercise it is just a matter of realizing that the series is a Leibniz' series for each fixed $x \in \mathbb{R}$ and that in order to establish uniform convergence in $\mathbb{R}$ it is enough to apply the very definition, namely

$$\lim_N \sup_{x \in \mathbb{R}} |S(x) - S_N(x)| = 0$$

which is Definition 6.4 made explicit by Definition 6.2. Here Leibniz' criterion is particularly handy because of the estimate $|S(x) - S_N(x)| \leq f_{N+1}(x)$

6.19 Determine where the series $\sum_{n=1}^{\infty}\left(e^{x/n^2} - \cos(x/n)\right)$ converges pointwise and where it converges uniformly.

6.5 Guided Exercises

Answer. Fix $x \in \mathbb{R}$ and observe that

$$\left| e^{x/n^2} - \cos(x/n) \right| \leq \left| e^{x/n^2} - 1 \right| + |1 - \cos(x/n)|.$$

In order to estimate each summand it is useful to apply Lagrange's theorem, first to $\varphi_n(x) = \exp(x/n^2)$ and then to $\psi_n(x) = \cos(x/n)$. From $\varphi_n(x) - \varphi_n(0) = \varphi'_n(c)x$ for some c between 0 and x, namely $0 < |c| < |x|$, it follows that

$$e^{x/n^2} - 1 = \frac{1}{n^2} e^{c/n^2} x \quad \Longrightarrow \quad \left| e^{x/n^2} - 1 \right| \leq \frac{|x|}{n^2} e^{c/n^2} \leq \frac{|x| e^{|x|}}{n^2},$$

and similarly from $\psi_n(x) - \psi_n(0) = \psi'_n(d)x$ for some d between 0 and x, namely $0 < |d| < |x|$, it follows that

$$1 - \cos(x/n) = \frac{x}{n} \sin(d/n) \quad \Longrightarrow \quad |1 - \cos(x/n)| = \left| \frac{x}{n} \sin(d/n) \right| \leq \frac{x^2}{n^2}.$$

Therefore the function $f_n(x) = e^{x/n^2} - \cos(x/n)$ satisfies

$$|f_n(x)| = \left| e^{x/n^2} - \cos(x/n) \right| \leq \frac{x^2 + |x| e^{|x|}}{n^2},$$

and since the series $\sum n^{-2}$ converges, $\sum f_n$ converges pointwise on $\mathbb{R}$. Take now a compact interval $[-a, a]$ for some $a > 0$. The same estimate achieved above yields

$$|f_n(x)| \leq \frac{a^2 + a e^a}{n^2},$$

thereby implying that $\sum f_n$ converges totally, hence uniformly, in each compact subset of $\mathbb{R}$. It cannot, however, converge uniformly on the whole $\mathbb{R}$ because the necessary condition

$$\lim_n \sup_{x \in \mathbb{R}} |f_n(x)| = 0$$

does not hold true. Indeed, for every integer $n \geq 1$

$$\sup_{x \in \mathbb{R}} |f_n(x)| = \sup_{x \in \mathbb{R}} f_n(x) \geq \sup_{x \in \mathbb{R}} \left(e^{x/n^2} - 1 \right) = +\infty.$$

This exercise is about estimating how fast the sequence $(f_n)_{n \geq 1}$ tends to zero, pointwise and uniformly. Now, both $\varphi_n(x) = \exp(x/n^2)$ and $\psi_n(x) = \cos(x/n)$ tend of course to 1 as $n \to \infty$, so it is natural to evaluate separately their distances from 1, which in both cases is dominated by small constants on compact sets but in neither case it is uniformly small on the whole $\mathbb{R}$. Hence $(f_n)_{n \geq 1}$ tends to 0 uniformly on every compact where it is nicely dominated, but this doesn't happen globally.

6.20 Determine where the series $\sum_{n=1}^{\infty} (e^{x/n} - 1) \log(1 + \frac{x^2}{n})$ converges pointwise and where it converges uniformly.

Answer. The function $f_n(x) = (e^{x/n} - 1) \log(1 + \frac{x^2}{n})$ vanishes at $x_0 = 0$ and hence the series $\sum f_n(0)$ converges. For any fixed $x \neq 0$ both sequences $(e^{x/n} - 1)_{n \geq 1}$ and $(\log(1 + \frac{x^2}{n}))_{n \geq 1}$ are positive and tend to 0 as $n \to \infty$ with order 1. Hence their product vanishes with order 2 and the series $\sum f_n(x)$ converges by the order criterion (Theorem 12.9 in [1]). Thus the set of pointwise convergence is $\mathbb{R}$. Fix next a compact interval $[a, b]$. From Lagrange's theorem it follows that for any $t \in \mathbb{R}$ there exists ξ with $0 < |\xi| < |t|$ such that $e^t = 1 + te^\xi$. Therefore, if $x \in [a, b]$ and $c = \max\{|a|, |b|\}$ for some ξ with $0 < |\xi| < |x|/n$ it holds

$$\left| e^{x/n} - 1 \right| = \frac{|x|}{n} e^\xi \leq \frac{c}{n} e^{|x|/n} \leq \frac{c}{n} e^c.$$

Finally, since $\log(1 + t) \leq t$ for every $t > 0$ it follows that for every $x \in [a, b]$ it is

$$|f_n(x)| \leq \frac{c}{n} e^c \frac{x^2}{n} \leq \frac{c^3 e^c}{n^2}.$$

Since the series $\sum n^{-2}$ converges, the series $\sum f_n$ converges totally hence uniformly in every interval $[a, b]$. The convergence is however not uniform on any unbounded interval J, either of the form $(-\infty, a]$ or $[a, +\infty)$ for some $a \in \mathbb{R}$. Indeed

$$\lim_{x \to \pm\infty} f_n(x) = \pm\infty \quad \Longrightarrow \quad \sup_{x \in J} f_n(x) = +\infty,$$

so that the sequence $(f_n)_{n \geq 1}$ does not converge uniformly to 0 on J.

This exercise is very similar to Exercise 6.19. Again everything is nicely dominated on compact intervals but not globally.

6.21 Consider the series $\sum_{n=1}^{\infty} (\cos(x/n) - 1)$.

(a) Find where the series converges pointwise and where it converges uniformly.
(b) Study the pointwise sum in $[-\pi, \pi]$ and draw a graph of it.

Answer. (a) Write $f_n(x) = (\cos(x/n) - 1)$ Using the second order McLaurin expansion, for any $t \in \mathbb{R}$ there is ξ with $0 < |\xi| < |t|$ such that

$$\cos t = 1 - \frac{t^2}{2} \cos \xi,$$

whence the estimate

$$|f_n(x)| \leq \frac{1}{2} \frac{c^2}{n^2}$$

6.5 Guided Exercises

for every $x \in I = [a, b]$, where $c = \max\{|a|, |b|\}$. Since $\sum n^{-2}$ converges, the series $\sum f_n$ converges totally hence uniformly in every bounded interval in $\mathbb{R}$. It does not converge uniformly in any unbounded interval $J = [a, +\infty)$ because any such interval contains infinitely many values of $n \in \mathbb{N}$ such that $n\pi \in J$, so that

$$\sup_{x \in J} |f_n(x))| = |f_n(n\pi))| = 2$$

and $(f_n)_{n \geq 1}$ does not converge uniformly to 0. A similar argument shows that $\sum f_n$ does not converge uniformly in any unbounded interval $(-\infty, a]$.

(b) Denote now by S the pointwise sum. From what has been established in (a), $S \in C^0(\mathbb{R})$. Furthermore, each f_n is differentiable and

$$f_n'(x) = -\frac{1}{n} \sin(x/n).$$

Thus,

$$|f_n'(x)| \leq \frac{|x|}{n^2}$$

so that in any interval $[a, b]$ the estimate

$$|f_n'(x)| \leq \frac{c}{n^2}$$

holds with $c = \max\{|a|, |b|\}$. Since $\sum n^{-2}$ converges, the formal derivative converges totally in any $[a, b]$ and $S \in C^1(\mathbb{R})$, with

$$S'(x) = -\sum_{n=1}^{\infty} \frac{1}{n} \sin(x/n).$$

Evidently, $x/n \in (0, \pi)$, hence $\sin(x/n) > 0$ if $x \in (0, \pi)$, and similarly $\sin(x/n) < 0$ if $x \in (-\pi, 0)$. Therefore $S'(x) > 0$ in $(-\pi, 0)$ and $S'(x) < 0$ in $(0, \pi)$. Thus S is strictly increasing in $[-\pi, 0]$ and strictly decreasing in $[0, \pi]$. Therefore $x_0 = 0$ is a local maximum. Actually, it is a global maximum because $f_n(x) \leq 0$ for every $x \in \mathbb{R}$, so that of course

$$S(x) = \sum_{n=1}^{\infty} f_n(x) \leq 0 = S(0).$$

Furthermore,

$$f_n''(x) = -\frac{1}{n^2} \cos(x/n)$$

and the obvious estimate

$$|f_n''(x)| \le \frac{1}{n^2}$$

shows that $S \in C^2(\mathbb{R})$ with

$$S''(x) = -\sum_{n=1}^{\infty} \frac{1}{n^2} \cos(x/n).$$

A straightforward analysis reveals that S is concave in $[-\pi/2, \pi/2]$. The graph of S is depicted in Fig. 6.9.

This exercise involves no particular difficulty, it just requires to handle carefully the necessary (but easy) estimates that are needed in order to show total convergence on compact intervals of $\sum f_n$, $\sum f_n'$ and $\sum f_n''$. The behaviour of S gets much more complicated away from the origin, as it appears from the bottom picture in Fig. 6.9.

6.22 Consider the function f defined by the series $\sum_{n=1}^{\infty} \dfrac{e^{-nx}}{n^2+1}$.

(a) Find $\mathrm{Dom}(f)$.
(b) Study continuity and differentiability of f.
(c) Compute $f(1)$ with an error no larger than 10^{-2}.

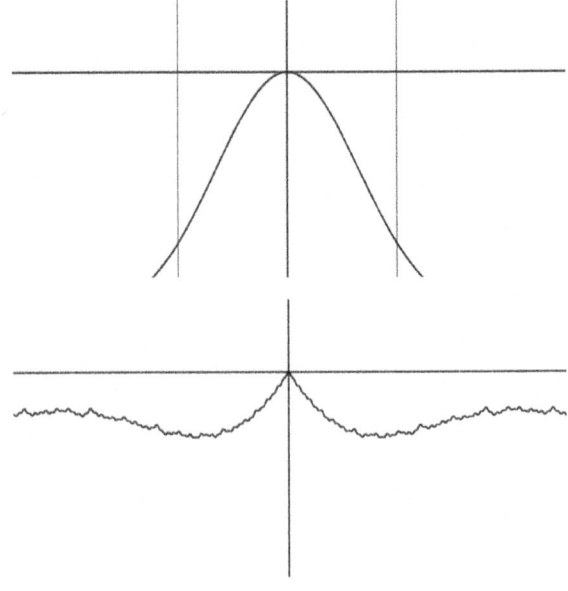

Fig. 6.9 The graph of the sum S in Exercise 6.21, restricted to $[-\pi, \pi]$ on top, where it appears quite nice; zooming out (bottom) reveals the more complicated behaviour of the sum S away from the origin, although $S \in C^2(\mathbb{R})$

6.5 Guided Exercises

Answer. (a) The functions $f_n(x) = \dfrac{e^{-nx}}{n^2+1}$ are positive on $\mathbb{R}$ and

$$\lim_n f_n(x) = \begin{cases} 0 & x \geq 0 \\ +\infty & x < 0, \end{cases}$$

so that necessarily $\mathrm{Dom}(f) \subseteq [0, +\infty)$. Further,

$$\sup_{x \geq 0} f_n(x) = \sup_{x \geq 0} \frac{e^{-nx}}{n^2+1} = \frac{1}{n^2+1}$$

and the series $\sum (n^2+1)^{-1}$ converges. Thus, by comparison, the series defining f converges uniformly, hence pointwise, on $[0, +\infty)$ and $\mathrm{Dom}(f) = [0, +\infty)$.

(b) Since each f_n is of class C^∞ (in particular continuous), the uniform convergence on $[0, +\infty)$ implies the continuity of f on its domain. Next, look at the formal derivative

$$\sum_{n=1}^\infty \frac{-ne^{-nx}}{n^2+1}$$

and fix $\delta > 0$. Since

$$\sup_{x \geq \delta} \left| \frac{-ne^{-nx}}{n^2+1} \right| = \frac{ne^{-n\delta}}{n^2+1}$$

and $\sum \dfrac{ne^{-n\delta}}{n^2+1}$ obviously converges, it follows that f is differentiable in $[\delta, +\infty)$ for every $\delta > 0$. Hence f is differentiable in $(0, +\infty)$ and for every $x \in (0, +\infty)$ it is

$$f'(x) = -\sum_{n=1}^\infty \frac{ne^{-nx}}{n^2+1} < 0.$$

The above formula shows that for any N it is $f'(x) < \sum_{n=1}^N \dfrac{-ne^{-nx}}{n^2+1}$, so that

$$\lim_{x \to 0^+} f'(x) < \lim_{x \to 0^+} \sum_{n=1}^N \frac{-ne^{-nx}}{n^2+1}$$

$$= \sum_{n=1}^N \lim_{x \to 0^+} \frac{-ne^{-nx}}{n^2+1}$$

$$= \sum_{n=1}^N \frac{-n}{n^2+1}.$$

Since the series in the last line diverges to $-\infty$, it follows that $\lim_{x \to 0^+} f'(x) = -\infty$ and so, applying de l'Hôpital's theorem, f is not differentiable at $x_0 = 0$.

(c) Evidently,
$$f(1) = \sum_{k=1}^{\infty} \frac{e^{-k}}{k^2+1} = \lim_n \sum_{k=1}^{n} \frac{e^{-k}}{k^2+1} =: \lim_n s_n$$

and hence
$$0 < f(1) - s_n = \sum_{k=n+1}^{\infty} \frac{e^{-k}}{k^2+1}$$
$$< \frac{1}{(n+1)^2+1} \sum_{k=n+1}^{\infty} e^{-k}$$
$$= \frac{1}{(n+1)^2+1} \left(\sum_{k=0}^{\infty} (e^{-1})^k - \sum_{k=0}^{n} (e^{-1})^k \right)$$
$$= \frac{1}{(n+1)^2+1} \left(\frac{1}{1-e^{-1}} - \frac{1-(e^{-1})^{n+1}}{1-e^{-1}} \right)$$
$$= \frac{1}{(n+1)^2+1} \frac{1}{e^{n+1} - e^n}$$

and this is $< 10^{-2}$ if $n = 2$. Hence, $s_2 < f(1) = f(1) - s_2 + s_2 < s_2 + 10^{-2}$, that is
$$\frac{1}{2e} + \frac{1}{5e^2} < f(1) < \frac{1}{2e} + \frac{1}{5e^2} + 10^{-2}.$$

This is essentially a standard exercise. The only slightly delicate issue is differentiability at the origin. Here, and elsewhere, it does not harm to recall that by de l'Hôpital's theorem if $f'(x) \to \pm\infty$ as $x \to 0^+$, then
$$\lim_{x \to 0^+} \frac{f(x) - f(0)}{x} = \lim_{x \to 0^+} f'(x) = \pm\infty,$$
so that $f'_+(0)$ does not exist. The use of partial sums to show that $f'(x) \to \pm\infty$ is a simple technique.

6.23 Consider the function f defined by the series $\sum_{n=1}^{\infty} \frac{n^x}{n+x^2}$.

(a) Find Dom(f).

(b) Find an upper bound for the integral $\int_{-2}^{-1} f(x)\, dx$.

Answer. (a) Observe that:

6.5 Guided Exercises

$$\lim_n f_n(x) := \lim_n \frac{n^x}{n+x^2} = \begin{cases} +\infty & x > 1 \\ 1 & x = 1 \\ 0 & x < 1 \end{cases}$$

and that when $x < 1$ the order with which $f_n(x)$ vanishes as $n \to \infty$ is exactly $1 - x$. It follows that the series converges if $1 - x > 1$, so that $\mathrm{Dom}(f) = (-\infty, 0)$.

(b) For any $x \in [-2, -1]$ it is $-2 \log n \le x \log n \le -\log n$, so that

$$\frac{n^x}{n+x^2} = \frac{e^{x \log n}}{n+x^2} \le \frac{e^{-\log n}}{n} = \frac{1}{n^2}.$$

This in particular implies that the series is totally convergent in $[-2, -1]$ and since the functions f_n are continuous in $[-2, -1]$, such is also f, hence integrable. Furthermore

$$\int_{-2}^{-1} f(x)\,dx = \sum_{n=1}^{\infty} \int_{-2}^{-1} f_n(x)\,dx \le \sum_{n=1}^{\infty} \int_{-2}^{-1} \frac{1}{n^2}\,dx = \sum_{n=1}^{\infty} \frac{1}{n^2}.$$

Therefore the upper bound is the sum of the series $\sum_{n=1}^{\infty} n^{-2}$, which is actually $\pi^2/6$ (see formula (6.28) in Exercise 6.42).

Here it is a matter of finding a good upper bound for f_n in the given interval, that is, showing that the series is totally convergent in $[-2, -1]$, with explicit upper bounds $|f_n(x)| \le c_n$, to wit $c_n = n^{-2}$.

6.24 Consider the series $\displaystyle\sum_{n=1}^{\infty} \frac{\cos(2^n x)}{2^n}$.

(a) Find the set I where the series converges pointwise and show that the series converges uniformly in I.
(b) Determine if the sum of the series is differentiable at the origin.

Answer. (a) Evidently,

$$\left|\frac{\cos(2^n x)}{2^n}\right| \le \frac{1}{2^n}$$

and the series $\sum 2^{-n}$ converges. Thus, the series converges totally, hence uniformly and absolutely, on $\mathbb{R}$.

(b) Recall that the derivative of an even differentiable function on $\mathbb{R}$ is an odd differentiable function on $\mathbb{R}$. Hence, since the sum S of the series is an even function, because such are all the summands, if S were differentiable at the origin it would be $S'(0) = 0$. Now, $S(0)$ is the sum of a geometric series, namely

$$S(0) = \sum_{n=1}^{\infty} \frac{1}{2^n} = 1.$$

Therefore the difference quotient at the origin is

$$R(x) := \frac{S(x) - 1}{x} = \frac{1}{x} \sum_{n=1}^{\infty} \left(\frac{\cos(2^n x)}{2^n} - \frac{1}{2^n} \right) = \sum_{n=1}^{\infty} \frac{\cos(2^n x) - 1}{2^n x}.$$

Consider now the sequence $(x_k)_{k \geq 0}$ with $x_k = 2^{-k}$. Evaluating the difference quotient at x_k yields

$$R(x_k) = (\cos 1 - 1) + \sum_{n \neq k} \frac{\cos(2^n x_k) - 1}{2^n x_k} \leq \cos 1 - 1 < 0.$$

Hence, if existing, it would be $S'(0) = \lim_k R(x_k) \leq \cos 1 - 1 < 0$, a contradiction. The series considered in this exercise has a very irregular behaviour, as illustrated in Fig. 6.10. It is clearly uniformly convergent, but the series of derivatives, namely $-\sum_{n=1}^{\infty} \sin(2^n x)$, is troublesome, as the easy estimate $|\sin(2^n x)| \leq 2^n |x|$ is not of any help and it is not even clear for which values of x one may safely say that $\lim_n \sin(2^n x) = 0$. So one has a reasonable suspect that differentiability is really not a trivial feature to detect, not even at the origin. Thus, a cautious attempt to make is to look at the difference quotient (at the origin). This has the opposite sign of that of x and the feeling grows that things might go wrong, in the sense that the behaviour for positive and negative (small) values of x seems quite different.

The conclusive argument is a combination of a symmetry observation (if existing $S'(0) = 0$) and an estimate on a suitable sequence. This is not a trivial exercise, it requires some ingenuity.

6.25 Given the series $\sum_{n=1}^{\infty} f_n$ where

$$f_n(x) = \int_{nx}^{+\infty} |t| \sin\left((t^2 + 1)^{-2}\right) dt,$$

determine the intervals in which it converges pointwise, where it converges absolutely and where it converges uniformly.

Answer. Put

$$g(t) = |t| \sin\left((t^2 + 1)^{-2}\right).$$

Fig. 6.10 The rather wild graph of the sum S in Exercise 6.24. The behaviour at the origin indicates a possible singularity

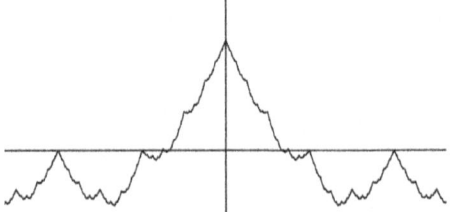

Clearly, g is continuous on $\mathbb{R}$ and $g(t) \to 0$ as $t \to +\infty$ with order 3. Therefore the improper integral $\int_a^{+\infty} g(t)\,dt$ converges for every $a \in \mathbb{R}$. Observe that

$$\lim_n f_n(x) = \begin{cases} 0 & x > 0 \\ \int_0^{+\infty} g(t)\,dt & x = 0 \\ 2\int_0^{+\infty} g(t)\,dt & x < 0, \end{cases}$$

since g is an even function, whence $\int_{-\infty}^{+\infty} g(t)\,dt = 2\int_0^{+\infty} g(t)\,dt$. Observe also that $g(t) > 0$ for $t > 0$ and hence $\int_0^{+\infty} g(t)\,dt > 0$. Therefore, if $x \leq 0$ the series diverges to $+\infty$. Next, if $x > 0$, then

$$0 \leq f_n(x) = \int_{nx}^{+\infty} |t| \sin\left((t^2+1)^{-2}\right) dt \leq \int_{nx}^{+\infty} \frac{t}{(t^2+1)^2}\,dt = \frac{1}{2(n^2x^2+1)}.$$

Since the series $\sum (n^2 x^2 + 1)^{-1}$ converges in $(0, +\infty)$, so does the series $\sum f_n(x)$. Furthermore, since $f_n(x) \geq 0$, the convergence is absolute in $(0, +\infty)$. As for uniform convergence, take $a > 0$. On the interval $[a, +\infty)$ the following simple estimate holds

$$\sup_{x \in [a,+\infty)} f_n(x) \leq \sup_{x \in [a,+\infty)} \frac{1}{2(n^2 x^2 + 1)} = \frac{1}{2(n^2 a^2 + 1)}$$

and since $\sum (n^2 a^2 + 1)^{-1}$ converges, the given series converges totally, hence uniformly, in every interval $[a, +\infty)$ with $a > 0$. Finally, the series does not converge uniformly in $(0, +\infty)$ because $(f_n)_{n \geq 1}$ does not converge uniformly to 0 in that set. Indeed,

$$\sup_{x \in (0,+\infty)} f_n(x) = \int_0^{+\infty} t \sin\left(\frac{1}{(t^2+1)^2}\right) dt > 0$$

and consequently, denoting by s the above supremum, which is independent of n,

$$\lim_n \sup_{x \in (0,+\infty)} f_n(x) = s > 0,$$

showing that the convergence $f_n \to 0$ is not uniform on the whole interval $(0, +\infty)$.

This exercise definitely requires attention because the functions f_n appearing in the series are integral functions of the positive function g, whereby the dependence on n and x is through the integration interval, which is $[nx, +\infty)$. The very first observation is thus that for fixed x this interval "tends" to the limiting interval

$$I_x = \begin{cases} \emptyset & x > 0 \\ [0, +\infty) & x = 0 \\ (-\infty, +\infty) & x < 0 \end{cases}$$

as $n \to +\infty$. Since the improper integral $\int_{I_x} g(t)\, dt$ converges to a positive value both when $I_x = [0, +\infty)$ and when $I_x = (-\infty, +\infty)$ it follows that $f_n(x) \not\to 0$ for $x \in (-\infty, 0]$. If $x > 0$, the integration interval progressively shrinks, so that the limit as $n \to +\infty$ (for fixed $x > 0$) of the improper integrals $f_n(x) = \int_{nx}^{+\infty} g(t)\, dt$, each of which is actually convergent, is therefore 0. This is when it makes sense to consider the series $\sum f_n(x)$. One then easily estimates $0 \le f_n(x) \le (n^2 x^2 + 1)^{-1}$ and this leads to absolute convergence. For total convergence on each interval $[a, +\infty)$ this estimate suffices to conclude. Global uniform convergence does not hold because the functions $f_n(x)$ all have the same limiting positive value for $x \to 0^+$, which violates uniform convergence to 0 on $(0, +\infty)$.

6.26 Establish where the function $f(x) = \sum_{n=0}^{+\infty} \dfrac{n^2 e^{-2n}}{n^2 + 1} x^n$ is differentiable.

Answer. This power series has convergence radius $r = e^2$ because

$$\lim_n \frac{a_{n+1}}{a_n} = \lim_n \frac{(n+1)^2 e^{-2(n+1)}}{(n+1)^2 + 1} \frac{n^2 + 1}{n^2 e^{-2n}} = \lim_n e^{-2} \frac{(n+1)^2}{(n+1)^2 + 1} \frac{n^2 + 1}{n^2} = e^{-2}$$

and formula (6.5) gives $r = e^2$. Hence the series certainly converges in $(-e^2, e^2)$. As for the boundary points, at $x = -e^2$ the numerical series

$$\sum_{n=0}^{+\infty} \frac{n^2 e^{-2n}}{n^2 + 1} (-1)^n e^{2n} = \sum_{n=0}^{+\infty} (-1)^n \frac{n^2}{n^2 + 1}$$

does not converge because its general term does not tend to zero. At $x = e^2$ the series diverges because

$$\sum_{n=0}^{+\infty} \frac{n^2 e^{-2n}}{n^2 + 1} e^{2n} = \sum_{n=0}^{+\infty} \frac{n^2}{n^2 + 1}.$$

Hence f is defined exactly in $(-e^2, e^2)$, where f is differentiable by Theorem 6.12.

This is a straightforward application of the basic techniques for power series, that is, evaluating the radius and then the endpoint behaviours. Notice that differentiability in the convergence interval is guaranteed, but in general one has to check what happens at its endpoints in case the power series converges in either of them. But this does not happen here.

6.27 Determine the convergence interval of the power series $\sum_{n=0}^{+\infty} \dfrac{(2x-1)^n}{7^n + 1}$ and discuss the behaviour at the endpoints.

6.5 Guided Exercises

Answer. Upon rewriting

$$\sum_{n=0}^{+\infty} \frac{(2x-1)^n}{7^n+1} = \sum_{n=0}^{+\infty} \frac{2^n(x-\frac{1}{2})^n}{7^n+1},$$

the given power series is centered at $x_0 = 1/2$. Applying the root test (6.5),

$$\lim_n \sqrt[n]{|a_n|} = \lim_n \sqrt[n]{\frac{2^n}{7^n+1}} = \lim_n \frac{2}{7}\sqrt[n]{\frac{1}{1+7^{-n}}} = \frac{2}{7},$$

so that the convergence interval is $(-3, 4)$. At $x = -3$ the series becomes

$$\sum_{n=0}^{+\infty} \frac{(-1)^n 7^n}{7^n+1},$$

which does not converge because $7^n/(1+7^n) \to 1$. At the other endpoint $x = 4$, the series becomes

$$\sum_{n=0}^{+\infty} \frac{7^n}{7^n+1},$$

and this diverges as just observed.

A straightforward exercise, which only requires to write the given series properly in the form $\sum a_n(x-x_0)^n$. Also, here the root test is preferable to the ratio test.

6.28 Find the McLaurin expansion of the function

$$f(x) = \begin{cases} \dfrac{x - \sin x}{x^3} & x \neq 0 \\ 1/6 & x = 0 \end{cases}$$

and compute $f^{(27)}(0)$ and $f^{(28)}(0)$.

Answer. It is well known that

$$\sin x = \sum_{n=0}^{\infty} (-1)^n \frac{x^{2n+1}}{(2n+1)!} = x + \sum_{n=1}^{\infty} (-1)^n \frac{x^{2n+1}}{(2n+1)!}.$$

This entails

$$x - \sin x = \sum_{n=1}^{\infty} (-1)^{n+1} \frac{x^{2n+1}}{(2n+1)!} = \frac{x^3}{6} + \sum_{n=2}^{\infty} (-1)^{n+1} \frac{x^{2n+1}}{(2n+1)!}.$$

Furthermore, dividing by x^3 and setting $m = n - 1$ yields

$$\frac{x - \sin x}{x^3} = \frac{1}{6} + \sum_{n=2}^{\infty} (-1)^{n+1} \frac{x^{2n-2}}{(2n+1)!}$$

$$= \frac{1}{6} + \sum_{m=1}^{\infty} (-1)^m \frac{x^{2m}}{(2m+3)!}$$

$$= \sum_{m=0}^{\infty} (-1)^m \frac{x^{2m}}{(2m+3)!}$$

which in particular shows that $f(x) \to 1/6$ for $x \to 0$. Thus, $f(x) = \sum_{n=0}^{\infty} a_n x^n$ with

$$a_n = \begin{cases} 0 & n = 2m+1 \\ \dfrac{(-1)^m}{(2m+3)!} & n = 2m. \end{cases}$$

Formula (6.8), namely $f^{(n)}(0) = n! a_n$, gives then

$$f^{(27)}(0) = 0, \qquad f^{(28)}(0) = 28! \frac{1}{31!} = \frac{1}{31 \cdot 30 \cdot 29} = \frac{1}{26970}.$$

This exercise is about manipulation of power series, based on the observation that

$$x - \sin x = \frac{x^3}{6} - \frac{x^5}{5!} + \frac{x^7}{7!} - \frac{x^9}{9!} \cdots$$

so that

$$\frac{x - \sin x}{x^3} = \frac{1}{6} - \frac{x^2}{5!} + \frac{x^4}{7!} - \frac{x^6}{9!} \cdots$$

and the right hand side looks really like a power series. Keeping track of the complete expression for the McLaurin series of the sine function yields the desired result.

6.29 Compute the sum of the series $\sum_{n=1}^{+\infty} n x^n$.

Answer. Clearly, the radius of the power series $\sum n x^n$ is 1 because

$$\frac{a_{n+1}}{a_n} = \frac{n+1}{n} \to 1.$$

6.5 Guided Exercises

Furthermore, the series does not converge a the endpoipnts ± 1 because the general term does not tend to 0 in either case. If f denotes the sum of the series in $(-1, 1)$, then

$$f(x) = x \sum_{n=1}^{+\infty} nx^{n-1} = x \sum_{n=1}^{+\infty} \frac{d}{dx} x^n = x \frac{d}{dx}\left(\sum_{n=1}^{+\infty} x^n\right) = x \frac{d}{dx}\left(\frac{x}{1-x}\right) = \frac{x}{(1-x)^2}.$$

All these formal manipulations are justified by Theorem 6.12.

Here the idea is to write $nx^n = x(nx^{n-1})$ because nx^{n-1} is the derivative of x^n. The point is that with power series, derivation under the integral sign is perfectly legitimate as long as one works inside the convergence interval, so that formal manipulations are actually rigorous. A last bit of information is given by the fundamental geometric series, which is the most basic power series with known sum.

6.30 Compute $f^{(7)}(0)$ and $f^{(10)}(0)$ of the function $f(x) = \int_0^{x^2} \frac{e^{2t}-1}{t}\, dt$.

Answer. The function $g(t) = (e^{2t}-1)/t$ is defined and continuous in $\mathbb{R} \setminus \{0\}$, and it may be extended to a continuous function $\tilde{g}$ at $x_0 = 0$ by setting $\tilde{g}(0) = 2$. Clearly,

$$\int_0^{x^2} g(t)\, dt = \int_0^{x^2} \tilde{g}(t)\, dt.$$

It follows that $\text{Dom}(f) = \mathbb{R}$. It is well-known that

$$e^t = \sum_{n=0}^{\infty} \frac{t^n}{n!},$$

and hence

$$\frac{e^{2t}-1}{t} = \sum_{n=1}^{\infty} \frac{2^n t^{n-1}}{n!} = \sum_{n=0}^{\infty} \frac{2^{n+1} t^n}{(n+1)!}$$

with uniform convergence in every compact interval. Therefore

$$f(x) = \int_0^{x^2} \tilde{g}(t)\, dt = \sum_{n=0}^{\infty} \frac{2^{n+1} x^{2(n+1)}}{(n+1)!(n+1)}.$$

Since all the coefficients of the odd powers of x vanish, $f^{(7)}(0) = 0$. Furthermore, for $n = 4$ it is

$$\frac{2^5}{5!} \frac{x^{10}}{5} = \frac{f^{(10)}(0)}{10!} x^{10}$$

and so
$$f^{(10)}(0) = \frac{10!}{5!}\frac{32}{5} = 193536.$$

This is straightforward manipulation starting from the exponential series.

6.31 Consider the function $f(x) = \int_0^x \frac{\cos t - e^t}{t}\, dt$.

(a) Find the McLaurin expansion of f, specifying its convergence interval.
(b) Compute $f(1/3)$ with an error not larger than 10^{-2}.
(c) Compute $f^{(3)}(0)$.

Answer. (a) Combining the series expansions of the exponential and cosine functions,

$$\cos t - e^t = \sum_{n=0}^{\infty}(-1)^n \frac{t^{2n}}{(2n)!} - \sum_{n=0}^{\infty}\frac{t^n}{n!}$$

$$= \sum_{m=0}^{\infty}\left[(-1)^m \frac{1}{(2m)!} - \frac{1}{(2m)!}\right]t^{2m} - \sum_{m=0}^{\infty}\frac{t^{2m+1}}{(2m+1)!}.$$

Hence, upon setting

$$\cos t - e^t = \sum_{n=0}^{\infty} a_n t^n$$

it follows

$$a_n = \begin{cases} -\dfrac{2}{n!} & n = 2+4m,\quad m \geq 0 \\[4pt] -\dfrac{1}{n!} & n = 2m+1,\quad m \geq 0 \\[4pt] 0 & \text{else.} \end{cases}$$

Notice that $a_0 = 0$ (so that it may by assumed that $n \geq 1$ in the series above) and that for $t \neq 0$

$$g(t) := \frac{\cos t - e^t}{t} = \sum_{n=1}^{\infty} a_n t^{n-1}.$$

This power series, however, has infinite radius of convergence and converges at $t_0 = 0$ to the value -1. Hence

$$\sum_{n=1}^{\infty} a_n t^{n-1} = \tilde{g}(t) := \begin{cases} g(t) & t \neq 0 \\ -1 & t = 0. \end{cases}$$

Clearly, $\tilde{g}$ is the continuous extension of g at $t_0 = 0$. The radius of convergence is $+\infty$, so the convergence interval is $(-\infty, +\infty)$. Using term by term integration,

6.5 Guided Exercises

$$f(x) = \int_0^x \tilde{g}(t)\,dt = \sum_{n=1}^\infty \frac{a_n}{n} x^n$$

for every $x \in \mathbb{R}$. This is the required McLaurin expansion.

(b) Evidently,

$$f(1/3) = \sum_{n=1}^\infty \frac{a_n}{n 3^n}$$

and if S_k denotes the kth partial sum, then observing that $|a_n| \leq 2/n!$ for every n

$$|f(1/3) - S_k| = \left| \sum_{n=k+1}^\infty \frac{a_n}{n 3^n} \right|$$

$$\leq \sum_{n=k+1}^\infty \frac{|a_n|}{n 3^n}$$

$$\leq 2 \sum_{n=k+1}^\infty \frac{1}{n} \frac{1}{n! 3^n}$$

$$\leq \frac{2}{k+1} \sum_{m=0}^\infty \frac{1}{(m+k+1)!} \frac{1}{3^{m+k+1}}$$

$$\leq \frac{2}{(k+1)3^{k+1}(k+1)!} \sum_{m=0}^\infty \frac{1}{m!} \frac{1}{3^m}$$

$$= \frac{2}{(k+1)3^{k+1}(k+1)!} e^{1/3},$$

where the last inequality holds because $(m+k+1)! \geq m!(k+1)!$. This is a particular case of the general inequality:

$$(p+q)! \geq p!q!$$

for every non negative integers p and q. Therefore,

$$|f(1/3) - S_k| \leq \frac{2}{(k+1)3^{k+1}(k+1)!} e^{1/3} \leq 10^{-2}$$

if $k \geq 2$. This gives

$$S_2 = -\frac{1}{3} - \frac{1}{2 \cdot 9} = -\frac{7}{18}.$$

(c) Finally,

$$\frac{f^{(3)}(0)}{3!} = \frac{a_3}{3} = -\frac{1}{3 \cdot 3!}$$

implies that $f^{(3)}(0) = -1/3$.

The first part of this exercise goes along the same lines as the previous one. The only slightly serious issue is the estimate of the series, which requires some ingenuity. The main idea is to shift the summation indices and then reduce the matter to an exponential series. This is because the latter is of course directly computable. It is worthwhile observing that the inequality $(p+q)! \geq p!q!$ is simply the statement

$$\binom{p+q}{p} = \binom{p+q}{q} = \frac{(p+q)!}{p!q!} \geq 1$$

6.32 Compute the Taylor series expansion of the function $f(x) = 1/x$ centered at $x_0 = 2$, specifying the convergence radius.

Answer. Using the expansion of the geometric power series, it is possible to write

$$\frac{1}{x} = \frac{1}{2-2+x} = \frac{1}{2} \cdot \frac{1}{1-(1-\frac{x}{2})} = \frac{1}{2} \sum_{n=0}^{\infty} \left(1 - \frac{x}{2}\right)^n$$

for $|1 - x/2| < 1$, namely for $|x - 2| < 2$. Therefore

$$\frac{1}{x} = \frac{1}{2} \sum_{n=0}^{\infty} \frac{(-1)^n (x-2)^n}{2^n} = \sum_{n=0}^{\infty} \frac{(-1)^n (x-2)^n}{2^{n+1}}$$

with convergence radius 2.

The idea of using the geometric power series is completely natural because $1/x$ is a translation of $1/(1-x)$, which is among the most fundamental power series. Then it is essentially algebraic manipulation.

6.33 Consider the function $f(x) = \sum_{n=1}^{\infty} \frac{(x^2-1)^{2n-1}}{n - \sqrt{n} + 1}$.

(a) Find the domain of f.
(b) Find the second partial sum of the Taylor expansion of f centered at $x_0 = 1$.
(c) Compute $\int_1^{4/3} x f(x) \, dx$ with an error not larger than 10^{-2}.

Answer. (a) Put for simplicity

$$f_n(x) = \frac{(x^2-1)^{2n-1}}{n - \sqrt{n} + 1}.$$

Now,

6.5 Guided Exercises

$$\lim_n f_n(x) = \begin{cases} +\infty & x^2 - 1 > 1 \\ 0 & x^2 - 1 = 1 \\ 0 & 0 \le |x^2 - 1| < 1, \end{cases}$$

where the convergence to 0 is of order exactly 1 when $x^2 - 1 = 1$ and of order greater than any $\alpha > 0$ when $0 < |x^2 - 1| < 1$. It follows that if $|x| \ge \sqrt{2}$ the series does not converge, whereas if $0 < |x| < \sqrt{2}$ the series converges. Finally,

$$f_n(0) = \frac{-1}{n - \sqrt{n} + 1}.$$

tends to 0 with order exactly 1 so that the series $\sum f_n(0)$ diverges to $-\infty$. Therefore $\mathrm{Dom}(f) = (-\sqrt{2}, 0) \cup (0, \sqrt{2})$.

(b) Observe that $f(x) = h \circ g(x)$ where

$$g(x) = x^2 - 1, \qquad h(y) = \sum_{n=1}^{\infty} \frac{y^{2n-1}}{n - \sqrt{n} + 1}$$

are both analytic functions, g on the whole $\mathbb{R}$ and h in $(-1, 1)$. Hence f is analytic for the values of x for which $g(x) \in (-1, 1)$, thus in $(-\sqrt{2}, 0) \cup (0, \sqrt{2})$. Notice also that the series defining f converges totally in every interval $[a, b] \subset (0, \sqrt{2})$. Being analytic, the function f admits a power series expansion at any point. Now,

$$f'(x) = h'(g(x))g'(x), \qquad f''(x) = h''(g(x))(g'(x))^2 + h'(g(x))g''(x).$$

Furthermore, since h is defined by a convergent power series, and since $g(1) = 0$

$$h'(y) = \sum_{n=1}^{\infty} \frac{(2n-1)y^{2n-2}}{n - \sqrt{n} + 1} \qquad \Longrightarrow \qquad h'(g(1)) = h'(0) = 1$$

and also

$$h''(y) = \sum_{n=2}^{\infty} \frac{(2n-1)(2n-2)y^{2n-3}}{n - \sqrt{n} + 1} \qquad \Longrightarrow \qquad h''(g(1)) = h''(0) = 0.$$

Finally, $g'(1) = 2$ and $g''(1) = 2$. It follows that $f(1) = 0$, $f'(1) = 2$ and $f''(1) = 2$. The second partial sum of the Taylor expansion of f centered at $x_0 = 1$ is

$$S_2(x) = f(1) + f'(1)(x - 1) + \frac{1}{2}f''(1)(x - 1)^2 = 2(x - 1) + (x - 1)^2.$$

(c) As established in (a) and also in (b), the series on the right hand side of

$$xf(x) = \sum_{n=1}^{\infty} \frac{x(x^2-1)^{2n-1}}{n - \sqrt{n+1}}$$

converges totally in the interval $[1, 4/3] \subset (0, \sqrt{2})$ and hence it is legitimate to integrate it term by term:

$$\int_1^{4/3} xf(x)\,dx = \sum_{n=1}^{\infty} \int_1^{4/3} \frac{x(x^2-1)^{2n-1}}{n - \sqrt{n+1}}\,dx$$

$$= \sum_{n=1}^{\infty} \frac{1}{4n} \frac{(x^2-1)^{2n}}{(n - \sqrt{n+1})} \bigg|_1^{4/3}$$

$$= \sum_{n=1}^{\infty} \frac{1}{4n} \frac{1}{(n - \sqrt{n+1})} \left(\frac{7}{9}\right)^{2n},$$

because, as easily checked,

$$\int x(x^2-1)^{2n-1}\,dx = \frac{(x^2-1)^{2n}}{4n} + c.$$

It follows that if S_k is the kth partial sum of the obtained numerical series, then

$$\left|\int_1^{4/3} xf(x)\,dx - S_k\right| = \sum_{n=k+1}^{\infty} \frac{1}{4n} \frac{1}{(n-\sqrt{n+1})} \left(\frac{7}{9}\right)^{2n}$$

$$\leq \frac{1}{4(k+1)} \frac{1}{(k+2-\sqrt{k+1})} \sum_{n=k+1}^{\infty} \left(\frac{49}{81}\right)^n.$$

Finally, using that

$$\sum_{n=k+1}^{\infty} y^n = \frac{y^{k+1}}{1-y}$$

and that with $y = 49/81$ it is $1/(1-y) = 81/32$ the resulting estimate is

$$\left|\int_1^{4/3} xf(x)\,dx - S_k\right| \leq \frac{1}{4(k+1)} \frac{1}{(k+2-\sqrt{k+1})} \left(\frac{49}{81}\right)^{k+1} \frac{81}{32}$$

and the right hand side is less than 10^{-2} provided that $k \geq 3$. The required approximation (from below) is thus

$$S_3 = \sum_{n=1}^{3} \frac{1}{4n} \frac{1}{(n-\sqrt{n+1})} \left(\frac{7}{9}\right)^{2n}.$$

6.5 Guided Exercises

The domain of f is found either directly by checking the convergence of the series or arguing as implicitly done in item (b), namely by considering the composition $f(x) = h \circ g(x)$. The computation of the requested partial sum, which is exactly the second Taylor polynomial of f at $x_0 = 1$, is computed using the chain rule applied to $f = h \circ g$ and the fact that h can be differentiated term by term. The idea for item (c) is that of integrating (again) term by term and then estimating the resulting numerical series by the appropriate geometric sum. The only serious issue is showing that integration term by term is legitimate, and this comes either from general results on power series or from direct observation on the order of convergence.

6.34 Consider the function $f(x) = \sum_{n=0}^{\infty} \dfrac{(3x^2 + 2x)^n}{(n+1)^2}$.

(a) Find the domain of f.
(b) Find where f is differentiable and compute $f^{(3)}(0)$.

Answer. (a) Put
$$f_n(x) = \frac{(3x^2 + 2x)^n}{(n+1)^2}.$$

Observe that
$$\left| \frac{(3x^2 + 2x)^n}{(n+1)^2} \right| \leq \frac{1}{(n+1)^2}$$

provided that $|3x^2 + 2x| \leq 1$, and that the series $\sum (n+1)^{-2}$ converges. Therefore the given series converges totally in the interval $[-1, 1/3]$, which is where $|3x^2 + 2x| \leq 1$ holds true. Observe that

$$\lim_n f_n(x) = \begin{cases} +\infty & 3x^2 + 2x > 1 \\ \nexists & 3x^2 + 2x < -1, \end{cases}$$

so that outside $[-1, 1/3]$ the series does not converge. Therefore $\text{Dom}(f) = [-1, 1/3]$.

(b) Every f_n is indefinitely differentiable in $\mathbb{R}$ with

$$f'_n(x) = \frac{n(3x^2 + 2x)^{n-1}(6x + 2)}{(n+1)^2},$$

so that

$$|f'_n(x)| \leq \frac{n \cdot C^{n-1} \cdot 4}{(n+1)^2} =: a_n$$

for every $x \in [a, b] \subset (-1, 1/3)$, where

$$C = \max_{[a,b]} |3x^2 + 2x| < 1.$$

Since the series $\sum a_n$ converges, the series of derivatives converges totally and hence uniformly in every compact interval contained in $(-1, 1/3)$, so the sum f is differentiable in $(-1, 1/3)$. It is, however, not differentiable at -1, as it is shown below. As just argued, in $(-1, 1/3)$,

$$f'(x) = \sum_{n=1}^{\infty} \frac{n}{(n+1)^2}(3x^2 + 2x)^{n-1}(6x + 2).$$

From the series expansion of the logarithm, for every $y \in [-1, 1)$

$$-\log(1-y) = \sum_{n=1}^{\infty} \frac{y^n}{n}$$

and hence for $x \in (-1, 1/3)$,

$$-\log(1 - 3x^2 - 2x) = \sum_{n=1}^{\infty} \frac{(3x^2 + 2x)^n}{n},$$

whence for $x \in (-1, -2/3)$

$$-\frac{\log(1 - 3x^2 - 2x)}{3x^2 + 2x} = \sum_{n=1}^{\infty} \frac{(3x^2 + 2x)^{n-1}}{n}.$$

Now, from the inequality

$$\frac{1}{n} \leq \frac{4n}{(n+1)^2},$$

which holds true for every integer $n \geq 1$, it follows that

$$-\frac{1}{4}\frac{\log(1 - 3x^2 - 2x)}{3x^2 + 2x} \leq \sum_{n=1}^{\infty} \frac{n}{(n+1)^2}(3x^2 + 2x)^{n-1}$$

and consequently if $x \in (-1, -2/3)$, then $6x + 2 < -2 < 0$ and

$$-\frac{1}{4}\frac{6x+2}{3x^2+2x}\log(1-3x^2-2x) \geq \sum_{n=1}^{\infty} \frac{n}{(n+1)^2}(3x^2+2x)^{n-1}(6x+2) = f'(x).$$

Observe that the left hand side diverges to $-\infty$ for $x \to (-1)^+$. Therefore

$$\lim_{x \to -1^+} f'(x) = -\infty.$$

6.5 Guided Exercises

Since f is continuous in -1 because it is the sum of a uniformly convergent series of continuous functions in $[-1, 1/3]$, the theorem of the limit of the derivative (Corollary 7.6 in [1]), implies that f is not differentiable at -1. A similar argument shows that f is not differentiable at $1/3$.

Finally, it is clear that f is analytic in its domain, because it may be written as the composition of the analytic functions $3x^2 + 2x$ and $\sum (n+1)^{-2} y^n$. It follows that the given convergent power series coincides with the McLaurin series expansion of f in a suitably small neighborhood of 0 contained in $(-1, 1/3)$, namely

$$f(x) = \sum_{n=0}^{\infty} \frac{(3x^2 + 2x)^n}{(n+1)^2} = \sum_{n=0}^{\infty} \frac{f^{(n)}(0)}{n!} x^n,$$

for x small enough. The first terms in the leftmost series give rise to

$$f(x) = 1 + \left(\frac{2}{4}x + \frac{3}{4}x^2\right) + \left(\frac{4}{9}x^2 + \frac{12}{9}x^3 + \frac{9}{9}x^4\right) + \frac{8}{16}x^3 + \left(\text{terms of order} \geq 4\right)$$

Therefore
$$\frac{f^{(3)}(0)}{3!} = \frac{12}{9} + \frac{8}{16} \quad \Longrightarrow \quad f^{(3)}(0) = 11.$$

This is a non-trivial exercise. Question (a) is actually standard. It can either be treated directly, as done here, or as a composition of a power series with a polynomial, which requires handling the endpoints as usual. This means that one first finds the convergence radius of $\sum (n+1)^{-2} y^n$, which is 1, then determines for which values of x it is $3x^2 + 2x \in (-1, 1)$, which happens exactly if $x \in (-1, 1/3)$, and finally studies the cases $x = -1$ and $x = 1/3$. Part (b) deserves comments. That everything is fine in $(-1, 1/3)$ is rather straightforward. But how does one come up with the logarithm? The main idea is that since formally

$$f'(x) = (6x+2) \sum_{n=1}^{\infty} \frac{n}{(n+1)^2} (3x^2 + 2x)^{n-1}.$$

one looks at the series on the right, without the factor $6x + 2$. Apart from composition with $3x^2 + 2x$, and also division by $3x^2 + 2x$, the "core" series to understand is

$$\sum_{n=1}^{\infty} \frac{n}{(n+1)^2} y^n$$

whose coefficient $n/(n+1)^2$ behaves like $1/n$. More precisely, the estimates from above and below

$$\frac{1}{4n} \leq \frac{n}{(n+1)^2} \leq \frac{1}{n},$$

indicate that what one really has to understand is $\sum n^{-1} y^n$, the well-known logarithmic series. Once this insight is achieved, then one works backwards.

6.35 Consider, for every positive integer p, the power series

$$f_p(x) = \sum_{n=1}^{\infty} \frac{n^p x^n}{n!}.$$

(a) Find where the series converges.
(b) Show that $f_p(x) e^{-x}$ is a polynomial of degree p and compute it for $p = 3$.

Answer. (a) Using the ratio test,

$$\frac{(n+1)^p \, n!}{(n+1)! \, n^p} = \left(1 + \frac{1}{n}\right)^p \frac{1}{n+1} \to 0$$

it is seen that the given series converges in $\mathbb{R}$ for every p.

(b) For $p = 1$

$$f_1(x) = \sum_{n=1}^{\infty} \frac{n x^n}{n!} = \sum_{n=1}^{\infty} \frac{x^n}{(n-1)!} = \sum_{n=0}^{\infty} \frac{x^{n+1}}{n!} = x \sum_{n=0}^{\infty} \frac{x^n}{n!} = x e^x,$$

so that $Q_1(x) = x$ satisfies $f_1(x) e^{-x} = Q_1(x)$. Next observe that

$$f_2(x) = \sum_{n=1}^{\infty} \frac{n^2 x^n}{n!}$$

and that differentiating term by term

$$f_1'(x) = \sum_{n=1}^{\infty} \frac{n^2 x^{n-1}}{n!}$$

so that manifestly $x f_1'(x) = f_2(x)$. Therefore, recalling that $f_1(x) = e^x Q_1(x) = e^x x$,

$$f_2(x) = x f_1'(x) = x (e^x x)' = x (e^x x + e^x) = e^x (x^2 + x).$$

Thus, $f_2(x) = e^x Q_2(x)$ with $Q_2(x) = x^2 + x$, a polynomial of degree 2. In general, the derivative of f_p can be computed via term by term differentiation, namely

$$f_p'(x) = \sum_{n=1}^{\infty} \frac{n^p n x^{n-1}}{n!} = \sum_{n=1}^{\infty} n^{p+1} \frac{x^{n-1}}{n!}.$$

6.5 Guided Exercises

Furthermore,

$$xf'_p(x) = \sum_{n=1}^{\infty} n^{p+1}\frac{x^n}{n!} = f_{p+1}(x).$$

Therefore, assuming inductively that $f_p(x) = e^x Q_p(x)$ with Q_p a polynomial of degree p, it follows that

$$f_{p+1}(x) = x\frac{d}{dx}\left(e^x Q_p(x)\right) = xe^x\left(Q_p(x) + Q'_p(x)\right) = e^x Q_{p+1}(x)$$

where

$$Q_{p+1}(x) = x(Q_p(x) + Q'_p(x))$$

is of course a polynomial of degree $p+1$. This proves the statement. In particular, since $Q_2(x) = x^2 + x$ and $Q'_2(x) = 2x + 1$, it follows

$$Q_3(x) = xQ_2(x) + xQ'_2(x) = \left(x^3 + x^2\right) + x(2x + 1) = x^3 + 3x^2 + x.$$

This exercise requires first making a couple of experiments, namely looking at f_1 and f_2 closely. The case f_1 is essentially trivial, and comparing f_1 with f_2 already exhibits the main point:

$$f_1(x) = \sum_{n=1}^{\infty}\frac{nx^n}{n!}, \quad f_2(x) = \sum_{n=1}^{\infty}\frac{n^2 x^n}{n!} = x\sum_{n=1}^{\infty}\frac{n(nx^{n-1})}{n!}$$

The function f_2 is obtained by two consecutive operations from f_1, that is, first differentiate and then multiply by x. Arguing inductively gives the result that

$$xf'_p(x) = f_{p+1}(x),$$

which is the key. This is not trivial to see, but fiddling with the first few values of p does give some good intuition.

6.36 Find the convergence radius and the sum of the series

$$1 + \sum_{n=1}^{\infty}\frac{x^n}{n(n+1)}.$$

Answer. By the ratio test, since

$$\lim_n \frac{|a_{n+1}|}{|a_n|} = \lim_n \frac{n(n+1)}{(n+1)(n+2)} = 1$$

the convergence radius is 1. Notice that both the numerical series that arise by evaluating at 1 and -1 the given power series, namely

$$\sum_{n=1}^{\infty} \frac{1}{n(n+1)}, \quad \sum_{n=1}^{\infty} \frac{(-1)^n}{n(n+1)}$$

converge. Hence the power series converges in $[-1, 1]$. By Abel's theorem, i.e. Theorem 6.11, the convergence is uniform in $[-1, 1]$, so that the sum $f(x)$ is continuous on $[-1, 1]$, and certainly differentiable in $(-1, 1)$. Differentiating term by term gives

$$f'(x) = \sum_{n=1}^{\infty} \frac{x^{n-1}}{n+1} = \sum_{n=0}^{\infty} \frac{x^n}{n+2}.$$

In particular it follows that $f'(0) = 1/2$. Multiplying by x^2 the above series yields the expression

$$x^2 f'(x) = \sum_{n=0}^{\infty} \frac{x^{n+2}}{n+2}$$

which allows a further simple observation, that is

$$\frac{d}{dx}(x^2 f'(x)) = \sum_{n=0}^{\infty} x^{n+1} = \sum_{n=1}^{\infty} x^n = \frac{x}{1-x}.$$

This is useful because by the fundamental theorem of Calculus it then follows that

$$x^2 f'(x) = \int_0^x \frac{t}{1-t} \, dt = \int_0^x \frac{t-1+1}{1-t} \, dt = -\int_0^x dt + \int_0^x \frac{1}{1-t} \, dt = -x - \log(1-x).$$

Therefore

$$f'(x) = \begin{cases} \dfrac{-x - \log(1-x)}{x^2} & x \neq 0 \\ 1/2 & x = 0. \end{cases}$$

Using again the fundamental theorem of Calculus, for every $x \in (0, 1)$ it is

$$f(x) - f(0) = \int_0^x f'(t) \, dt = \lim_{y \to 0^+} \int_y^x \left(\frac{-t - \log(1-t)}{t^2} \right) dt.$$

6.5 Guided Exercises

Integrating by parts gives

$$\int_y^x \left(\frac{-t - \log(1-t)}{t^2}\right) dt = \frac{t + \log(1-t)}{t}\bigg|_y^x + \int_y^x \frac{1}{t}\frac{t}{1-t} dt$$

$$= \frac{\log(1-x)}{x} - \log(1-x) - \frac{\log(1-y)}{y} + \log(1-y).$$

and since

$$\lim_{y \to 0^+} -\frac{\log(1-y)}{y} + \log(1-y) = 1,$$

it follows that for every $x \in (0, 1)$

$$f(x) - f(0) = \frac{\log(1-x)}{x} - \log(1-x) + 1 \implies f(x) = \frac{(1-x)\log(1-x)}{x} + 2$$

where it was taken into account that $f(0) = 1$. The argument is similar for the case where $x \in (-1, 0)$. Finally

$$\lim_{x \to -1^+} \frac{(1-x)\log(1-x)}{x} + 2 = 2 - 2\log 2$$

$$\lim_{x \to 1^-} \frac{(1-x)\log(1-x)}{x} + 2 = 2$$

and hence the complete answer is

$$f(x) = \begin{cases} 2 + \dfrac{(1-x)\log(1-x)}{x} & x \in (-1, 0) \cup (0, 1) \\ 2 - 2\log 2 & x = -1 \\ 1 & x = 0 \\ 2 & x = 1. \end{cases}$$

This exercise is slightly demanding because it requires an idea, basically to look at the derived series, and, in fact, to do that twice. In broad terms, the first idea comes from the following observation. Upon differentiation the general term of the given series is mapped into a simpler one, because indeed

$$\frac{d}{dx}\left(\frac{x^n}{n(n+1)}\right) = \frac{x^{n-1}}{n+1}.$$

One can then multiply by x^2 and repeat the trick:

$$\frac{d}{dx}\left(x^2 \frac{x^{n-1}}{n+1}\right) = \frac{d}{dx}\left(\frac{x^{n+1}}{n+1}\right) = x^n.$$

Hence, by a sequence of simple operations (differentiate, multiply by x^2 and differentiate again) one gets the geometric series, whose sum $g(x)$ is known. Then one needs to proceed backwards on $g(x)$: integrate, divide by x^2 and integrate again. This is the core of the method, whose details are worked out above.

6.37 Draw the graph of $f(x) = \sum_{n=0}^{\infty} \frac{(\sqrt{x+1})^n}{n+1}$.

Answer. (a) Put $F(x) = \sum_{n=0}^{\infty} \frac{x^n}{n+1}$, so that $f(x) = F(\sqrt{x+1})$. Therefore

$$\text{Dom}(f) = \{x \in \mathbb{R} : x \geq -1, \sqrt{x+1} \in \text{Dom}(F)\}.$$

Now, F is the sum of a power series with convergence radius equal to 1. Furthermore, the series does not converge in $x = 1$ while it does converge at $x = -1$. Hence $\text{Dom}(F) = [-1, 1)$. Consequently,

$$\sqrt{x+1} < 1 \implies -1 \leq x < 0 \implies \text{Dom}(f) = [-1, 0).$$

By Abel's theorem, $F \in C^0([-1, 1))$ and since f is a composition of continuous functions it follows that $f \in C^0([-1, 0))$. Furthermore, F is differentiable in $(-1, 1)$ and

$$F'(x) = \sum_{n=1}^{\infty} \frac{n}{n+1} x^{n-1} = \sum_{n=0}^{\infty} \frac{n+1}{n+2} x^n.$$

By the chain rule, f is differentiable in $(-1, 0)$ and

$$f'(x) = \frac{1}{2\sqrt{x+1}} \sum_{n=0}^{\infty} \frac{n+1}{n+2} (x+1)^{n/2}.$$

Clearly, $f'(x) > 0$ for every $x \in (-1, 0)$, so that f is increasing in its domain. By estimating the sum from below with the term corresponding to $n = 0$, which is $1/2$, it follows that

$$f'(x) \geq \frac{1}{4\sqrt{x+1}}.$$

Therefore

$$\lim_{x \to -1^+} f'(x) = +\infty$$

and hence f is not differentiable at $x_0 = -1$ Finally, $f(-1) = F(0) = 1$ and a qualitative graph of F is as in Fig. 6.11.

The main step here is to understand that it is convenient to write $f(x) = F(\sqrt{x+1})$ where F is a tractable power series. Simple tasks such as finding the domain and establishing continuity and the sign of the derivative are well within reach by direct manipulation of the power series.

Fig. 6.11 The graph of the function in Exercise 6.37

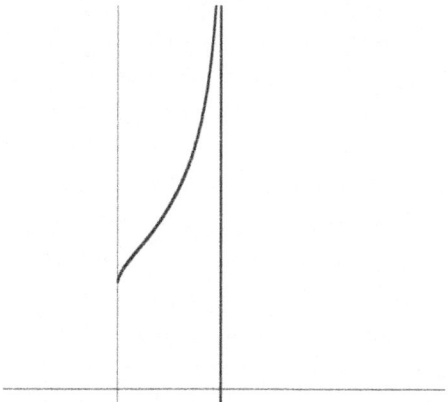

6.38 Consider the function f which is the 2π-periodic extension of the function whose value in $[-\pi, \pi)$ is $|x|$.

(a) Find the Fourier series of f.
(b) Discuss the pointwise and uniform convergence of the Fourier series.
(c) Write Parseval's formula in this case.
(d) Prove that $\sum_{n=0}^{\infty}(2n+1)^{-2} = \pi^2/8$.

Answer. (a) The function f is continuous on $\mathbb{R}$. Clearly, for every positive integer k

$$b_k = \frac{1}{\pi}\int_0^{2\pi} f(x)\sin(kx)\,dx = 0$$

because f is an even function and $x \mapsto \sin(kx)$ is odd, so that $x \mapsto f(x)\sin(kx)$ is odd. Furthermore,

$$a_0 = \frac{1}{\pi}\int_{-\pi}^{\pi} f(x)\,dx = \frac{2}{\pi}\int_0^{\pi} f(x)\,dx = \frac{2}{\pi}\int_0^{\pi} x\,dx = \pi,$$

whereas for $k \geq 1$, arguing as above

$$\begin{aligned}
a_k &= \frac{2}{\pi}\int_0^{\pi} x\cos(kx)\,dx \\
&= \frac{2}{\pi}\left[\frac{x}{k}\sin(kx)\Big|_0^{\pi} - \int_0^{\pi}\frac{1}{k}\sin(kx)\,dx\right] \\
&= -\frac{2}{k\pi}\left[-\frac{\cos(kx)}{k}\Big|_0^{\pi}\right] \\
&= \frac{2}{\pi k^2}\Big[\cos(k\pi) - 1\Big] \\
&= \frac{2}{\pi k^2}\Big[(-1)^k - 1\Big]
\end{aligned}$$

$$= \begin{cases} 0 & k \text{ even} \\ \dfrac{-4}{\pi k^2} & k \text{ odd.} \end{cases}$$

Therefore, the Fourier series of f is

$$\frac{\pi}{2} - \frac{4}{\pi} \sum_{n=0}^{\infty} \frac{1}{(2n+1)^2} \cos((2n+1)x).$$

(b) Since $f \in PS_{2\pi}$ and it is also continuous, its Fourier series of converges absolutely and uniformly to f in $\mathbb{R}$.

(c) Since the b_n's vanish, Parseval equality is:

$$\frac{1}{\pi} \int_{-\pi}^{\pi} |f(x)|^2 \, dx = \frac{a_0^2}{2} + \sum_{n=1}^{\infty} a_n^2.$$

Now, the left hand side is

$$\frac{1}{\pi} \int_{-\pi}^{\pi} |f(x)|^2 \, dx = \frac{2}{\pi} \int_{0}^{\pi} x^2 \, dx = \frac{2\pi^2}{3}$$

and the right hand side is

$$\frac{a_0^2}{2} + \sum_{n=1}^{\infty} a_n^2 = \frac{\pi^2}{2} + \frac{16}{\pi^2} \sum_{n=0}^{\infty} \frac{1}{(2n+1)^4}.$$

Equating

$$\frac{2\pi^2}{3} = \frac{\pi^2}{2} + \frac{16}{\pi^2} \sum_{n=0}^{\infty} \frac{1}{(2n+1)^4} \quad \Longrightarrow \quad \sum_{n=0}^{\infty} \frac{1}{(2n+1)^4} = \frac{\pi^4}{96}.$$

(d) Since $f(0) = 0$ and the Fourier series converges to f everywhere,

$$0 = \frac{\pi}{2} - \frac{4}{\pi} \sum_{n=0}^{\infty} \frac{1}{(2n+1)^2} \quad \Longrightarrow \quad \sum_{n=0}^{\infty} \frac{1}{(2n+1)^2} = \frac{\pi^2}{8}.$$

This is a basic exercise, where the periodic function is even, continuous and piecewise smooth. Thus one computes the a_n coefficients via the defining integrals that is formula (6.20) which for $T = 2\pi$ reads

$$a_n = \frac{2}{\pi} \int_0^{\pi} f(x) \cos(nx) \, dx,$$

and attains absolute and uniform convergence on $\mathbb{R}$ of the Fourier series. Parseval's identity yields the precise evaluation of an interesting series, namely:

$$1 + \frac{1}{3^4} + \frac{1}{5^4} + \frac{1}{7^4} + \cdots = \frac{\pi^4}{96}.$$

Evaluation at the origin of the convergent Fourier series yields the precise evaluation of a closely related, equally interesting, series, namely:

$$1 + \frac{1}{3^2} + \frac{1}{5^2} + \frac{1}{7^2} + \cdots = \frac{\pi^2}{8}.$$

6.39 Consider the function

$$f(x) = \begin{cases} \sin x & x \in (0, \pi/2] \\ 1 & x \in (\pi/2, 2]. \end{cases}$$

(a) Draw the graph of the sum of the Fourier series of the 2-periodic extension of f.
(b) Find the 2-periodic a_n Fourier coefficients of the 2-periodic extension of f.
(c) Compute the sum $\sum_{n=1}^{+\infty}(a_n^2 + b_n^2)$, where $(a_n)_{n\geq 0}$ and $(b_n)_{n\geq 1}$ are the Fourier coefficients of the 2-periodic extension of f.

Answer. (a) Upon drawing the graph of f, as in Fig. 6.12, it is immediate to realize that the 2-periodic extension of f is in PS_2, it is continuous at $\{\pi/2 + 2k : k \in \mathbb{Z}\}$ but it is not continuous at the endpoints, namely at $\{2k : k \in \mathbb{Z}\}$. Since

$$\lim_{x \to 0^+} f(x) = 0, \qquad \lim_{x \to 2\pi^-} f(x) = 1$$

the Fourier series of f converges to the value $1/2$ at all points $\{2k : k \in \mathbb{Z}\}$ and to the 2-periodic extension of f at all other points. Thus, the graph is as in Fig. 6.12.
(b) Evidently,

$$a_n = \int_0^2 f(x)\cos(\pi n x)\,dx = \int_0^{\pi/2} \sin x \cos(\pi n x)\,dx + \int_{\pi/2}^2 \cos(\pi n x)\,dx.$$

Now

$$\int_0^{\pi/2} \sin x \cos(\pi n x)\,dx = \frac{1}{2}\int_0^{\pi/2}[\sin(1+\pi n)x + \sin(1-\pi n)x]\,dx$$

$$= \frac{1}{2}\left[-\frac{\cos\frac{\pi}{2}(1+\pi n)}{1+\pi n} - \frac{\cos\frac{\pi}{2}(1-\pi n)}{1-\pi n} + \frac{1}{1+\pi n} + \frac{1}{1-\pi n}\right]$$

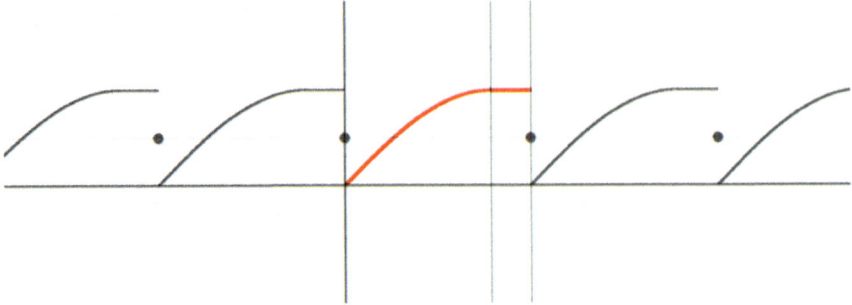

Fig. 6.12 In red the graph of the function f in Exercise 6.39, and in black the graph of the sum of the Fourier series of its 2-periodic extension

$$= \frac{\pi n}{\pi^2 n^2 - 1} \sin \frac{\pi^2 n}{2} + \frac{1}{1 - \pi^2 n^2}$$

and

$$\int_{\pi/2}^{2} \cos(\pi n x)\, dx = \begin{cases} -\dfrac{\sin(\pi^2 n/2)}{\pi n} & n \neq 0 \\ 2 - \pi/2 & n = 0. \end{cases}$$

Notice that summing the two above integral in the particular case $n = 0$ yields

$$a_0 = 3 - \frac{\pi}{2}.$$

(c) From Parseval's identity

$$\sum_{n=1}^{\infty}(a_n^2 + b_n^2) = \int_0^2 f^2(x)\, dx - \frac{a_0^2}{2} = \int_0^{\pi/2} \sin^2(x)\, dx + \int_{\pi/2}^{2} dx - \frac{1}{2}\left(3 - \frac{\pi}{2}\right)^2$$

$$= \frac{\pi}{4} + 2 - \frac{\pi}{2} - \frac{1}{2}\left(3 - \frac{\pi}{2}\right)^2 = -\frac{5}{2} + \frac{5\pi}{4} - \frac{\pi^2}{8}.$$

The first issue is a straightforward application of Dirichlet theorem, whereby the only discontinuities of the periodic extension occur at the endpoints, namely at the even integers $\{2k : k \in \mathbb{Z}\}$. Part (b) is an integration interlude that produces as byproduct the value of a_0, which is what is needed to give a numerical answer to item (c) via Parseval's identity.

6.40 Consider the function

$$f(x) = \begin{cases} x & x \in (0, 1) \\ 3 & x \in (1, 2]. \end{cases}$$

6.5 Guided Exercises

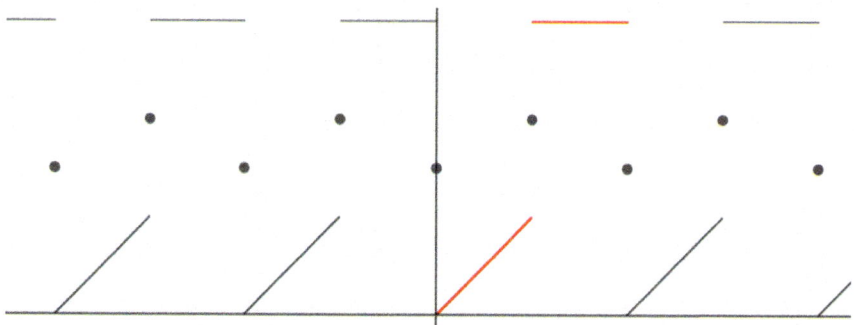

Fig. 6.13 In red the graph of the function f in Exercise 6.40, and in black the graph of the sum of the Fourier series of its 2-periodic extension

(a) Draw the graph of the sum of the Fourier series of the 2-periodic extension of f.
(b) Find the 4-periodic cosine Fourier series of f.

Answer. (a) The 2-periodic extension of f is in PS_2, it is not continuous neither at the points $\{1 + 2k : k \in \mathbb{Z}\}$ nor at the "endpoints", namely at $\{2k : k \in \mathbb{Z}\}$. The Fourier series of f converges to the average of the left and right limits, namely to the value 2 at the points $\{1 + 2k : k \in \mathbb{Z}\}$, and, similarly, to the value $3/2$ at the points $\{2k : k \in \mathbb{Z}\}$, which are the jump discontinuities. Finally, it converges to the 2-periodic extension of f at all other points. Thus, the graph is as in Fig. 6.13.

(b) The 4-periodic extension of the even extension f_e of f to the interval $[-2, 2]$ has cosine Fourier series given by formula (6.21), which in the case $T = 4$ is

$$\frac{a_0}{2} + \sum_{n=1}^{\infty} a_n \cos\left(\frac{\pi}{2} n x\right) \qquad (6.26)$$

and where the coefficients a_n are given by (6.20), which for $T = 4$ becomes

$$a_n = \frac{1}{2} \int_{-2}^{2} f_e(x) \cos\left(\frac{\pi}{2} n x\right) dx = \int_{0}^{2} f(x) \cos\left(\frac{\pi}{2} n x\right) dx. \qquad (6.27)$$

Computing,

$$a_0 = \int_{0}^{2} f(x) \, dx = \int_{0}^{1} x \, dx + \int_{1}^{2} 3 \, dx = \frac{1}{2} + 3 = \frac{7}{2},$$

and for $n \geq 1$, an integration by parts yields

$$a_n = \int_{0}^{1} x \cos\left(\frac{\pi}{2} n x\right) dx + 3 \int_{1}^{2} \cos\left(\frac{\pi}{2} n x\right) dx$$

$$= \frac{2}{n\pi}\sin(\frac{\pi}{2}n) + \frac{4}{n^2\pi^2}\left(\cos(\frac{\pi}{2}n) - 1\right) - \frac{6}{\pi n}\sin(\frac{\pi}{2}n)$$

$$= \begin{cases} \dfrac{1}{m^2\pi^2}((-1)^m - 1) & n = 2m \\[2ex] (-1)^m\dfrac{4}{(2m-1)\pi} - \dfrac{4}{(2m-1)^2\pi^2} & n = 2m - 1. \end{cases}$$

Therefore, the series is

$$\frac{7}{4} + \sum_{m=1}^{\infty}\left[\frac{1}{m^2\pi^2}((-1)^m - 1)\cos(\pi m x)\right.$$
$$\left. + \left((-1)^m\frac{4}{(2m-1)\pi} - \frac{4}{(2m-1)^2\pi^2}\right)\cos(\frac{\pi}{2}(2m-1)x)\right].$$

This is a standard exercise, where the behaviour of the sum of the Fourier series can be derived from the properties of the given function (Fig. 6.14).

6.41 Consider the function

$$f(x) = \begin{cases} 0 & x \in (0, 1) \\ x & x \in (1, 2]. \end{cases}$$

(a) Determine the 4-periodic cosine series of f and draw the graph of its sum.
(b) Find an interval in which the above series converges totally.

Answer. (a) Let f_e denote the even extension of f to the interval $(-2, 2)$. The 4-periodic cosine series of f is then as (6.26) where a_n is given by (6.27), namely

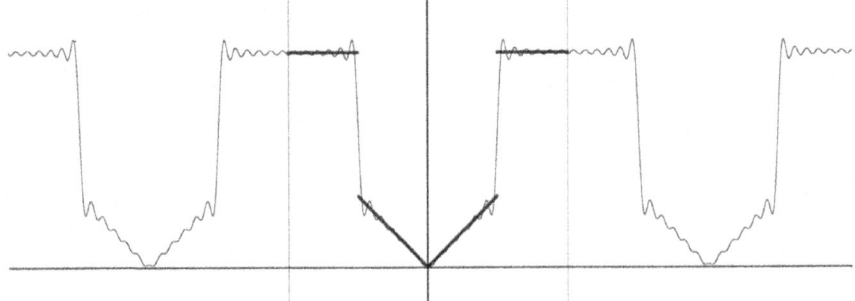

Fig. 6.14 In black the graph of the function f_e in Exercise 6.40, and in grey the graph of the partial sum $S_n(x)$ with $n = 12$ of the Fourier series of its 4-periodic extension, a continuous function with a visible "overshoot" near the discontinuities known as the Gibbs' phenomenon

6.5 Guided Exercises

$$a_n = \int_0^2 f(x) \cos\left(\frac{\pi}{2}nx\right) dx = \int_1^2 x \cos\left(\frac{\pi}{2}nx\right) dx.$$

If $n \neq 0$, then

$$a_n = \frac{4}{\pi^2 n^2}(\cos(n\pi) - \cos(n\pi/2)) - \frac{2}{\pi n} \sin(n\pi/2)$$

$$= \begin{cases} 0 & n = 4k \\ \dfrac{8}{\pi^2(2k+1)^2} & n = 2(2k+1) \\ -\dfrac{4}{\pi^2(2k+1)^2} + (-1)^{k+1}\dfrac{2}{\pi(2k+1)} & n = 2k+1 \end{cases}$$

and $a_0 = 3/2$. Therefore, the cosine Fourier series is

$$\frac{3}{4} + \sum_{k=0}^{\infty}\left[\left(-\frac{4}{\pi^2(2k+1)^2} + (-1)^{k+1}\frac{2}{\pi(2k+1)}\right) \cos\left(\frac{\pi}{2}(2k+1)x\right) \right.$$
$$\left. + \frac{8}{\pi^2(2k+1)^2} \cos(\pi(2k+1)x)\right].$$

The 4-periodic extension of f_e is in PS$_4$ and therefore the Fourier series converges to it at all points of continuity, and to the average of the left and right limits, namely $1/2$, at the jump discontinuities, namely on the set $\{2k+1 : k \in \mathbb{Z}\}$. Hence the graph is as in Fig. 6.15.

(b) In every closed interval of the form $[a, b]$ with $2k + 1 < a < b < 2k + 3$ and $k \in \mathbb{Z}$ the function is continuous, hence the series converges totally on $[a, b]$.

This is a completely standard exercise, where the only task is that of considering the even extension and hence to use the correct version of formulae (6.20) for the cosine coefficients. The convergence properties of the series are inferred directly from the elementary graph.

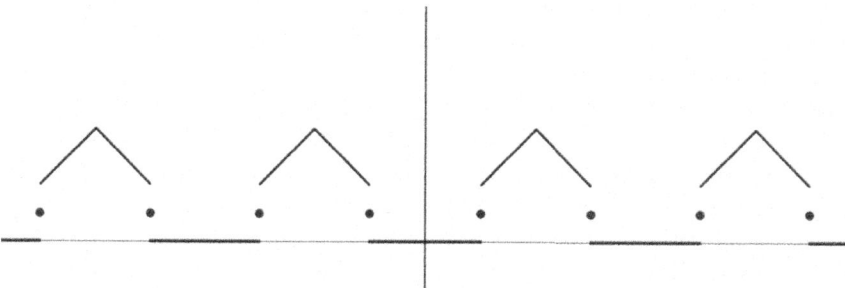

Fig. 6.15 The graph of the 4-periodic function in Exercise 6.41

6.42 Consider the function $f(x) = \dfrac{x^2}{4} - \dfrac{\pi x}{2} + \dfrac{\pi^2}{6}$, with $x \in [0, \pi)$.

(a) Determine the 2π-periodic cosine series of f.
(b) Show that $\sum_{n=1}^{\infty} n^{-2} = \pi^2/6$.
(c) Show that $\sum_{n=1}^{\infty} (-1)^n (2n+1)^{-3} = \pi^3/32$.

Answer. (a) Let f_e denote the even extension of f to the interval $(-\pi, \pi)$, which is actually given by

$$f_e(x) = \frac{x^2}{4} - \frac{\pi |x|}{2} + \frac{\pi^2}{6}.$$

The 2π-periodic series of f_e is then as (6.21) where a_n is given by (6.20), namely

$$a_n = \frac{2}{\pi} \int_0^\pi f(x) \cos(nx)\, dx = \frac{2}{\pi} \int_0^\pi \left(\frac{x^2}{4} - \frac{\pi x}{2} + \frac{\pi^2}{6} \right) \cos(nx)\, dx$$

If $n = 0$, then

$$a_0 = \frac{2}{\pi} \int_0^\pi \left(\frac{x^2}{4} - \frac{\pi x}{2} + \frac{\pi^2}{6} \right) dx = \frac{2}{\pi} \left[\frac{x^3}{12} - \frac{\pi}{4} x^2 + \frac{\pi^2}{6} x \right]_0^\pi = 0.$$

If $n \neq 0$, then

$$a_n = \frac{2}{\pi} \int_0^\pi \left(\frac{x^2}{4} - \frac{\pi x}{2} + \frac{\pi^2}{6} \right) \cos(nx)\, dx$$

$$= \frac{2}{\pi} \left\{ \left[\left(\frac{x^2}{4} - \frac{\pi x}{2} + \frac{\pi^2}{6} \right) \frac{\sin(nx)}{n} \right]_0^\pi - \frac{1}{2n} \int_0^\pi (x - \pi) \sin(nx)\, dx \right\}$$

$$= -\frac{1}{\pi n} \int_0^\pi (x - \pi) \sin(nx)\, dx$$

$$= \frac{1}{\pi n} \left\{ \left[(x - \pi) \frac{\cos nx}{n} \right]_0^\pi - \frac{1}{n} \int_0^\pi \cos(nx)\, dx \right\}$$

$$= \frac{1}{n^2}.$$

Therefore, the 2π-periodic cosine series of f is

$$\sum_{n=1}^{\infty} \frac{\cos(nx)}{n^2}.$$

(b) The 2π-periodic extension g of f_e is actually continuous on $\mathbb{R}$ because

$$\lim_{x \to -\pi^+} f_e(x) = \lim_{x \to \pi^-} f_e(x) = -\frac{\pi^2}{12}$$

6.5 Guided Exercises

and it is differentiable everywhere except at the integer multiples of 2π, as it is easily established at the origin because of the presence of $|x|$. Since $g \in PS_{2\pi} \cap C(\mathbb{R})$, the Fourier series of g converges to g uniformly in $\mathbb{R}$ and in particular, since $g(0) = \pi^2/6$ it follows

$$\sum_{n=1}^{\infty} \frac{1}{n^2} = \frac{\pi^2}{6}, \tag{6.28}$$

as desired.

(c) Observe that, proceeding as in the computation of a_0,

$$\int_0^{\pi/2} f(x)\,dx = \left[\frac{x^3}{12} - \frac{\pi}{4}x^2 + \frac{\pi^2}{6}x\right]_0^{\pi/2} = \frac{\pi^3}{32}.$$

By the uniform convergence of the Fourier series of g on $[0, \pi/2]$, where obviously $f = g$, it is legitimate to integrate term by term, obtaining

$$\frac{\pi^3}{32} = \int_0^{\pi/2} g(x)\,dx$$

$$= \int_0^{\pi/2} \sum_{n=1}^{\infty} \frac{\cos(nx)}{n^2}\,dx$$

$$= \sum_{n=1}^{\infty} \int_0^{\pi/2} \frac{\cos(nx)}{n^2}\,dx$$

$$= \sum_{n=1}^{\infty} \left[\frac{\sin(nx)}{n^3}\right]_0^{\pi/2}$$

$$= \sum_{n=0}^{\infty} \frac{(-1)^n}{(2n+1)^3},$$

whereby it was used that

$$\sin\left(\frac{n\pi}{2}\right) = \begin{cases} 0 & n = 2k \\ (-1)^k & n = 2k+1 \end{cases}$$

as k varies in the non negative integers.

This is a standard exercise, with a little twist. Part (a) is about writing correctly the cosine series, and it contains a small but important point when computing a_0, namely the fact that the (indefinite) integral of f is a polynomial of third degree in x, which happens to vanish when evaluated at π. Part (b) requires to properly justify that $g(0)$ is the Fourier series at zero, which leads to the very famous formula (6.28). This formula can actually be derived in a number of different ways, very often related to Fourier series. Part (c) is where the twist occurs. One naturally wonders what to do in order to obtain the formula, but a minute pause reveals that the hint comes from the third power of π. Indeed, as already mentioned, when computing a_0 one finds a polynomial of third degree in x which actually attains the value $\pi^3/32$ when evaluated at $\pi/2$. This suggests to compute the integral of f over the interval $[0, \pi/2]$, and to do this term by term.

6.43 Let $A \in \mathbb{R}$ be a fixed real number and consider the functions

$$f(x) = \begin{cases} A & x \in (0, \pi) \\ -A & x \in (-\pi, 0), \end{cases}$$

$$g(x) = \begin{cases} 1 & x \in (0, \frac{\pi}{2}) \\ -1 & x \in (\frac{\pi}{2}, \pi), \end{cases} \qquad h(x) = \begin{cases} 1 & x \in (-\frac{\pi}{2}, \frac{\pi}{2}) \\ 0 & x \in (-\pi, -\frac{\pi}{2}) \cup (\frac{\pi}{2}, \pi). \end{cases}$$

(a) Find the Fourier series of the 2π-periodic extension of f and find its sum, specifying in which intervals it converges uniformly.
(b) Show that $\sum_{n=0}^{\infty}(-1)^n(2n+1)^{-1} = \pi/4$.
(c) Find the 2π-periodic cosine series of g.
(d) Find the Fourier series of the 2π-periodic extension of h.

Answer. (a) Since f and its 2π-periodic extension are odd functions, $a_n = 0$ for every non-negative integer n. Further,

$$\begin{aligned} b_n &= \frac{2}{\pi} \int_0^\pi f(x) \sin(nx) \, dx \\ &= \frac{2A}{\pi} \left[-\frac{\cos(nx)}{n} \right]_0^\pi \\ &= \frac{2A}{\pi} \left[\frac{1-(-1)^n}{n} \right] \\ &= \begin{cases} 0 & n \text{ even} \\ \frac{4A}{n\pi} & n \text{ odd.} \end{cases} \end{aligned}$$

Therefore the Fourier series of the 2π-periodic extension of f is

$$\frac{4A}{\pi} \sum_{n=0}^{\infty} \frac{1}{2n+1} \sin((2n+1)x).$$

It converges uniformly to the 2π-periodic extension of f in every compact interval of the form $[a, b]$ with $k\pi < a < b < (k+1)\pi$ for every integer k, and to 0 at every integer multiple of π.

6.5 Guided Exercises

(b) For $A = 1$ and $x = \pi/2$ the series converges to $f(\pi/2) = 1$ and hence

$$1 = \frac{4}{\pi} \sum_{n=0}^{\infty} \frac{1}{2n+1} \sin((2n+1)\frac{\pi}{2}) = \frac{4}{\pi} \sum_{n=0}^{\infty} \frac{1}{2n+1} \cos(n\pi) = \frac{4}{\pi} \sum_{n=0}^{\infty} \frac{(-1)^n}{2n+1},$$

which implies $\sum_{n=0}^{\infty}(-1)^n(2n+1)^{-1} = \pi/4$.

(c) Observe that for every $x \in (-\pi, \pi)$ it holds that $g_e(x) = \tilde{f}(x + \frac{\pi}{2})$, where $\tilde{f}$ is the 2π-periodic extension of f when $A = 1$. Hence the cosine series of g is the Fourier series of $\tilde{f}(x + \frac{\pi}{2})$, namely

$$\frac{4}{\pi} \sum_{n=0}^{\infty} \frac{1}{2n+1} \sin\left((2n+1)(x + \frac{\pi}{2})\right) = \frac{4}{\pi} \sum_{n=0}^{\infty} \frac{1}{2n+1} \sin\left((2n+1)x + n\pi + \frac{\pi}{2}\right)$$

$$= \frac{4}{\pi} \sum_{n=0}^{\infty} \frac{1}{2n+1} \cos\left((2n+1)x + n\pi\right)$$

$$= \frac{4}{\pi} \sum_{n=0}^{\infty} \frac{1}{2n+1} \cos\left((2n+1)x\right) \cos\left(n\pi\right)$$

$$= \frac{4}{\pi} \sum_{n=0}^{\infty} \frac{(-1)^n}{2n+1} \cos\left((2n+1)x\right).$$

(d) Evidently, $h(x) = (g_e(x) + 1)/2$, and hence the Fourier series of the 2π-periodic extension of h is

$$\frac{1}{2}\left(\frac{4}{\pi} \sum_{n=0}^{\infty} \frac{(-1)^n}{2n+1} \cos\left((2n+1)x\right) + 1\right) = \frac{1}{2} + \frac{2}{\pi} \sum_{n=0}^{\infty} \frac{(-1)^n}{2n+1} \cos\left((2n+1)x\right).$$

This exercise is simple but instructive. In (a) it is required to find the Fourier series of the so-called square wave, which is quite important in applications. Notice that the Fourier coefficients b_n decay as $(2n+1)^{-1}$, that is, slowly, but have alternating sign. The functions g_e and h are obtained from f by setting $A = 1$ and then by simple manipulations. Indeed, g_e is just a translation of f to the left by $\pi/2$ and h is just a translation upward by 1 followed by multiplication by $1/2$. Notice that in the former case the translation changes the parity of the function (from odd to even) which is reflected in the fact that a sine series becomes a cosine series, whereas in the latter case the parity is unchanged because shifting and shrinking (or dilating) vertically does not affect it. The graph of some partial Fourier sums are as in Fig. 6.16.

6.44 Consider the function $f(x) = \sin x$, with $x \in [0, \pi/2)$.

(a) Determine the π-periodic cosine series of f.
(b) Show that $\sum_{n=1}^{\infty}(4n^2 - 1)^{-1} = 1/2$.
(c) Show that $\sum_{n=1}^{\infty}(4n^2 - 1)^{-2} = (\pi^2/16) - 1/2$.

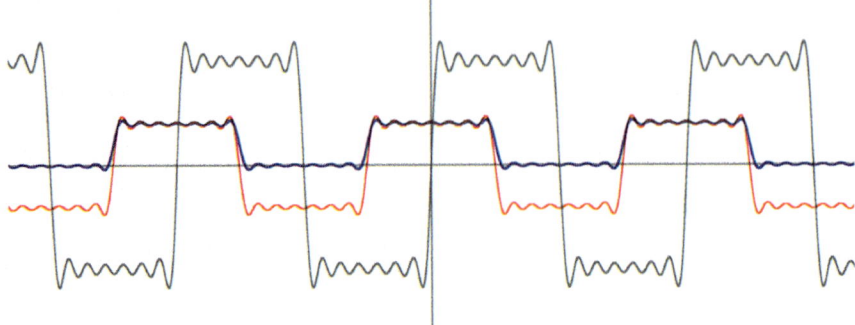

Fig. 6.16 The graph of the Fourier partial sums of the 2π-periodic extensions of the various functions in Exercise 6.43; precisely, in black those of f (with $A = 2.5$), in red those of g and in blue those of h

Answer. (a) It is clear that the even extension of f to the interval $I = (-\pi/2, \pi/2)$ is piecewise smooth and continuous, and in fact it coincides on I with the function $g(x) = |\sin x|$. The Fourier coefficients of the cosine series are therefore

$$a_0 = \frac{2}{\pi}\int_{-\pi/2}^{\pi/2} |\sin x|\,dx = \frac{4}{\pi}\int_0^{\pi/2} \sin x\,dx = \frac{4}{\pi},$$

and, observing that

$$\sin(2n+1)x = \sin(2nx + x) = \cos(2nx)\sin x + \cos x \sin(2nx)$$
$$\sin(2n-1)x = \sin(2nx - x) = -\cos(2nx)\sin x + \cos x \sin(2nx)$$

it follows that, subtracting the above equalities, for every integer $n \geq 1$

$$\begin{aligned}
a_n &= \frac{2}{\pi}\int_{-\pi/2}^{\pi/2} |\sin x|\cos(2nx)\,dx \\
&= \frac{4}{\pi}\int_0^{\pi/2} \sin x \cos(2nx)\,dx \\
&= \frac{4}{\pi}\int_0^{\pi/2} \frac{1}{2}\Big[\sin(2n+1)x - \sin(2n-1)x\Big]\,dx \\
&= \frac{2}{\pi}\int_0^{\pi/2} \sin(2n+1)x\,dx + \frac{2}{\pi}\int_0^{\pi/2} \sin(1-2n)x\,dx \\
&= \frac{4}{\pi(1-4n^2)},
\end{aligned}$$

where the final result easily follows by evaluating, and then summing, the two indicated elementary integrals. Therefore, the requested cosine series is

6.5 Guided Exercises

$$\frac{2}{\pi} + \frac{4}{\pi} \sum_{n=1}^{\infty} \frac{\cos(2nx)}{1 - 4n^2}.$$

(b) The Fourier series found above converges uniformly to the π-periodic extension of g to the real line, namely to $|\sin x|$, because the latter is piecewise smooth and continuous. In particular,

$$0 = g(0) = \frac{2}{\pi} + \frac{4}{\pi} \sum_{n=1}^{\infty} \frac{1}{1 - 4n^2},$$

whence

$$\sum_{n=1}^{\infty} \frac{1}{4n^2 - 1} = \frac{1}{2},$$

as desired.

(c) Consider next Parseval's identity (6.24), namely, since $b_n = 0$ for all n

$$\frac{2}{\pi} \int_{-\pi/2}^{\pi/2} (|\sin x|)^2 \, dx = \frac{a_0^2}{2} + \sum_{n=1}^{\infty} a_n^2.$$

The left hand side is equal to

$$\frac{1}{\pi} \int_{-\pi/2}^{\pi/2} (1 - \cos 2x) \, dx = 1 - 0 = 1$$

and the right hand side to

$$\frac{8}{\pi^2} + \frac{16}{\pi^2} \sum_{n=1}^{\infty} \frac{1}{(1 - 4n^2)^2}.$$

Equating them yields precisely

$$\sum_{n=1}^{\infty} \frac{1}{(4n^2 - 1)^2} = \frac{\pi^2}{16} - \frac{1}{2}.$$

This is a direct exercise, in the sense that all it is required is to apply the defining formulas. Here some trigonometric identities were used to speed up the calculations, but a direct and straightforward approach would have led to the same answers with just about the same effort.

6.45 Consider the function $f(x) = x^4$, with $x \in (-\pi, \pi)$.

(a) Determine the 2π-periodic Fourier series of f and study its convergence properties.
(b) Show that $\sum_{n=1}^{\infty} n^{-4} = \pi^4/90$.

Answer. (a) Clearly, the 2π-periodic extension g of f is continuous and piecewise smooth, hence the Fourier series converges uniformly to g in $\mathbb{R}$. As for the explicit calculation of the series, observe that f is even and hence $b_n = 0$ for all n. Next,

$$a_0 = \frac{2}{2\pi} \int_{-\pi}^{\pi} x^4 \, dx = \frac{2}{5}\pi^4$$

$$\begin{aligned}
a_n &= \frac{1}{\pi} \int_{-\pi}^{\pi} x^4 \cos(nx) \, dx \\
&= \frac{2}{\pi} \int_{0}^{\pi} x^4 \cos(nx) \, dx \\
&= \frac{2}{\pi} \left\{ \left[\frac{x^4 \sin(nx)}{n}\right]_0^{\pi} - \frac{4}{n} \int_0^{\pi} x^3 \sin(nx) \, dx \right\} \\
&= -\frac{8}{\pi n} \left\{ \left[\frac{-x^3 \cos(nx)}{n}\right]_0^{\pi} + \frac{3}{n} \int_0^{\pi} x^2 \cos(nx) \, dx \right\} \\
&= -\frac{8}{\pi n^2}(-\pi^3(-1)^n) - \frac{24}{\pi n^2} \left\{ \left[\frac{x^2 \sin(nx)}{n}\right]_0^{\pi} - \frac{2}{n} \int_0^{\pi} x \sin(nx) \, dx \right\} \\
&= (-1)^n \frac{8\pi^2}{n^2} + \frac{48}{\pi n^3} \left\{ \left[\frac{-x \cos(nx)}{n}\right]_0^{\pi} + \frac{1}{n} \int_0^{\pi} \cos(nx) \, dx \right\} \\
&= (-1)^n \left(\frac{8\pi^2}{n^2} - \frac{48}{n^4}\right).
\end{aligned}$$

Therefore, the Fourier series is

$$\frac{\pi^4}{5} + \sum_{n=1}^{\infty} (-1)^n \left(\frac{8\pi^2}{n^2} - \frac{48}{n^4}\right) \cos(nx).$$

(b) Evaluating the above series at $x = \pi$, where it converges, the equality

$$\pi^4 = \frac{\pi^4}{5} + \sum_{n=1}^{\infty} \left(\frac{8\pi^2}{n^2} - \frac{48}{n^4}\right)$$

6.5 Guided Exercises

holds. Using next that $\sum_1^\infty n^{-2} = \pi^2/6$, see formula (6.28), it follows that

$$\frac{4}{5}\pi^4 = 8\pi^2 \sum_{n=1}^{\infty} \frac{1}{n^2} - 48 \sum_{n=1}^{\infty} \frac{1}{n^4} = \frac{4}{3}\pi^4 - 48 \sum_{n=1}^{\infty} \frac{1}{n^4},$$

whence

$$\sum_{n=1}^{\infty} \frac{1}{n^4} = \frac{\pi^4}{90}.$$

This exercise only requires some attention in performing integration by parts a number of times and then observing that the series converges pointwise at π. Knowledge of $\sum_1^\infty n^{-2} = \pi^2/6$, namely formula (6.28), is necessary.

6.46 Consider the function $f(x) = |\cos x|$, with $x \in (-\pi, \pi)$.

(a) Determine the 2π-periodic Fourier series of f and study its convergence properties.
(b) Show that $\sum_{n=1}^{\infty} (-1)^{n+1} (4n^2 - 1)^{-1} = (\pi - 2)/4$.

Answer. (a) Denote by g the 2π-periodic extension of f. Observe that f, hence g, is even and hence $b_n = 0$ for all n. Next,

$$a_0 = \frac{1}{\pi} \int_{-\pi}^{\pi} |\cos x| \, dx = \frac{4}{\pi} \int_0^{\pi/2} \cos x \, dx = \frac{4}{\pi}$$

$$a_1 = \frac{1}{\pi} \int_{-\pi}^{\pi} |\cos x| \cos x \, dx = 0$$

and for every integer $n > 1$

$$a_n = \frac{1}{\pi} \int_{-\pi}^{\pi} |\cos x| \cos(nx) \, dx$$

$$= \frac{2}{\pi} \int_0^{\pi} |\cos x| \cos(nx) \, dx$$

$$= \frac{2}{\pi} \left(\int_0^{\pi/2} \cos x \cos(nx) \, dx - \int_{\pi/2}^{\pi} \cos x \cos(nx) \, dx \right)$$

$$= \frac{1}{\pi} \int_0^{\pi/2} \left\{ \cos((n+1)x) + \cos((n-1)x) \right\} dx$$

$$\quad - \frac{1}{\pi} \int_{\pi/2}^{\pi} \left\{ \cos((n+1)x) + \cos((n-1)x) \right\} dx$$

$$= \frac{2}{\pi(n+1)} \sin\left(\frac{n+1}{2}\pi\right) + \frac{2}{\pi(n-1)} \sin\left(\frac{n-1}{2}\pi\right)$$

$$= \left(\frac{2}{\pi(n+1)} - \frac{2}{\pi(n-1)} \right) \sin\left(\frac{n+1}{2}\pi\right)$$

$$= \frac{4}{\pi(1-n^2)} \sin\left(\frac{n+1}{2}\pi\right).$$

Now, observing that for $m \geq 1$

$$\sin\left(\frac{n+1}{2}\pi\right) = \cos\left(n\frac{\pi}{2}\right) = \begin{cases} (-1)^m & n = 2m \\ 0 & n = 2m-1 \end{cases}$$

it follows that

$$a_n = \begin{cases} (-1)^m \dfrac{4}{\pi(1-(2m)^2)} & n = 2m \\ 0 & n = 2m-1. \end{cases}$$

Consequently, the Fourier series is

$$\frac{2}{\pi} + \frac{4}{\pi} \sum_{m=1}^{\infty} \frac{(-1)^m}{1-4m^2} \cos(2mx).$$

(b) Clearly, the 2π-periodic extension g of f is continuous and piecewise smooth, hence the Fourier series converges uniformly to g in $\mathbb{R}$. Therefore,

$$1 = f(0) = \frac{2}{\pi} + \frac{4}{\pi} \sum_{m=1}^{\infty} \frac{(-1)^m}{1-4m^2} \quad \Longrightarrow \quad \frac{\pi-2}{4} = \sum_{m=1}^{\infty} \frac{(-1)^{m+1}}{1-4m^2}.$$

This exercise requires a careful examination of the various different cases that occur when computing a_n because the formula depends on the value of "n modulo 4" due to the possible different values of

$$\sin\left(\frac{n+1}{2}\pi\right),$$

which need to be analyzed. Indeed, as n varies, $\pi(n+1)/2$ runs over the integer multiples of $\pi/2$, which take up the values $1, 0, -1, 0, \ldots$. This care is somewhat typically needed when dealing with sufficiently complicated calculations of Fourier coefficients. The second half amounts to properly justify the operation of assigning the value zero to the variable x inside the series.

6.6 Problems

6.47 Consider the sequence $(f_n)_{n\geq 1}$, where

$$f_n(x) = \begin{cases} n & |x| < \frac{1}{n} \\ \dfrac{n}{n|x|+1} & |x| \geq \frac{1}{n}. \end{cases}$$

(a) Describe the pointwise limit f of $(f_n)_{n\geq 1}$, specifying the domain I of f.
(b) Does $f_n \rightrightarrows f$ on I?
(c) Compute, if existing, the limit of the numerical sequence $\int_{-1}^{1} f_n(x)\,dx$.

6.48 Consider the sequence $(f_n)_{n\geq 1}$ defined by

$$f_n(x) = \begin{cases} 0 & x \leq 0 \\ x/n & 0 < x \leq n \\ 1 & x > n. \end{cases}$$

(a) Describe the pointwise limit f of $(f_n)_{n\geq 1}$.
(b) Discuss the uniform convergence.

6.49 Determine the pointwise limit and decide where the convergence is uniform for the sequence $(f_n)_{n\geq 1}$ where $f_n(x) = \dfrac{x+n}{ne^x+1}$.

6.50 Consider the sequence $(f_n)_{n\geq 1}$ defined by

$$f_n(x) = \begin{cases} n(x^2-1) & 0 \leq x \leq 1 \\ 0 & 1 < x \leq 1+1/n \\ \exp\left(\dfrac{2-x}{n}\right)\arctan x & x > 1+1/n. \end{cases}$$

(a) Describe the pointwise limit f of $(f_n)_{n\geq 1}$.
(b) Is the convergence uniform on the intervals $[2, b]$ for every $b > 2$?
(c) Is the convergence uniform on the intervals $[2, +\infty]$?

6.51 Consider the sequence $(f_n)_{n\geq 1}$ defined by $f_n(x) = \dfrac{2+\sin(nx)}{1+(n^2x^2-1)^2}$.

(a) Describe the pointwise limit f of $(f_n)_{n\geq 1}$.
(b) Is the convergence uniform on $[-1, 1]$?
(c) Is the convergence uniform on $[1, +\infty]$?

6.52 Consider the sequence $(f_n)_{n\geq 1}$ defined by $f_n(x) = \dfrac{ne^{-nx}}{n^2 + e^{-2nx}}$.
(a) Describe the pointwise limit f of $(f_n)_{n\geq 1}$.
(b) Is the convergence uniform on $[-e, e]$?
(c) Is the convergence uniform on $[0, +\infty]$?

6.53 Consider the sequence $(f_n)_{n\geq 1}$ defined by $f_n(x) = \dfrac{x}{1 + nx^2}$.
(a) Establish what is the pointwise limit of $(f_n)_{n\geq 1}$ and discuss where the uniform convergence occurs.
(b) Establish what is the pointwise limit of $(f'_n)_{n\geq 1}$ and discuss where the uniform convergence occurs.

6.54 Consider the sequence $(f_n)_{n\geq 1}$ defined on $[0, +\infty)$ by $f_n(x) = \left(1 + \dfrac{x}{n}\right)^n$. Establish what is the pointwise limit of $(f_n)_{n\geq 1}$ and discuss where the uniform convergence occurs.

6.55 Consider the sequence $(f_n)_{n\geq 1}$ defined on $[0, +\infty)$ by $f_n(x) = \dfrac{1}{1 + (x - n)^2}$.
(a) Establish what is the pointwise limit f of $(f_n)_{n\geq 1}$ and discuss where the uniform convergence occurs.
(b) Is it true that $f'_n(x) \to f'(x)$?

6.56 Consider the sequence $(f_n)_{n\geq 1}$ defined by

$$f_n(x) = \begin{cases} 0 & x \in (-\infty, 0] \cup [\frac{1}{n}, +\infty) \\ 2n^{a+1}x & x \in (0, \frac{1}{2n}) \\ 2n^a - 2n^{a+1}x & x \in [\frac{1}{2n}, \frac{1}{n}), \end{cases}$$

where $a > 0$ is a parameter.
(a) Establish what is the pointwise limit f of $(f_n)_{n\geq 1}$ and discuss where the uniform convergence occurs.
(b) Compute $\lim\limits_n \int_0^{1/n} f_n(x)\, dx$ as $a > 0$ varies.

6.57 Consider the sequence $(f_n)_{n\geq 1}$ defined by $f_n(x) = \dfrac{n}{nx^2 - x + a_n}$ where $(a_n)_{n\geq 1}$ admits limit.
(a) Find the conditions on $(a_n)_{n\geq 1}$ for which the domain of f_n is $\mathbb{R}$ for every $n \geq 1$.
(b) Assume that the domain of f_n is $\mathbb{R}$ for every $n \geq 1$ and compute the limit as $(a_n)_{n\geq 1}$ varies.
(c) Find a sequence $(a_n)_{n\geq 1}$ such that $(f_n)_{n\geq 1}$ converges uniformly on $\mathbb{R}$ to the null function.

6.6 Problems

6.58 Consider the sequence $(f_n)_{n\geq 1}$ defined by $f_n(x) = \dfrac{\arctan(x+n)}{x^2+n}$.
(a) Find the pointwise limit.
(b) Compute, if existing, $\lim_n \int_0^2 f_n(x)\,dx$.
(c) Compute, if existing, $\lim_n \int_0^{+\infty} f_n(x)\,dx$.

6.59 Consider the sequence $(f_n)_{n\geq 1}$ defined by $f_n(x) = \dfrac{n^k x}{n^3 x^2 + nx + 1}$, where $k \in \mathbb{N}$ is a parameter.
(a) Discuss pointwise and uniform convergence as k varies.
(b) Fix $k=3$. Compute, if existing, $\lim_n \int_{1/n}^n f_n(x)\,dx$.

6.60 Consider the sequence $(f_n)_{n\geq 1}$ defined recursively by

$$\begin{cases} f_0(x) = \sin x \\ f_{n+1}(x) = \sin\left(\dfrac{\pi}{2} f_n(x)\right). \end{cases}$$

(a) Describe the pointwise limit f of $(f_n)_{n\geq 1}$, specifying the domain I of f.
(b) Discuss the uniform convergence within I.

6.61 Consider the series $\displaystyle\sum_{n=1}^{\infty}\left(\dfrac{\sin^n x}{n^2} - \dfrac{\sin^{n+1} x}{(n+1)^2}\right)$.
(a) Find the set I of pointwise convergence.
(b) Does the series converge uniformly on I?
(c) Establish if the sum is differentiable on I.

6.62 Establish where the following series converge pointwise and where they converge uniformly.

(a) $\displaystyle\sum_{n=1}^{\infty} 3^n \sin\dfrac{x}{5^n}$

(b) $\displaystyle\sum_{n=1}^{\infty} (-1)^n e^{-n\cos x}$

(c) $\displaystyle\sum_{n=1}^{\infty} (-1)^n \dfrac{x^4}{1+n^4 x^8}$

(d) $\displaystyle\sum_{n=1}^{\infty} nxe^{-nx}$.

6.63 Establish where $\displaystyle\sum_{n=1}^{\infty} \dfrac{|e^{nx} - e^{-nx} + 1|}{n^4 + 1}$ converges pointwise.

6.64 Establish where $\displaystyle\sum_{n=1}^{\infty} \dfrac{\log(1+x^n)}{n^{x+1}}$ converges pointwise.

6.65 Consider the series $\sum_{n=1}^{\infty} \dfrac{\sqrt{n+1}-\sqrt{n}}{n} \sin(x^n)$.
(a) Find the set I of pointwise convergence to a sum S.
(b) Where is S continuous?

6.66 Consider the series $\sum_{n=1}^{\infty} f_n(x)$ where

$$f_n(x) = \begin{cases} 0 & x \in (-\infty, 0] \cup [\tfrac{2}{n}, +\infty) \\ nx & x \in (0, \tfrac{1}{n}] \\ 2 - nx & x \in (\tfrac{1}{n}, \tfrac{2}{n}). \end{cases}$$

Determine the interval I where the series converges pointwise and establish if the convergence is absolute and uniform on I.

6.67 Consider the series $\sum_{n=1}^{\infty} \dfrac{\log(1+nx)}{nx^n}$ with $x > 0$.
(a) Find the set I of pointwise convergence to a sum S.
(b) Discuss the uniform convergence within I.

6.68 Determine where the series $\sum_{n=1}^{\infty} \left(\left(1+\dfrac{1}{n}\right)^{n^x} - 1\right)$ converges.

6.69 Consider the function $S(x) = \sum_{n=1}^{\infty} \log(1+|x|^n) \sin(e^{-n|x|})$.

(a) Is S continuous on $\mathbb{R}$?
(b) Is S differentiable on $\mathbb{R}$?

6.70 Consider the series $\sum_{n=1}^{\infty} \left(1 - \dfrac{n}{x^4+x^2+1}\right) \sin\left(\dfrac{x^4+x^2+1}{n}\right)$.
(a) Discuss the pointwise convergence of the series.
(b) Discuss the uniform convergence of the series.

6.71 Consider the function $S(x) = \sum_{n=1}^{\infty} \dfrac{\sqrt{x^2 n^2 + 1} - n|x|}{n}$.

(a) Find the domain of S.
(b) Find where S is continuous.

6.6 Problems

6.72 Consider the function $S(x) = \sum_{n=1}^{\infty} n^x \log(1 + \frac{1}{n})$.
(a) Find the domain of S.
(b) Find where S is continuous.

6.73 Consider the series $\sum_{n=1}^{\infty} n^x (e^{nx} - n)$.
(a) Find where the series converges totally.
(b) Study the convergence of $\int_{-3}^{-2} S(x)\, dx$, where S is the pointwise sum of the series.

6.74 Establish where the series $\sum_{n=1}^{\infty} \frac{\log(1 + (1-x)x^n)}{n^{4/3}}$ converges pointwise and where it converges totally.

6.75 Establish where $S(x) = \sum_{n=0}^{\infty} \frac{n!}{e^{n^2}} x^n$ is continuous and where it is differentiable.

6.76 Consider $S(x) = \sum_{n=1}^{\infty} \frac{n}{(n+2)^3} x^n$.
(a) Find the domain of S.
(b) Compute $S'''(0)$.

6.77 Express the function $f(x) = \int_0^x \frac{\cos(t\sqrt{t}) - 1}{t^2}\, dt$, with $x > 0$, as a power series.

6.78 Express the integral $\int_0^1 \frac{\log(1 + x^2)}{\sqrt{x}}\, dx$ as a convergent series.

6.79 Evaluate the integral $\int_0^1 \frac{e^{2x^2} - 1}{x}\, dx$ with an error smaller than 10^{-1}.

6.80 Consider the power series $\sum_{n=1}^{\infty} (x^n - x^{n-1})$. Establish where the series converges pointwise, absolutely, uniformly or totally.

6.81 Consider the function $f(x) = \int_{-1}^{x} \frac{t\log(2+t) - 1}{t}\, dt$.
(a) Establish if there exists an expansion of f as a convergent power series centered at $x_0 = -1$ and, if so, specify the convergence radius.
(b) Compute, if existing, $f^{(5)}(-1)$.
(c) Compute $f(-1/2)$ with an error smaller than 10^{-2}.

6.82 Consider the function:
$$f(x) = \begin{cases} \dfrac{\sqrt[3]{(x+1)^2} - 1}{x} & x \neq 0 \\ 2/3 & x = 0. \end{cases}$$

(a) Does f admit infinitely many derivatives at $x_0 = 0$?
(b) Compute, if existing, $f''(0)$ and $f'''(0)$.

6.83 Consider the function: $f(x) = \dfrac{x}{(x^2+1)^2}$.
(a) Find the McLaurin series expansion of f and specify its convergence radius.
(b) Compute, if existing, $f^{(9)}(0)$.

6.84 Consider the function: $f(x) = \sqrt{1+x}$.
(a) Find the McLaurin series expansion of f and specify its convergence radius.
(b) Compute, $\sqrt{1.2}$ with an error smaller than $5 \cdot 10^{-6}$.

6.85 Find the Taylor series expansion of $f(x) = x^{-2}$ centered at $x_0 = -2$ and specify the convergence interval.

6.86 Find the sum of the series $\sum_{n=1}^{\infty} (-1)^n \dfrac{(e^x - 1)^n}{n}$ and specify the convergence interval.

6.87 Consider the function $f(x) = \dfrac{1}{\sqrt{1+e^{-x}}}$.
(a) Find a series of functions converging to f somewhere, and specify where.
(b) Compute $f(1)$ with an error not greater than 10^{-2}.

6.88 (a) Find the McLaurin series of $\log\left(\dfrac{1+x}{1-x}\right)$, specifying the convergence interval.
(b) Compute $\log 2$ with an error not larger than $5 \cdot 10^{-5}$.

6.89 (a) Find where the series $\sum_{n=1}^{\infty} n^2 x^n$, convergences.
(b) Determine the explicit analytic expression of the sum.

6.90 Compute, with an error not larger than $2 \cdot 10^{-3}$, the integral
$$\int_0^{1/3} \dfrac{\arctan x - \log(1+x^2)^{1/x}}{x^3} \, dx.$$

6.6 Problems

6.91 Consider the series $\sum_{n=2}^{\infty} \dfrac{(-1)^n(n+1)}{n^3-1} x^{2n}$.

(a) Determine where the series converges.
(b) Determine where the series converges uniformly.
(c) Determine where the pointwise limit S is differentiable.
(d) Show that the integral $\int_2^{+\infty} S(1/x)\,dx$ converges and compute an approximate value with an error no greater than 10^{-3}.

6.92 Consider the series $\sum_{n=1}^{\infty} \dfrac{n}{n^2+3n+2} x^n$.

(a) Determine where the series converges.
(b) Determine the pointwise limit S.

6.93 Consider the function

$$f(x) = \begin{cases} -1 & x \in (-2, -1] \\ x & x \in (-1, 1) \\ 1 & x \in [1, 2) \end{cases}$$

(a) Find the Fourier coefficients of f.
(b) Find the sum of the series in -2 and 2.

6.94 Consider the function $f(x) = \cos x$ for $x \in (0, \pi)$.
(a) Find the sine series of f in $[-\pi, \pi]$.
(b) Find the sum of the series in $\pm\pi$ and 0.

6.95 Consider the function

$$f(x) = \begin{cases} \cos x & x \in (0, \pi] \\ 1 & x \in (\pi, 2\pi). \end{cases}$$

(a) Find b_1, b_2, b_3 and b_4 in the sine series of f in $[-2\pi, 2\pi]$.
(b) Find the sum of the series in $9\pi/4$.

6.96 Find the sine series of $f(x) = \cos(6x)$ for $x \in [0, \pi/3]$ with period $2\pi/3$

6.97 Determine the 2π-periodic Fourier series of

$$f(x) = \begin{cases} 0 & x \in (-\pi, 0) \\ x & x \in [0, \pi). \end{cases}$$

and find its sum.

6.98 (a) Determine the 2π-periodic Fourier series of $f(x) = x^2 + x$ in $(-\pi, \pi)$ and find its sum.

(b) Compute the sum of $\sum_{n=1}^{\infty} \dfrac{n^2 + 4}{n^4}$.

6.99 Determine the 2π-periodic Fourier series of $f(x) = x \sin x$ in $[-\pi, \pi]$ and find its sum.

6.100 Determine the 2π-periodic Fourier series of $f(x) = x \sin(2x)$ in $(0, 2\pi)$ and find its sum.

6.101 Let $f \in PC([0, T])$ and denote by a_n and b_n its Fourier coefficients, and let g be the sum of the series.
(a) Suppose that $|a_n| \leq 1/n^4$ and $|b_n| \leq 1/n^4$ for $n \geq 1$. Is it true that $g \in C^2(\mathbb{R})$?
(b) Establish if it is possible that $|a_n| = 1/n^{1/2}$.

Reference

1. Baronti, M., De Mari, F., van der Putten, R., Venturi, I.: Calculus problems. Unitext, 101, La Matematica per il 3+2. Springer (2016)

Solutions

Problems of Chapter 1

1.16 $\text{Dom}(f_1) = \mathbb{R}^2 \setminus \{\text{axes}\}$. $\text{Dom}(f_2) = \mathbb{R}^2 \setminus \{(0,0)\}$. $\text{Dom}(f_3) = \mathbb{R}^2 \setminus \{(x,y) : y > 1 - x\}$. For f_4 and f_5 see Fig. A.1. $\text{Dom}(f_6) = \{(x,y) \in \mathbb{R}^2 : x \geq 0\} \cup \{(x,0) \in \mathbb{R}^2 : x < 0\}$.

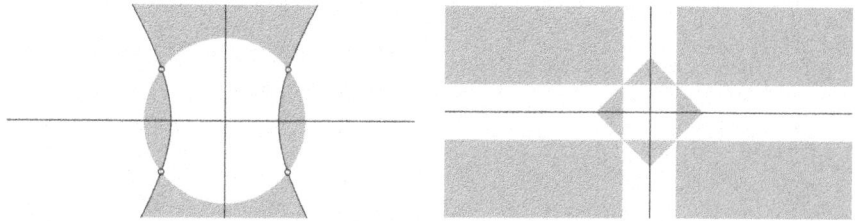

Fig. A.1 Problem 1.16: in grey the domains of f_4 (left) and of f_5 (right)

1.17 The level sets of f_1 they are the hyperbolae $xy = c$ with $c \neq 0$; those of f_2 the union of the coordinate axes ($c = 0$) and the graphs of the functions $x \mapsto c/x^2$ with $c \neq 0$; those of f_3 are the ellipses $x^2 + 2y^2 = c$ with $c > 0$ and those of f_4 are the lines $y = [(1 - \sin c)/(1 + \sin c)]x$ with $c \in (-\pi/2, \pi/2]$, so in fact the lines $y = kx$ with $k > 0$, and the y-axis, both without the origin.

1.18 One example is $A = (0,1) \cup [2,3)$ and $B = [1,2)$, with $\overline{A} = [0,1] \cup [2,3]$ and $\overline{B} = [1,2]$; the four sets are $\{2\}, \{1\}, \{1,2\}$ and $\emptyset$.

1.19 $\overline{E} = E$.

1.20 A is compact, hence closed and bounded, and not open. See Fig. A.2.

1.21 B is the infinite closed strip $(-\infty, \alpha] \times [0,1]$, where $\alpha \in (1/4, 1/2)$ is the only solution of $e^x + x - 2 = 0$.

Fig. A.2 The set A of Problem 1.20; the larger circle has equation $(x-1)^2 + (y+2)^2 = 5$

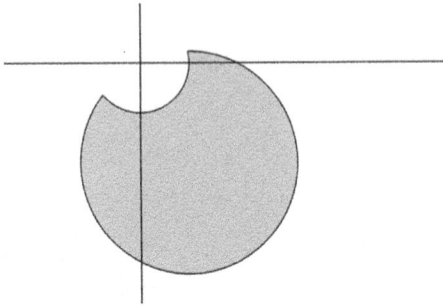

1.22 A is bounded but neither open nor closed; hence not compact. See Fig. A.3.

Fig. A.3 In grey the set A of Problem 1.22; the smaller circle has equation $x^2 + y^2 = 1$, the bigger has equation $(x-2)^2 + (y-2)^2 = 8$

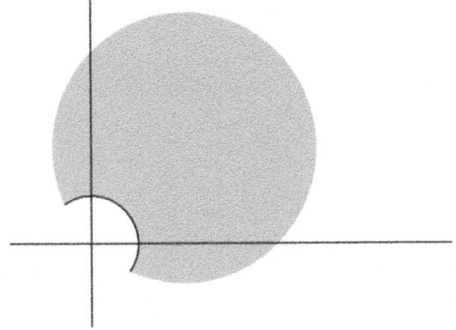

1.23 It is closed and unbounded, see Fig. 1.24.

1.24 Points in A have x-coordinate either ≤ -3 or ≥ 3, and the corresponding subsets of A are closed and disjoint; see Fig. A.4.

Fig. A.4 The set A of Problem 1.24

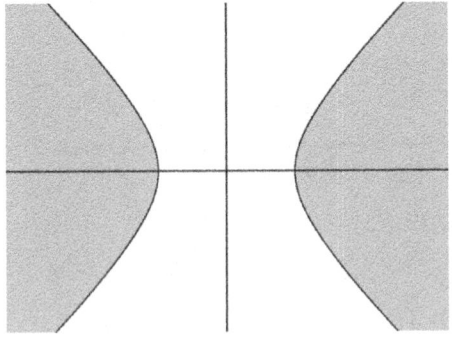

1.25 (a) No: not closed. (b) No: not bounded. (c) No: not closed. (d) Yes. (e) Yes. (f) No: not closed, open. (g) No: not bounded.

1.26 (a) Yes. (b) Yes. (c) No: not closed. (d) No: not bounded.

1.27 $W = \{(x, y) \in \mathbb{R}^2 : x > 0, \ y \geq 1/x\}$ has $p(W) = (0, +\infty)$; see Fig. A.5.

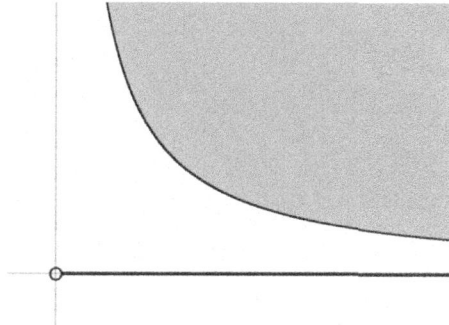

Fig. A.5 The sets $W \subset \mathbb{R}^2$ and $p(W) \subset \mathbb{R}$ of Problem 1.27

1.28 The set A is bounded but is neither closed nor open; see Fig. A.6.

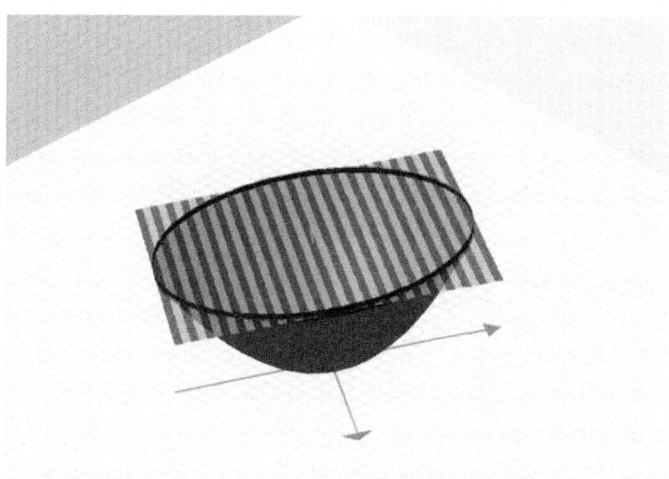

Fig. A.6 The region A of Problem 1.28, inside the elliptic paraboloid and below the plane

1.29 The sets A, B and $A \cup B$ are pathwise connected; see Fig. A.7.

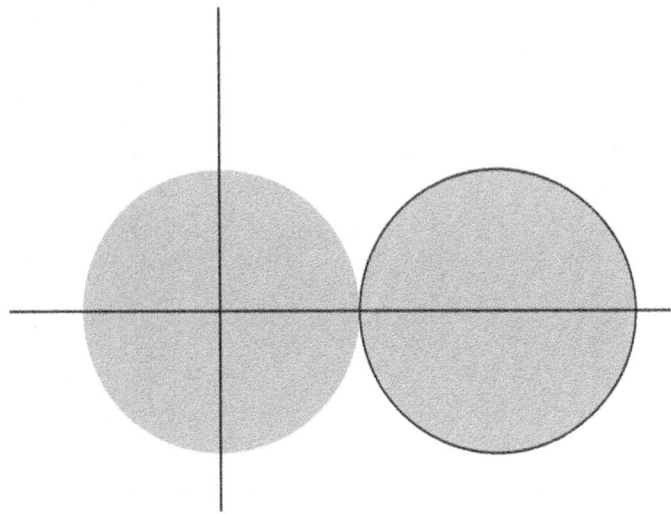

Fig. A.7 The sets A (left, without its boundary), B (right, with boundary) and $A \cup B$ (with half of its boundary) of Problem 1.29

Problems of Chapter 2

2.28 (a) 0. (b) 0. (c) It does not exist. (d) 0. (e) It does not exist. (f) It does not exist.

2.29 (a) 0. (b) 0.

2.30 (a) It does not exist. . (b) It does not exist. (c) 0. (d) It does not exist.

2.31 (a) It does not exist. (b) It does not exist. (c) It does not exist. (d) It does not exist. (e) $+\infty$. (f) It does not exist. (g) It does not exist. (h) 0.

2.32 (a) $+\infty$ if $\alpha < 4$, it does not exist otherwise. (b) It does not exist. (c) 0 if $\alpha < 3/2$, it does not exist otherwise. (d) It does not exist. (e) 0 if $\alpha < 2$, it does not exist otherwise. (f) 0 if $\alpha < 1$, 1 if $\alpha = 1$ and $+\infty$ if $\alpha > 1$. (g) 0. (h) 0.

2.33 0.

2.34 A is bounded, open and convex. It is neither closed nor compact.

2.35 A is both bounded and connected.

2.36 The domain is closed, bounded (hence compact) and connected. The set of level $c \in [-\pi/2, \pi/2]$ is the circle centered in $(1, 2)$ and radius $\sqrt{5 + \sin c}$.

2.37 A is connected but it is neither convex nor compact.

2.38 $k \neq 0$.

2.39 A is open, connected, convex and bounded. The level set is the upper half of the circle centered at $(2, 0)$ and radius 2.

2.40 A is closed and bounded (hence compact) connected and convex. The level set is the half below the line $x = y$ of the circle centered at $(0, 2)$ and radius 2.

2.41 A is closed and bounded (hence compact) and connected. It is not convex but *star-shaped*, where this means that there exists a point in the set which can be joined to every other point in the set with a segment lying in the set.

2.42 The set A is neither open nor closed, nor bounded, nor connected. The level set through the point $(-3, 1)$ is the parabola $x = 2y^2 - 2y - 3$ except for the points $(9, 3)$ and $(1, -1)$.

2.43 The set A_a is open for every $a \in \mathbb{R}$; the set B_a is closed if and only if $a \leq 1$.

Problems of Chapter 3

3.28 (a) $\mathbb{R}^2 \setminus \{(0, 0)\}$. (b) $\mathbb{R}^2 \setminus \{(0, 0)\}$. (c) $\mathbb{R}^2 \setminus \{(0, 0)\}$. (d) $\mathbb{R}^2$. (e) $\mathbb{R}^2 \setminus \{(0, 0)\}$.

3.29 (a) $f_x = \dfrac{1 + \tan^2(x/y)}{y \tan(x/y)}$, $f_y = -\dfrac{x}{y^2} \dfrac{1 + \tan^2(x/y)}{\tan(x/y)}$.

(b) $f_x = yx^{y-1}$, $f_y = x^y \log x$.

(c) $f_x = 3x^2 y^2 z + 2$, $f_y = 2x^3 yz - 3$, $f_z = x^3 y^2 + 1$.

(d) $f_x = -\dfrac{y}{x^2} e^{\sin(y/x)} \cos(y/x)$, $f_y = \dfrac{1}{x} e^{\sin(y/x)} \cos(y/x)$.

(e) $f_x = \dfrac{-y}{x^2 + y^2}$, $f_y = \dfrac{x}{x^2 + y^2}$.

(f) $f_x = \dfrac{1}{\sqrt{x^2 + y^2}}$, $f_y = \dfrac{y}{x^2 + y^2 + x\sqrt{x^2 + y^2}}$.

(g) $f_x = \dfrac{1}{\sqrt{y} \tan(x/\sqrt{y})}$, $f_y = -\dfrac{x}{2\sqrt[3]{y^2} \tan(x/\sqrt{y})}$.

(h) $f_x = y^z x^{y^z - 1}$, $f_y = x^{(y^z)} z y^{z-1} \log x$, $f_z = x^{(y^z)} y^z \log x \log y$.

(i) if $xy \neq 0$, then $f_x = \dfrac{xy\sqrt{|y|}}{2|x|\sqrt{|x|}}$, $f_y = \dfrac{3y^2\sqrt{|x|}}{2|y|\sqrt{|y|}}$; if $y = 0$, then $f_x = f_y = 0$.

(l) $f_x = \dfrac{2x}{x^2 + y^2}$, $f_y = \dfrac{2y}{x^2 + y^2}$.

(m) $\nabla f = (a_1, \ldots, a_n)$.

3.30 (a) f is not continuous, it has partial derivatives, it is not differentiable.
(b) f is not continuous, it does not have partial derivatives, it is not differentiable.
(c) f is continuous, it has partial derivatives, it is differentiable.

3.31 (a) It does not exist. (b) 3/5. (c) 0. (d) $-2/\sqrt{6}$. (e) $1/\sqrt{6}$. (f) $\sqrt{5}/2$.

3.32 (a) Yes. (b) No. (c) No. (d) For no unit vector Q.

3.33 (a) Yes. (b) Yes, and $\nabla f(0,0) = (0,0)$. (c) No. (d) Only for the unit vectors that correspond to the angles $\theta = 0, \pi/2, \pi, 3\pi/2$.

3.34 The function f is continuous for $\alpha > 2$ and differentiable for $\alpha > 3$. The function g is continuous for $\beta \leq 0$ and $\gamma = 0$ and differentiable for $\beta \leq 0$ and $\gamma = 0$.

3.35 (a) For $0 < \alpha < 3$. (b) For $0 < \alpha < 5/2$.

3.36 (a) For $\alpha > 1$. (b) For $\alpha > 2$.

3.37 (a) For $\alpha > 0$. (b) For $\alpha > 1/2$. (c) For $\alpha > 1/2$.

3.38 (a) For $\alpha > 1/3$. (b) $(\partial f/\partial Q)(0,1)$ exists only for $Q = (\pm 1, 0)$, $Q = (0, \pm 1)$ and for these Q they vanish.

3.39 The function vanishes on the axes, hence $\nabla f(0,0) = (0,0)$. The limit (3.7) in polar coordinates is $\rho^{2/3} \cos^{2/3} \theta \sin \theta$, whence differentiability. The partial f_x exists only away from the y-axis, where $f_x(x,y) = (2/3)x^{-1/3}y$, and at the origin, where it vanishes. It is manifestly not continuous: take for example the curve $y = x^{1/4}$, along which $f_x(x, x^{1/4}) = (2/3)x^{-1/12}$; clearly, $x^{-1/12}$ does not tend to 0 as $x \to 0^+$.

3.40 The partial f_y vanishes along the x-axis and at the origin, and equals $f_y(0,y) = 2y\sin(1/y) - \cos(1/y)$ on the punctured y-axis, hence it is obviously not continuous. The second order mixed partials trivially vanish everywhere in $\mathbb{R}^2$.

3.41 (a) Yes. (b) Yes. (c) No. (d) Yes.

3.42 (a) Yes. (b) Yes. (c) Yes. (d) No.

3.43 (a) 0. (b) It is not continuous.

3.44 (a) Yes. (b) $L_{(1,1)}(x,y) = \sqrt{2}(2(x+y), (x-y))$. (c) $\det JF(x,y) = -4(x^2 + y^2)$ is differentiable on $\mathbb{R}^2$.

3.45 (a) $G \circ F(x,y) = (\sin(xe^y), \cos(ye^x))$. (b) $F \circ G(x,y) = (e^{\cos y}\sin x, e^{\sin x}\cos y)$.

(c) $J(G \circ F)(x,y) = \begin{bmatrix} \cos(xe^y)e^y & \cos(xe^y)xe^y \\ -\sin(ye^x)ye^x & -\sin(ye^x)e^x \end{bmatrix}$.

(d) $J(F \circ G)(x,y) = \begin{bmatrix} e^{\cos y}\cos x & -e^{\cos y}\sin x \sin y \\ e^{\sin x}\cos x \cos y & -e^{\sin x}\sin y \end{bmatrix}$.

3.46 F vanishes with order 1/4.

3.47 The McLaurin polynomial of order 5 of F is $p_5(x) = x + \frac{x^5}{4}$.

3.48 (a) $\mathrm{Dom}(F) = [-1, +\infty)$.
(b) Yes, because $F'(0) = \int_0^1 \frac{dy}{(1+y)\sqrt{y}} > 0$.

3.49 (a) $\text{Dom}(F) = (-\infty, 1] \cup (2, +\infty)$.
(b) $F'(0) = -\log 2 + \dfrac{e^y}{1+y^2} dy$.

3.50 After differentiation under the integral sign and integration by parts, one finds $H'(t) = -1/(1+t^2)$. Hence $H(t) = \dfrac{\pi}{2} - \arctan t$ and thus $\int_{-\infty}^{+\infty} \dfrac{\sin x}{x} dx = \pi$.

Problems of Chapter 4

4.20 There are no local extrema.

4.21 The minimum is at $(-1, 0)$ and the maximum at $(1, 1)$.

4.22 The minima are at $\pm(\sqrt{2}/2, -\sqrt{2})$ and at the maxima at $\pm(1, 1)$.

4.23 The maxima are at $\{(x, 0) : x \in [0, 1]\} \cup \{(0, y) : y \in [0, 1]\}$ and the minimum is at $(1/2, 1/2)$.

4.24 The maximum is at $(1, 0)$ and the minimum at $(1, 1)$.

4.25 The maximum is at $(0, 0)$ and the minima at $(1, 0)$, $(-1, 0)$ and $(0, -1)$.

4.26 The maximum is at $(0, -\sqrt{2})$ and the minimum at $(0, \sqrt{2})$.

4.27 There are no maxima. The minima are at $\left(\pm\sqrt{5/2}, \tfrac{1}{2}\right)$.

4.28 The maximum is at $(0, -1, 2\sqrt{2})$ and the minimum at $(0, -1, -2\sqrt{2})$.

4.29 If $k \in (-2, 2)$, then $(0, 0)$ is a global minimum. If $k \in (-\infty, -2) \cup (2, +\infty)$, there are no global minima and $(0, 0)$ is a saddle point. If $k = 2$, then the line $x + y = 0$ consists of global minima and if $k = -2$, then the line $x - y = 0$ consists of global minima. In any case there are no global maxima.

4.30 There are no global extrema in $\mathbb{R}^2$; $(1/2, 1/3)$ is a local maximum; the points of the form $(x, 0)$ are local minima if $x \in (0, 1)$ and local maxima if either $x < 0$ or $x > 1$. On Q the unique global minimum is at $(-3/5, -2/5)$ and the unique global maximum is at $(1/2, -1/2)$.

4.31 If $k > 0$, then $(0, 0)$ is the unique critical point and it is a global minimum. If $k < 0$, then $(0, 0)$ is the unique critical point and it is a saddle point. If $k = 0$, then all the points of the y-axis are critical and they are all global minima. In any case, there are no global maxima.

4.32 If $k \neq -1$, then $(0, 0)$ is the unique critical point and it is a saddle point. If $k = -1$, then all the points of the y-axis are critical and they are all global minima. In any case there are no global maxima.

4.33 Since $x'(0) = 1$ and $x''(0) = -7$, the local graph is as in Fig. A.8, left.

4.34 Since $x'(0) = 1/2$ and $x''(0) = 15/8$, the local graph is as in Fig. A.8, right.

Fig. A.8 Left: the full level set considered in Problem 4.33 depicted in grey, and the local graph of the implicit function $x = x(y)$ in black; right: a portion of the level set considered in Problem 4.34 depicted in grey, and the local graph of the implicit function $x = x(y)$ in black

4.35 $(1/4, 1)$ is a global minimum and $(1, 0)$ is a global maximum; the points of the form $(0, y)$ are local maxima if $y \in [0, 2)$ and local minima if $y \in (2, 3]$.

4.36 $(2, 3\pi/4)$ and $(2, \pi/4)$ are global maxima and all the points of the form $(x, \pi/2)$ with $1 \leq x \leq 2$ are global minima.

4.37 If $k \neq 0$, then the level set defines both a function of $y = y(x)$ and a function $x = x(y)$; if $k = 0$, then the level set defines only a function $x = x(y)$.

4.38 The expansion is $y(x) = \dfrac{1}{2}x - \dfrac{1}{16}x^2 + o\left(x^2\right)$, see Fig. A.9.

4.39 All points in $\{(x, y) \in \mathbb{R}^2 : x^2 = y^3, \ x \neq 0\}$ are global maxima and all points in $\{(x, y) \in \mathbb{R}^2 : x^2 = -y^3, \ x \neq 0\}$ are global minima; $(\pm 2^{-1/4}, 2^{-1/6})$ are global maxima and $(\pm 2^{-1/4}, -2^{-1/6})$ are global minima on $\{(x, y) \in \mathbb{R}^2 : x^4 + y^6 = 1\}$.

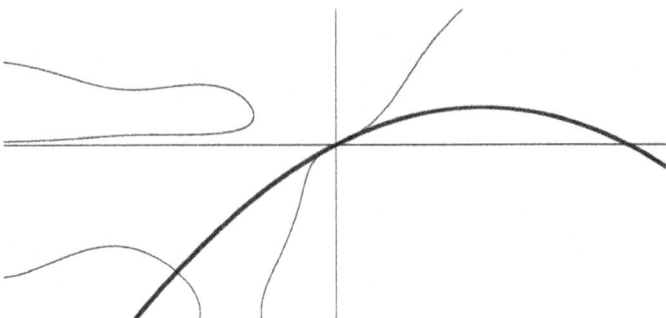

Fig. A.9 In grey the complicated zero-level set of the function in Problem 4.38 and in black the Taylor polynomial at the origin of the local function $x \mapsto y(x)$

4.40 All points on $\{(x, y, z) \in \mathbb{R}^3 : x^2 + (y - \frac{1}{2})^2 = \frac{5}{4} - \frac{1}{e}, z = y + 1\}$ are global minima, and $(0, 1/2, 5/4)$ is a global maximum.

4.41 The function $x = \varphi(y)$ is defined in a neighborhood of $y_0 = 0$ that contains the interval $[(1 - \sqrt{5})/2, (1 + \sqrt{5})/2]$ at which extreme points φ vanishes (see Fig. A.10).

Fig. A.10 In grey the set of level 1 of the function in Problem 4.41 and in black the portion of the graph of of the local function $\varphi(y)$ defined on $[(1 - \sqrt{5})/2, (1 + \sqrt{5})/2]$, together with its two zeroes

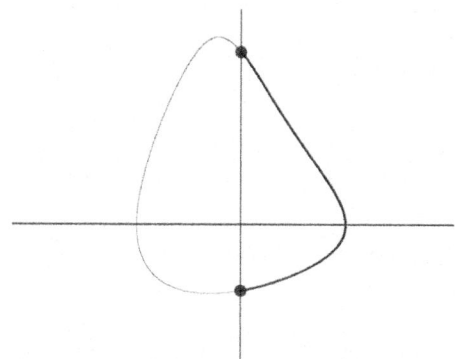

4.42 The point $(0, 0)$ is the minimum and $(\pm 1, 0)$ are the maxima.

4.43 A critical point is $(0, 0)$. An upper bound of $z(x, y)$ is $\log(1 + \pi/2)$ on its domain, which is $\{(x, y) \in \mathbb{R}^2 : x^2 + y^2 < \exp\left(\frac{\pi}{2} + 1 - 1\right)\}$).

Problems of Chapter 5

5.34 (a) $(e - 1)/3$. (b) $2 - \log 2$. (c) $3 - \frac{3\sqrt{2}}{4} \arctan(2\sqrt{2})$. (d) $77824/45$.
(e) $2\sqrt{3} - 2/3$. (f) $27/16$. (g) 0. (h) $1/4 + \log(3/4)$. (i) $-19/96$.
(j) $(-7\pi/24)(\cos 4 - \cos 1) + 3/2 \log \sqrt{2}/2$. (k) $(320\sqrt{2} - 512)/15$.
(l) $4/7(8\sqrt{2} - 1)(1 - 1/\sqrt[4]{2})$. (m) $17\pi/2 - 17 \arctan(3/\sqrt{7}) - 13\sqrt{7}/6$.
5.35 $9/2$.
5.36 (b) $4\left(2 + \sqrt{2}\right)/3$.
5.37 $12 \log(3/2) - 14/3$.
5.38 $\dfrac{125}{128} \pi^3 \left(\dfrac{5}{8}\pi^2 - \dfrac{1}{3}\right) + \dfrac{1}{15}$.
5.39 15.
5.40 $B = \left(0, \dfrac{8}{3} \dfrac{5\sqrt{2} - 2\sqrt{5}}{20 \arctan 2 - 5\pi}\right)$.
5.41 $B = \left(0, \dfrac{384}{105\pi}\right)$.

5.43 (a) $1/48$. (b) $\sqrt{2}/10$. (c) $27\pi/16$. (d) $9\pi/8$. (e) $195\pi/32$. (f) $192\sqrt{3}\pi/5$. (g) $(9\pi^2/2) - 4\pi$. (h) $2\pi(1 + \arctan 2 - \arctan 4) + \pi \log(\sqrt{17/5})/2$. (i) $64/3$.

5.44 $16/3$.
5.45 $5\pi/6$.
5.46 32π.
5.47 36π.
5.48 (a) $\pi[(1/\sqrt[4]{8}) - (1/h)]$. (b) $\pi/\sqrt[4]{8}$.
5.49 $\pi/4$.
5.50 $7\pi/6$.
5.51 $(0, 0, 1069/312)$.
5.52 $141\pi/4$.

Problems of Chapter 6

6.47 (a) $f(x) = 1/|x|$ on $I = \mathbb{R} \setminus \{0\}$; (b) $f_n \not\rightrightarrows f$ on I; (c) $\int_{-1}^{1} f_n(x)\,dx \to +\infty$.

6.48 (a) The sequence converges pointwise to the null function. (b) The convergence is not uniform on $\mathbb{R}$ but is uniform on every compact interval and in every interval of the form $(-\infty, a]$.

6.49 The sequence converges pointwise to e^{-x} on $\mathbb{R}$; the convergence is not uniform on $\mathbb{R}$ but is uniform on every compact interval.

6.50 (a) The sequence converges pointwise on $[1, +\infty)$ to the function

$$f(x) = \begin{cases} 0 & x = 1 \\ \arctan x & x > 1. \end{cases}$$

(b) Yes. (c) No.

6.51 (a) The pointwise limit is $f(x) = \begin{cases} 0 & x \neq 0 \\ 1 & x = 0 \end{cases}$. (b) No. (c) Yes.

6.52 (a) The pointwise limit is $f(x) = 0$ in $\mathbb{R}$. (b) No. (c) Yes.

6.53 (a) The pointwise and uniform limit is $f(x) = 0$ in $\mathbb{R}$. (b) The pointwise limit of $(f_n')_{n \geq 1}$ is $g(x) = \begin{cases} 0 & x \neq 0 \\ 1 & x = 0 \end{cases}$. The convergence is uniform in the closed intervals not containing the origin.

6.54 The pointwise limit is $f(x) = e^x$ on $[0, +\infty)$ and the convergence is uniform on all intervals $[0, a]$ for any $a > 0$.

6.55 (a) The pointwise limit is $f(x) = 0$ on $[0, +\infty)$ and the convergence is uniform on all compact intervals $[a, b]$. (b) Yes.

6.56 (a) The pointwise limit is $f(x) = 0$ on $\mathbb{R}$ but the convergence is not uniform; it is uniform in $(-\infty, 0] \cup [b, +\infty)$ for every $b > 0$. (b) The required limit is: 0 if $0 < a < 1$; $1/2$ if $a = 1$ and $+\infty$ if $a > 1$.

6.57 (a) $a_n > 1/4n$ for every $n \geq 1$. (b) The pointwise limit is $g(x) = 1/x^2$ if a_n converges or diverges with order less than 1; it is $h(x) = 1/(x^2 + \ell_1)$ if a_n diverges of order 1 and $a_n/n \to \ell_1$; it is the null function if a_n diverges of order greater than 1. (c) Any sequence diverging with order greater than 1.

6.58 (a) $\lim_n f_n(x) = 0$. (b) $\lim_n \int_0^2 f_n(x)\,dx = 0$. (c) By dominated convergence, that is Theorem 6.4, $\lim_n \int_0^{+\infty} f_n(x)\,dx = 0$.

6.59 (a) If $k > 3$ then the sequence converges only at $x = 0$ to 0; if $k = 3$ the pointwise limit is

$$f(x) = \begin{cases} 1/x & x \neq 0 \\ 0 & x = 0 \end{cases}$$

and the convergence is uniform in every closed set not containing the origin; if $k \in \{0, 1, 2\}$ the pointwise limit is the null function on $\mathbb{R}$; if $k \in \{0, 1\}$ the convergence is uniform on $\mathbb{R}$; if $k = 2$ the convergence is uniform in every closed set not containing the origin. (b) $+\infty$.

6.60 (a) The pointwise limit f is defined on $\mathbb{R}$, it is 2π periodic and on $(-\pi, \pi]$ it is

$$f(x) = \begin{cases} -1 & x \in (-\pi, 0) \\ 0 & x \in \{0, \pi\} \\ 1 & x \in (0, \pi); \end{cases}$$

(b) $f_n \rightrightarrows f$ on every compact subset of $\mathbb{R}$ which does not intersect $\{k\pi : k \in \mathbb{Z}\}$.

6.61 (a) $I = \mathbb{R}$. (b) Yes. (c) Yes.

6.62 (a) Pointwise convergent in $I = \mathbb{R}$ and uniformly in every compact subset of $\mathbb{R}$; (b) pointwise convergent in every interval $(-\frac{\pi}{2} + 2k\pi, \frac{\pi}{2} + 2k\pi)$ with $k \in \mathbb{Z}$ and uniformly in every compact subset thereof; (c) pointwise and uniformly convergent in $\mathbb{R}$; (d) pointwise convergent in $[0, +\infty)$ and uniformly in every $[a, +\infty)$ for every $a > 0$.

6.63 $\{0\}$.

6.64 $(-1, +\infty)$.

6.65 (a) $I = \mathbb{R}$. (b) S is continuous on $\mathbb{R}$.

6.66 $I = \mathbb{R}$; the convergence is absolute but not uniform on $\mathbb{R}$.

6.67 (a) $I = (1, +\infty)$. (b) Uniform convergence on all intervals $[a, +\infty)$ with $a > 1$.

6.68 $I = (-\infty, 0)$.

6.69 (a) Yes. (b) No, it is not differentiable at $x_0 = 0$.

6.70 (a) $\mathbb{R}$. (b) In every bounded interval.

6.71 (a) $\mathbb{R} \setminus \{0\}$. (b) It is continuous on its domain.

6.72 (a) $(-\infty, 0)$. (b) It is continuous on its domain.

6.73 (a) It converges totally in $(-\infty, a]$ for every $a < -2$ but not in $(-\infty, -2)$.
(b) The integral diverges to $-\infty$.

6.74 It converges pointwise $(1 - \sqrt{5}/2, 1]$ and totally in $[a, 1]$ for every $a \in (1 - \sqrt{5}/2, 1)$; it does not converge totally in $(1 - \sqrt{5}/2, b]$ for any $b \in (1 - \sqrt{5}/2, 1]$.

6.75 It is differentiable in $\mathbb{R}$.

6.76 (a) $[-1, 1]$. (b) $18/125$.

6.77 $f(x) = \sum_{n=1}^{\infty} \dfrac{(-1)^n \, x^{3n-1}}{(2n)! \, 3n - 1}$.

6.78 $2 \sum_{n=0}^{\infty} \dfrac{(-1)^n}{(n+1)(4n+5)}$.

6.79 $\sum_{n=1}^{5} \dfrac{2^n}{n! \, 2n}$.

6.80 The series converges pointwise and absolutely in $(-1, 1]$; it converges totally, hence uniformly, in any compact $K \subset (-1, 1)$; it converges neither uniformly nor totally in $(-1, 1]$

6.81 (a) $f(x) = (x+1) + \sum_{n=1}^{\infty} \left(1 + \dfrac{(-1)^{n+1}}{n}\right) \dfrac{(x+1)^{n+1}}{n+1}$, with radius 1. (b) 18.

(c) $\dfrac{1}{2} + \sum_{n=1}^{4} \left(1 + \dfrac{(-1)^{n+1}}{n}\right) \dfrac{1}{(n+1)2^{n+1}}$.

6.82 (a) Yes. (b) $f''(0) = 8/81$ and $f'''(0) = -14/81$.

6.83 (a) $f(x) = \sum_{n=1}^{\infty} (-1)^{n+1} n x^{2n-1}$, with radius 1. (b) $f^{(9)}(0) = 5 \cdot 9! = 1.814.400$.

6.84 (a) $f(x) = \sum_{n=0}^{\infty} \binom{1/2}{n} x^n$, with radius 1. (b) $\sum_{n=0}^{6} \binom{1/2}{n} \dfrac{1}{5^n}$.

6.85 $x^{-2} = \dfrac{1}{4} \sum_{n=0}^{\infty} \dfrac{n+1}{2^n} (x+2)^n$, with interval $(-4, 0)$.

6.86 $-x$, with interval $(-\infty, \log 2]$.

6.87 (a) $\sum_{n=0}^{\infty} \binom{-1/2}{n} e^{-nx}$, convergence in $[0, +\infty)$. (b) $\sum_{n=0}^{3} \binom{-1/2}{n} e^{-n}$.

6.88 (a) $2\sum_{n=1}^{\infty} x^{2n-1}/2n - 1$, convergence in $(-1, 1)$.
(b) $\log 2 \approx 2\sum_{n=1}^{4} 1/(3^{2n-1}(2n - 1))$.

6.89 (a) $(-1, 1)$ (b) $(x + x^2)/(1 - x)^3$.

6.90 $1/18$.

6.91 (a) $[-1, 1]$. (b) $[-1, 1]$. (c) $[-1, 1]$. (d) $1/56$.

6.92 (a) $[-1, 1)$. (b) $f(x) = \begin{cases} (x - 2)\log(1 - x) - 2x/x^2 & x \in [-1, 1) \setminus \{0\} \\ 0 & x = 0. \end{cases}$

6.93 (a) $a_n = 0$ and $b_n = [4\sin(n\pi/2) + (-1)^{n+1}2n\pi](n\pi)^{-2}$. (b) 0.

6.94 (a) $\pi^{-1}\sum_{n=1}^{\infty}(8n\sin 2nx)/(4n^2 - 1)$. (b) 0.

6.95 (a) $b_1 = 4/3\pi$, $b_2 = -2/\pi$, $b_3 = 28/15\pi$ $b_4 = 4/3\pi$. (b) -1.

6.96 $\sum_{n=1}^{\infty} \frac{4}{\pi} \frac{2n - 1}{(2n - 3)(2n + 1)} \sin((6n - 3)x)$.

6.97 The sum of

$$\frac{\pi}{4} - \frac{2}{\pi}\sum_{n=1}^{\infty} \frac{\cos((2n - 1)x)}{(2n - 1)^2} + \sum_{n=1}^{\infty} \frac{(-1)^{n+1}}{n} \sin(nx)$$

is the 2π-periodic extension of f except at $\pi + 2k\pi$ where it converges at $\pi/2$, for every $k \in \mathbb{Z}$.

6.98 (a) The sum of

$$\frac{\pi^2}{3} + 2\sum_{n=1}^{\infty}(-1)^n \left(\frac{2\cos(nx)}{n^2} - \frac{\sin(nx)}{n}\right)$$

is the 2π-periodic extension of f except at $\pi + 2k\pi$ where it converges at π^2, for every $k \in \mathbb{Z}$. (b) By Parseval's identity the value is $(4\pi^4 + 15\pi^2)/90$.

6.99 $a_0 = 2$, $a_1 = -1/2$, $a_n = (-1)^n 2/(1 - n^2)$ for $n \geq 2$, $b_n = 0$ for all $n \geq 1$. The sum is is the 2π-periodic extension of f, which is continuous.

6.100 $a_0 = -1$, $a_1 = -4/3$, $a_2 = -1/4$, $a_n = 4/n^2 - 4$ for $n \geq 3$, $b_2 = \pi$, $b_n = 0$ for all $n \neq 2$. The sum is is the 2π-periodic extension of f, which is continuous.

6.101 (a) Yes. (b) No, by Parseval equality.

Subject Index

B
Ball, 8
Barycenter, 194
Boundary, 10

C
Chain rule, 91
Constraint, 151
Coordinates
　angle, 22
　polar, 21
　pole, 23
　radius, 22
　regular transformation, 189
　spherical, 23
　system, 21
Cover
　open, 15
Curve
　closed, or loop, 57
　derivative, or velocity, 57
　image, 57
　in Euclidean space, 57
　Jordan, 57
　path or arc, 57
　simple, 57

D
Derivative
　directional, 86
　gradient, 86
　higher order partials, 91
　mixed partials, 92
　partial, 86
　under the integral sign, 95
Disk, 11
Distance
　Euclidean, 7
　normed space distance, 7
　or metric, 7
Domain
　normal, 188
Dot product, 4

F
Function
　analytic, 253
　characteristic, 185
　continuous, 53
　density, 194
　differentiable, 87
　differential, 87, 89
　graph, 18
　implicit, 154
　indicator, 185
　Lagrangian, 153
　of several variables, 17
　periodic, 254
　periodic extension, 256
　piecewise continuous, 255
　piecewise smooth, 257
　scalar field, 17
　step, 185
　trigonometric polynomial, 254

vector field, 17
vector valued map, 17

H
Hessian
  form, 150
  matrix, 92

I
Inequalities
  Cauchy-Schwarz, 5
  triangle, 5
Inner product, 1, 4
Integral
  change of variables, 190, 193
  depending on parameters, 95
  double, 183, 185
  reduction formulae, 188, 193
  Riemann, 186
  Riemann integrable, 186
  triple, 183, 191
Inverse image, 55

J
Jacobian
  matrix, 90

L
Lagrange multipliers, 153
Length
  vector, 1
Limit, 47
  at a point, 47
  at infinity, 49
  sequential, 50
Locally, 50

M
Map
  bilinear, 4
Mass
  center of mass, 194
  of a region, 195
Measure
  inner Peano-Jordan, 184
  of a n-rectangle, 183
  outer Peano-Jordan, 184
  Peano-Jordan measure, 184
Meridian section, 197

Moment of inertia, 195–197

N
Neighborhood
  of a point, 8
  punctured, 8
Norm, 1, 5, 6
  Eucledean, 5
n-rectangle, 183
  almost disjoint, 184

P
Period, 254
Point
  accumulation, 11
  critical, 148
  extreme
    constrained, 147, 152
    global, 148
    global maximum, 148
    global minimum, 148
    local, 148
    local maximum, 148
    local minimum, 148
    strong, 148
    unconstrained, 147
  interior, 10
  isolated, 12
  limit, 11
  saddle, 149

Q
Quadratic form, 94, 149
  indefinite, 149
  negative definite, 149
  negative semidefinite, 149
  positive definite, 149
  positive semidefinite, 149

S
Sequence
  Cauchy, 16
  converging, 16
  of functions, 243
  of points, 15
  pointwise convergence, 244
  subsequence, 16
  uniform convergence, 244
Series
  absolute convergence, 248
  continuity of the sum, 249

formal derivative, 249
formal integral, 250
Fourier
    coefficients, 256
    cosine expansion, 259
    series of a periodic function, 254
    sine expansion, 259
geometric, 247
partial sums, 246
pointwise convergence, 247
power series, 250
    coefficients, 250
    convergence interval, 250
    convergence radius, 250
    McLaurin series, 253
    Taylor series, 253
    Taylor series expansion, 253
sum, 247
telescopic, 247
term by term differentiation, 249
term by term integration, 250
total convergence, 248
trigonometric, 254
uniform convergence, 248, 260
Set
    bounded, 14
    closed, 8
        unit ball, 10
    closure, 10
    compact, 15
    connected, 13
        arcwise or path connected, 58
        polygonally, 13
    convex, 14
    dense, 13
    disconnected, 13
        disconnecting subsets, 13
    interior, 10
    level, 18
    open, 8
    Peano-Jordan measurable, 184
    star-shaped, 341
Simplex, 223
Solid of revolution, 197

Space
    complex vector space, 4
    Euclidean, 1
    inner product, 4
    metric, 7
    normed, 7
    vector, 1
Sphere
    circle, 11
    unit, 11

T
Tangent plane, 88
Taylor expansion
    first order Taylor polynomial, 89
    second order Taylor polynomial, 94
Theorem
    Abel, 251
    Bolzano-Weierstrass, 16
    Dini, 155
    Dirichlet, 258
    dominated convergence, 246
    Fermat, 148
    Guldino or Pappus, 198
    Heine-Borel, 15
    identity principle of power series, 252
    Lagrange multipliers, 152
    Parseval identity, 259
    permanence of sign, 50, 54
    Riemann-Lebesgue, 260
    root and ratio tests, 251
    Schwarz, 93
    squeeze, 50
    Weierstrass, 148

V
Vector
    basis, 3
    dimension, 3
    in Eucledean space, 1
    linear combination, 2
    of a vector space, 2
    scalar multiplication, 2

The manufacturer's authorised representative in the EU is Springer Nature Customer Service Centre GmbH, Europaplatz 3, 69115 Heidelberg, Germany. If you have any concerns regarding our products, please contact ProductSafety@springernature.com

Printed and bound by CPI Group (UK) Ltd, Croydon, CR0 4YY
26/03/2026
02078940-0001